Constitutive Models of Solid Materials

Constitutive Models of Solid Materials

From Mechanical Principles to Engineering Applications

Yao Yao, Hu Fang, Hongcun Guo, Tao Zeng, and Xu He

WILEY VCH

Authors

Professor Yao Yao
Northwestern Polytechnical University
No.1 DongXiang Rd
ChangAn District
Xi'an
CH, 710072

Dr. Hu Fang
Northwestern Polytechnical University
No. 1 Dongxiang Road
Chang'an District
Xi'an
CH, 710072

Dr. Hongcun Guo
Xi'an University of Architecture and Technology
No.13, Yanta Road
Beilin District
Xi'an
CH, 710055

Professor Tao Zeng
Xi'an University of Architecture and Technology
No.13, Yanta Road
Beilin District
Xi'an
CH, 710055

Professor Xu He
Nanjing University of Science and Technology
No. 200, Xiaolingwei Street
Xuanwu District
Nanjing
CH, 210094

Cover Design: Wiley
Cover Image: © PIRO4D/Pixabay

Typesetting: Lumina Datamatics,
Madurai, India

Printing and Binding: CPI Group (UK) Ltd,
Croydon, CR0 4YY

Library of Congress Card No.: applied for

British Library Cataloguing-in-Publication Data A catalogue record for this book is available from the British Library.

Bibliographic information published by the Deutsche Nationalbibliothek
The Deutsche Nationalbibliothek lists this publication in the Deutsche Nationalbibliografie; detailed bibliographic data are available on the Internet at http://dnb.d-nb.de.

Hardback ISBN: 9783527355532
ePDF ISBN: 9783527853502
ePub ISBN: 9783527853496
oBook ISBN: 9783527853519

C9783527355532_300326

Contents

Preface

Solid materials constitute one of the fundamental pillars of modern science and engineering. From aerospace to biomedicine, and from energy development to microelectronic devices, their mechanical behavior directly influences the performance and reliability of engineering structures. The constitutive relations of solid materials—that is, the laws governing their response to external loads and environmental conditions—serve as a bridge connecting microscopic material structures to macroscopic mechanical behavior. These relations form a critical foundation for solid mechanics, materials science, and engineering applications.

This book aims to provide a systematic introduction to the fundamental theories, modeling approaches, and recent advances in the constitutive behavior of solid materials, offering readers a comprehensive knowledge framework ranging from foundational concepts to cutting-edge developments. The content covers tensor analysis, elasticity theory, plasticity theory, damage mechanics, micromechanics, numerical implementation of constitutive models, and the emerging application of machine learning methods in fundamental mechanics and constitutive modeling. Additionally, constitutive model examples for several typical solid materials are presented. The book emphasizes both theoretical rigor and practical engineering relevance, striving to help readers gain a deep understanding of the intrinsic mechanical behavior of materials while mastering key techniques and methodologies in constitutive modeling.

The writing of this book is based on the author's years of teaching and research experience in solid mechanics, constitutive theory, and computational mechanics, incorporating the latest research developments from both domestic and international sources. In terms of organization, the book begins with mathematical foundations (tensor analysis) and progressively delves into constitutive theories at different scales, including continuum-based elastoplastic models, failure theories accounting for damage evolution, homogenization methods rooted in micromechanics, and data-driven machine learning modeling techniques. Furthermore, the book discusses the numerical implementation of constitutive models in finite element analysis and demonstrates practical applications through constitutive examples of typical materials such as electronic packaging solders, metals, rocks, and concrete.

The intended readership includes senior undergraduates, graduate students, and researchers in fields such as mechanics, materials science, mechanical engineering, and aerospace engineering, as well as practicing engineers engaged in related disciplines. It is hoped that this book will provide readers with a solid theoretical foundation while inspiring further exploration of novel constitutive modeling methods and technologies.

During the preparation of this book, the author has drawn upon numerous classical references and cutting-edge studies, and has received invaluable support and assistance from colleagues, peers, and students; to whom sincere gratitude is expressed. Given the vast and rapidly evolving nature of constitutive theory for solid materials, any omissions or inaccuracies in this work are solely the author's responsibility, and constructive feedback from readers is warmly welcomed.

Yao Yao
September 2025

1

Overview of Continuum Mechanics

1.1 Definition of Tensor

1.1.1 Vectors and Tensors

In physics and mechanics, certain physical quantities are inherently independent of the choice of coordinate systems. A typical example is the vector, which represents a geometric quantity possessing both magnitude and direction. Common vector quantities in mechanics include displacement, velocity, force, and momentum, which are conventionally represented as directed line segments.

Vectors play a fundamental role across various natural sciences, including mathematics, physics, and engineering. Geometrically, a vector is often visualized as an arrowed line segment, where the length denotes its magnitude and the arrow indicates its direction. In contrast to vectors, scalar quantities such as mass, temperature, and energy—possess magnitude but lack direction.

In mathematics, vectors are formally defined as elements of vector spaces (or linear spaces), which are abstract structures characterized by operations of addition and scalar multiplication. When interpreted physically, defining a vector requires the introduction of a norm and an inner product within the framework of Euclidean space, enabling the quantification of length and angle. For instance, the gradient of a scalar field is a vector field, while the derivative of a scalar with respect to another scalar remains a scalar.

Although a vector is an invariant geometric entity, its numerical components vary with the choice of coordinate system. To represent a vector, one must specify its components relative to a particular basis. These components transform according to specific rules–coordinate transformation laws—that preserve the underlying physical meaning of the vector. This transformation behavior underpins the vector's coordinate independence and leads naturally to the generalized concept of tensors.

Tensors generalize vectors and scalars, providing a powerful mathematical framework for formulating physical laws in a coordinate-independent manner. In many physical theories, it is necessary to express laws in terms of component quantities; however, the component values typically change under a transformation of the coordinate system. To ensure the validity of physical laws in any frame of reference, the transformation rules must apply consistently to all terms in an equation [1–3]. This principle guarantees the covariance of physical laws.

Constitutive Models of Solid Materials: From Mechanical Principles to Engineering Applications, First Edition.

From a physical standpoint, the components of a tensor correspond to measurable quantities in a given frame of reference, whereas the tensor itself represents the intrinsic physical quantity. Observations may differ between reference frames, but the form of the physical laws and equations remains invariant. To fully describe a tensorial physical quantity, one must understand not only its components in a particular frame but also how those components transform under changes of reference. It is this transformation behavior that preserves the form of physical equations and ensures their universal applicability.

1.1.2 Definition of Tensors

1. Definitions

The introduced coordinate system represents the physical quantity $\boldsymbol{T}$ with some directional combination. Let the physical quantity $\boldsymbol{T}$ be represented in different Cartesian coordinate systems as:

$$\begin{aligned} \boldsymbol{T} &= T_{ij\cdots k}\boldsymbol{e}_i\boldsymbol{e}_j\cdots\boldsymbol{e}_k \\ \boldsymbol{T}' &= T'_{rs\cdots t}\boldsymbol{e}'_r\boldsymbol{e}'_s\cdots\boldsymbol{e}'_t \end{aligned} \tag{1.1}$$

where $\boldsymbol{e}_i\,\boldsymbol{e}_j\ldots\boldsymbol{e}_k$ is the combined base vector required to represent the directionality of the physical quantity, and let the number of combined base vectors be n (i.e. n base vectors $\boldsymbol{e}$ multiplied). If its coordinate components meet the following coordinate conversion rules:

$$T_{ij\cdots k} = \beta_{ir'}\beta_{js'}\cdots\beta_{kt'}T'_{rs\cdots t} \tag{1.2}$$

Combined Eq. (1.1) has

$$T'_{rs\cdots t}\boldsymbol{e}'_r\boldsymbol{e}'_s\cdots\boldsymbol{e}'_t = \beta_{ir'}\beta_{js'}\cdots\beta_{kt'}T'_{rs\cdots t}\boldsymbol{e}_i\boldsymbol{e}_j\cdots\boldsymbol{e}_k = T_{ij\cdots k}\boldsymbol{e}_i\boldsymbol{e}_j\cdots\boldsymbol{e}_k \tag{1.3}$$

i.e. $\boldsymbol{T} = \boldsymbol{T}'$ such a physical quantity is called a tensor, and the number n of the combined basis vectors is called the order of this tensor. It can be seen that tensors represent physical quantities with a certain combination of directions, which are independent of the choice of coordinate system [4–7].

For example, physical quantities such as temperature or mass density of an object are functions of position, independent of direction, and are called zero-order tensors, which are expressed as scalars. Other physical quantities are expressed by vectors, such as the displacement, velocity, acceleration of a particle, and the force exerted on an object, which can be expressed as $\boldsymbol{T} = T_i\boldsymbol{e}_i$, the number of combined base vectors $n = 1$, i.e. the vector is a first-order tensor. When the number of combined basis vectors is $n = 2$, it is called a second-order tensor, such as the inertia tensor representing the mass distribution of the object, the strain tensor representing the deformation at a point, and the corresponding stress tensor. Second-order tensors can be expressed as $\boldsymbol{T} = T_{ij}\boldsymbol{e}_i\boldsymbol{e}_j$. A tensor can be represented in a variety of ways, such as in $\boldsymbol{T}$ or other capitalized italic, bold. In the case of a given coordinate system, it can also be expressed in the form of omitting the base vector, as in the case of a second-order tensor T_{ij}, which represents a set of tensor components, where i and j are free indicators. The advantage of this representation is that it is expressed in scalar form, and when performing operations, it conforms to the well-known scalar algorithm [1, 5, 7]. Sometimes, for clarity, the second-order tensor $\boldsymbol{T}$ can also be represented by a matrix of 3 × 3:

$$\begin{bmatrix} T_{11} & T_{12} & T_{13} \\ T_{21} & T_{22} & T_{23} \\ T_{31} & T_{32} & T_{33} \end{bmatrix} \tag{1.4}$$

Something like T_{12} represents a component of the tensor. Vectors can be represented in Ti or in arrays of 1 × 3:

$$\begin{bmatrix} T_1 \\ T_2 \\ T_3 \end{bmatrix} = \begin{bmatrix} T_1 & T_2 & T_3 \end{bmatrix}^T \tag{1.5}$$

2. Objectivity of Tensors
A tensor represents a physical quantity that is independent of the observer or the choice of reference frame. This invariance ensures that physical laws expressed in tensorial form remain valid in all coordinate systems, thereby satisfying the principle of objectivity required in physical formulations. Although the tensor itself is invariant, its component representation depends on the choice of basis vectors associated with a specific frame of reference. When the coordinate system changes, the basis vectors transform accordingly, and the tensor components adjust in such a way that the overall tensor remains unchanged. In this sense, a tensor is a covariant combination of basis and components. Thus, while the same tensor may have different component forms in different frames of reference, the underlying physical quantity it represents is preserved across all observers [1, 5, 7].
Therefore, it can be understood that if the tensor satisfies the corresponding coordinate transformation law, the tensor can be objective. For example, the scalar a, the first-order tensor (vector) $\boldsymbol{b}$, the second-order tensor $\boldsymbol{T}$ expressed in the reference line $Ox_1x_2x_3$ expressed under the reference line $Ox_1'x_2'\,x_3'$ need to satisfy the following relationship:

$$\boldsymbol{T}' = \boldsymbol{QTQ}^T \tag{1.6}$$

So, at the level of coordinate transformation, the objectivity of the tensor and the indistinguishable property of the reference frame of the tensor are equivalent.
However, it cannot be understood that if the tensor satisfies the coordinate transformation rules of the tensor, it is said that the tensor has objectivity, and there will be some special tensors. For example, the deformation gradient tensor $\boldsymbol{F}$ is a second-order tensor categorically speaking, but its coordinate transformation rule is not $\boldsymbol{F}' = \boldsymbol{QFQ}^T$, but $\boldsymbol{F}' = \boldsymbol{QF}$, this is because $\boldsymbol{F}$ is a special second-order tensor, also called a two-point tensor, because the base $\boldsymbol{e}_i \otimes \boldsymbol{E}_j$, the base $\boldsymbol{e}_i$ under the current configuration changes with time and is affected by the rotation tensor $\boldsymbol{Q}$; At the same time, the substrate $\boldsymbol{E}_j$ under the initial configuration is always unchanged, so it is not affected by $\boldsymbol{Q}$. Therefore, the two-point tensor $\boldsymbol{F}$ needs to satisfy the principle of objectivity, and the coordinate transformation law is the same as the first-order tensor, which is $\boldsymbol{F}' = \boldsymbol{QF}$.

At the same time, some tensors have objectivity, but their material time differentiation is not. For example, the Cauchy stress tensor σ has objectivity $\left(\boldsymbol{\sigma} = \boldsymbol{Q\sigma Q}^T\right)$, but the Cauchy stress rate tensor σ is not:

$$\sigma' = \left(\mathbf{Q}\sigma\mathbf{Q}^{\mathrm{T}}\right)' = \mathbf{Q}\sigma\mathbf{Q}^{\mathrm{T}} + \mathbf{Q}\sigma\mathbf{Q}^{\mathrm{T}} + \mathbf{Q}\sigma\mathbf{Q}^{\mathrm{T}} \neq \mathbf{Q}\sigma\mathbf{Q}^{\mathrm{T}} \tag{1.7}$$

Therefore, to ensure the objectivity of the Cauchy stress rate, several scholars have proposed objective stress rate formulations in tensorial form that are consistent with the transformation rules required for second-order tensors under a change of reference frame [8–10]. When establishing constitutive models—particularly those in rate form, such as hypoelastic models—it is essential to determine whether the employed tensor quantities are objective. If the material time derivative of a tensor is not objective, then any constitutive relation formulated using it may fail to be frame-indifferent, thereby violating a fundamental physical requirement.

Objective constitutive equations are typically expressed in rate form to capture time-dependent behavior, and ensuring the objectivity of stress and strain rates is crucial for the correctness of these models. If non-objective tensors are used directly in the formulation, it becomes challenging to construct frame-indifferent constitutive laws without introducing additional correction terms or structural complexity.

That said, it is not always necessary to use objective tensors in constitutive modeling. In certain cases, non-objective tensors can still yield physically meaningful results if the constitutive equations are constructed in terms of scalar invariants (e.g. norm or trace) or functions of tensors that remain invariant under observer transformations. In practice, many constitutive models adopt such an approach, leveraging observer-invariant quantities rather than explicitly ensuring the objectivity of every tensor involved.

For example, Green's elastic constructs use a tensor of the observer's invariant quantity, which is like:

$$S = \rho^{\,0} \frac{\partial \psi^{\mathrm{e}}}{\partial E} = 2\rho^{0} \frac{\partial \psi^{\mathrm{e}}}{\partial C} \tag{1.8}$$

In Eq. (1.8), S, E and C are all observer invariants, and can also be written as $S^* = \bar{\rho}\,\dfrac{\partial \psi^e\,(E^*)}{\partial E}$, for $S = S'$, this is a constitutive equation that satisfies the objectivity of physical laws.

1.2 Coordinate Transformations

1.2.1 Summation Convention

Consider the following sums:

$$s = a_1 x_1 + a_2 x_2 + \cdots + a_n x_n. \tag{1.9}$$

Equation (1.9) can be simplified using a summation notation:

$$s = \sum_{i=1}^{n} a_i x_i \tag{1.10}$$

At the same time, it is possible to use any symbol instead of i as a free indicator, such as

$$s = \sum_{j=1}^{n} a_j x_j, \quad s = \sum_{m=1}^{n} a_m x_m, \quad s = \sum_{k=1}^{n} a_k x_k \tag{1.11}$$

Whether it is i in the equation, or j, m, k, it is a dummy metric because the sum is independent of the letters used for indexing. The equation can be further simplified if the following convention is adopted: when an index is repeated once, it is a dummy metric that represents the sum of the index with the integers 1, 2, …, n. This convention is called Einstein's summation convention.

Using this convention, the equation can be simply written:

$$s = a_i x_i \text{ or } s = a_j x_j \text{ or } s = a_m x_m \cdots \tag{1.12}$$

It is important to emphasize that expressions such as $a_i b_i x_i$ or $a_m b_m x_m$ are not defined in this convention. This means that when summation conventions are used, indicator symbols should not be repeated more than once. Therefore, the expression of the summation form:

$$\sum_{i=1}^{n} a_i b_i x_i \tag{1.13}$$

must retain its sum symbol.

For ease of understanding, n in the sum formula is taken as 3 in the following text:

$$a_i x_i = a_1 x_1 + a_2 x_2 + a_3 x_3, \quad a_{ii} = a_{11} + a_{22} + a_{33}. \tag{1.14}$$

The summation convention can obviously be used to denote the sum of two or the sum of three things, etc. For example:

$$\alpha = \sum_{i=1}^{3} \sum_{j=1}^{3} a_{ij} x_i x_j \tag{1.15}$$

or

$$\alpha = a_{ij} x_i x_j. \tag{1.16}$$

Equation (1.16) can be fully expanded as:

$$\begin{aligned} \alpha = a_{ij} x_i x_j &= a_{11} x_1 x_1 + a_{12} x_1 x_2 + a_{13} x_1 x_3 + a_{21} x_2 x_1 + a_{22} x_2 x_2 \\ &\quad + a_{23} x_2 x_3 + a_{31} x_3 x_1 + a_{32} x_3 x_2 + a_{33} x_3 x_3 \end{aligned} \tag{1.17}$$

Similarly, the tag symbol $a_{ijk} x_i x_j x_k$ represents the triple sum of 27 items, i.e.

$$\sum_{i=1}^{3} \sum_{j=1}^{3} \sum_{k=1}^{3} a_{ijk} x_i x_j x_k = a_{ijk} x_i x_j x_k. \tag{1.18}$$

Consider the following set of equations:

$$
\begin{aligned}
x_1' &= a_{11}x_1 + a_{12}x_2 + a_{13}x_3, \\
x_2' &= a_{21}x_1 + a_{22}x_2 + a_{23}x_3, \\
x_3' &= a_{31}x_1 + a_{32}x_2 + a_{33}x_3,
\end{aligned} \tag{1.19}
$$

The use of summation conventions can be abbreviated as:

$$
\begin{aligned}
x_1' &= a_{1m}x_m, \\
x_2' &= a_{2m}x_m, \\
x_3' &= a_{3m}x_m,
\end{aligned} \tag{1.20}
$$

Further simplification can be obtained:

$$x_i' = a_{im}x_m, \quad i = 1, 2, 3 \tag{1.21}$$

An indicator that appears only once in each term of the equation, such as indicator i in Eq. (1.21), is called a free exponent. In general, a free exponent is an integer of 1, 2, or 3. Thus, Eq. (1.21) is an abbreviation of three equations, each with the sum of the three to the right.

1.2.2 Kronecker Increment

The Kronecker delta, denoted by $\boldsymbol{\delta}_{ij}$, can be defined as:

$$\delta_{ij} = \begin{cases} 1 & \text{if} : i = j \\ 0 & \text{if} : i \neq j \end{cases} \tag{1.22}$$

i.e.

$$\delta_{11} = \delta_{22} = \delta_{33} = 1, \delta_{12} = \delta_{13} = \delta_{21} = \delta_{23} = \delta_{31} = \delta_{32} = 0 \tag{1.23}$$

In other words, Kronecker's delta matrix is an identity matrix:

$$[\delta_{ij}] = \begin{bmatrix} \delta_{11} & \delta_{12} & \delta_{13} \\ \delta_{21} & \delta_{22} & \delta_{23} \\ \delta_{31} & \delta_{32} & \delta_{33} \end{bmatrix} = \begin{bmatrix} 1 & 0 & 0 \\ 0 & 1 & 0 \\ 0 & 0 & 1 \end{bmatrix} \tag{1.24}$$

According to the characteristics of the indicator and the summation convention, the Kronecker increment has the following properties.

Nature 1:

$$\delta_{ii} = \delta_{11} + \delta_{22} + \delta_{33} = 1 + 1 + 1 \tag{1.25}$$

i.e.

$$\delta_{ii} = 3 \tag{1.26}$$

Nature 2:

$$\delta_{im}a_m = a_i \tag{1.27}$$

Nature 3:

$$\begin{aligned}
\delta_{1m}T_{mj} &= \delta_{11}T_{1j} + \delta_{12}T_{2j} + \delta_{13}T_{3j} = T_{1j} \\
\delta_{2m}T_{mj} &= \delta_{21}T_{1j} + \delta_{22}T_{2j} + \delta_{23}T_{3j} = T_{2j} \\
\delta_{3m}T_{mj} &= \delta_{31}T_{1j} + \delta_{32}T_{2j} + \delta_{33}T_{3j} = T_{3j}
\end{aligned} \tag{1.28}$$

i.e.

$$\delta_{im}T_{mj} = T_{ij} \tag{1.29}$$

Nature 4: if $\boldsymbol{e}_1$, $\boldsymbol{e}_2$, and $\boldsymbol{e}_3$ are unit vectors perpendicular to each other, then there is

$$\boldsymbol{e}_i \cdot \boldsymbol{e}_j = \delta_{ij} \tag{1.30}$$

1.2.3 Coordinate Transformations

In three-dimensional space, three lines x_1, x_2 and x_3 orthogonal to each other at the origin O form a Cartesian coordinate system. $\boldsymbol{e}_1$, $\boldsymbol{e}_2$, and $\boldsymbol{e}_3$ represent the direction of these three lines, respectively, and their length is 1, which is called the base vector. It is customary to use the right-hand coordinate system, as shown in Figure 1.1. The dot product of any two of its base vectors is

$$\boldsymbol{e}_i \cdot \boldsymbol{e}_j = \delta_{ij} = \delta_{ji} = \begin{cases} 1, i = j \\ 0, i \neq j \end{cases} \tag{1.31}$$

where $\boldsymbol{\delta_{ij}}$ is the Kronecker operator, and the subscripts i and j can be freely selected from 1, 2, and 3, which is called the free indicator. If expressed in matrix form, there is

$$[\delta_{ij}] = \begin{bmatrix} 1 & 0 & 0 \\ 0 & 1 & 0 \\ 0 & 0 & 1 \end{bmatrix} \tag{1.32}$$

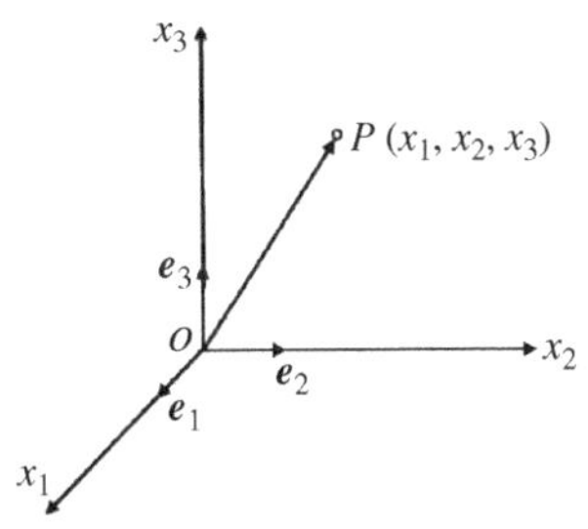

Figure 1.1 Cartesian coordinate system and basis vectors.

$$x = x_1\boldsymbol{e}_1 + x_2\boldsymbol{e}_2 + x_3\boldsymbol{e}_3 \tag{1.33}$$

According to Einstein's summation convention, Eq. (1.33) can be abbreviated as:

$$x = x_i\boldsymbol{e}_i \tag{1.34}$$

where the subscript i is repeated and is called a dumb mark. In a three-dimensional Euclidean space, the dumb mark i is summed by 1, 2, and 3, respectively, so that the dumb mark can be represented by any letter. For example

$$\delta_{ii} = \delta_{11} + \delta_{22} + \delta_{33} = 3 \tag{1.35}$$

Let another set of right-hand Cartesian coordinate systems $Ox_1'x_2'x_3'$ intersect at the origin O, and its base vectors are e_1', e_2', e_3'. The dot product of the two Cartesian coordinate systems base vectors e_i and e_i' is the direction cosine of the two rectilinear directions:

$$\beta_{ij'} = \boldsymbol{e}_i \cdot \boldsymbol{e}_j' = \cos\left(x_i, x_j'\right), \beta_{i'j} = \boldsymbol{e}_i' \cdot \boldsymbol{e}_j = \cos\left(x_i', x_j\right) \tag{1.36}$$

It is denoted by a matrix as

$$[\beta_{ij'}] = \begin{bmatrix} \cos(x_1, x_1') & \cos(x_1, x_2') & \cos(x_1, x_3') \\ \cos(x_2, x_1') & \cos(x_2, x_2') & \cos(x_2, x_3') \\ \cos(x_3, x_1') & \cos(x_3, x_2') & \cos(x_3, x_3') \end{bmatrix} \tag{1.37}$$

where $[\boldsymbol{\beta}_{i'j}]$ is the transpose matrix of $[\boldsymbol{\beta}_{ij'}]$, i.e. $[\boldsymbol{\beta}_{i'j}] = [\boldsymbol{\beta}_{ij'}]^T$, or $\boldsymbol{\beta}_{i'j} = \boldsymbol{\beta}_{j'i}$. Since $[\boldsymbol{\beta}_{ij'}]$ $[\boldsymbol{\beta}_{j'k}] == [\boldsymbol{\delta}_{ik}]$, it is shown that $[\boldsymbol{\beta}_{ij'}]$ is an orthogonal matrix, i.e. $[\boldsymbol{\beta}_{i'j}] = [\boldsymbol{\beta}_{ij'}]^{-1}$.

Thus, the transformation relationship between the two sets of basis vectors satisfies the transformation:

$$\boldsymbol{e}_i = \beta_{ij'}\boldsymbol{e}_j' \text{ or } \boldsymbol{e}_i' = \beta_{i'j}\boldsymbol{e}_j \tag{1.38}$$

The position of the P point can also be expressed by the vector diameter $\boldsymbol{x}'$, $\boldsymbol{x}' = \boldsymbol{x}$, by Eq. (1.38), there is

$$x_i\boldsymbol{e}_i = x_j'\boldsymbol{e}_j' = x_j'\beta_{j'i}\boldsymbol{e}_i = \beta_{ij'}x_j'\boldsymbol{e}_i \tag{1.39}$$

Comparison Eq. (1.45) is available on both sides

$$\begin{aligned} x_i &= \beta_{ij'}x_j', \text{ or} \\ x_i' &= \beta_{i'j}x_j \end{aligned} \tag{1.40}$$

Equation (1.46) is the coordinate transformation relationship between two sets of coordinate systems, $[\boldsymbol{\beta}_{i'j}]$ or $[\boldsymbol{\beta}_{ij'}]$ is called the coordinate transformation matrix.

1.2.4 Permutation Symbols

The permutation symbol represented by ε_{ijk} is defined as follows:

$$\varepsilon_{ijk} = \left\{ \begin{array}{|c|} 1 \\ -1 \\ 0 \end{array} \right\} \text{Traverse the order of 1, 2, 3 according to } i,\ j,\ k \tag{1.41}$$

i.e.

$$\begin{aligned} &\varepsilon_{123} = \varepsilon_{231} = \varepsilon_{312} = 1 \\ &\varepsilon_{213} = \varepsilon_{321} = \varepsilon_{132} = -1 \\ &\varepsilon_{111} = \varepsilon_{112} = \varepsilon_{222} = \ldots = 0 \end{aligned} \tag{1.42}$$

It can be recorded as:

$$\varepsilon_{ijk} = \varepsilon_{jki} = \varepsilon_{kij} = -\varepsilon_{jik} = -\varepsilon_{kji} - \varepsilon_{ikj} \tag{1.43}$$

From the vector mixture product, it can be seen that $(\boldsymbol{a}, \boldsymbol{b}, \boldsymbol{c})$ denotes the volume of a parallelepiped composed of $\boldsymbol{a}, \boldsymbol{b}, \boldsymbol{c}$. Thus, the displacement symbol can be represented by the mixed product of the Cartesian coordinate system base vector

$$\boldsymbol{e}_{ijk} = \left(\boldsymbol{e}_i, \boldsymbol{e}_j, \boldsymbol{e}_k\right) \tag{1.44}$$

If $\{\boldsymbol{e}_1, \ \boldsymbol{e}_2, \ \boldsymbol{e}_3\}$ is a set of vectors that follow the right-hand rule, then there is

$$\boldsymbol{e}_1 \times \boldsymbol{e}_2 = \boldsymbol{e}_3, \boldsymbol{e}_2 \times \boldsymbol{e}_1 = -\boldsymbol{e}_3, \boldsymbol{e}_2 \times \boldsymbol{e}_3 = \boldsymbol{e}_1, \boldsymbol{e}_3 \times \boldsymbol{e}_2 = -\boldsymbol{e}_1 \tag{1.45}$$

It can also be shortened as:

$$\boldsymbol{e}_i \times \boldsymbol{e}_j = \varepsilon_{ijk}\boldsymbol{e}_k = \varepsilon_{jki}\boldsymbol{e}_k = \varepsilon_{kij}\boldsymbol{e}_k \tag{1.46}$$

Now there are two certain vectors

$$\boldsymbol{a} = a_i\boldsymbol{e}_i \tag{1.47}$$

$$\boldsymbol{b} = b_i\boldsymbol{e}_i \tag{1.48}$$

Since the results of the vector cross product are in different directions, they can be obtained

$$\boldsymbol{a} \times \boldsymbol{b} = (a_i\boldsymbol{e}_i) \times \left(b_j\boldsymbol{e}_j\right) = a_ib_j\left(\boldsymbol{e}_i \times \boldsymbol{e}_j\right) = a_ib_j\varepsilon_{ijk}\boldsymbol{e}_k \tag{1.49}$$

1.3 Basic Operations of Tensors

1.3.1 Vector Operations

Vectors are physical quantities characterized by both magnitude and direction, and are commonly used to represent displacement, velocity, force, and other directional phenomena in physics. Geometrically, vectors are typically illustrated as directed line segments (arrows), where the length indicates magnitude and the orientation indicates direction.

Mathematically, vectors are elements of a vector space and can be manipulated through well-defined operations, including vector addition, scalar multiplication, dot product (also known as inner product), and cross product. These operations obey specific algebraic rules and are foundational to the formulation of physical laws and mathematical models in mechanics and engineering.

Vector Addition: Add the length and direction of the two vectors to get a new vector. For example:

$$\overrightarrow{AB} + \overrightarrow{BC} = \overrightarrow{AC} \tag{1.50}$$

where $\overrightarrow{AB}$, $\overrightarrow{BC}$, $\overrightarrow{AC}$ are all vectors, representing the displacement from point A to point B, from point B to point C, and from point A to point C, respectively. Add $\overrightarrow{AB}$ and $\overrightarrow{BC}$ to give $\overrightarrow{AC}$. This process can be represented by the law of triangles or the rule of horizontal quadrilaterals.

Vector Subtraction: Reverse the length and direction of one vector to the length and direction of another, and then do the addition operation. For example:

$$\overrightarrow{AB} - \overrightarrow{AC} = \overrightarrow{CB} \tag{1.51}$$

where $\overrightarrow{AB}$, $\overrightarrow{AC}$, $\overrightarrow{CB}$ represent the displacement from point A to point B, from point A to point C, and from point C to point B, respectively. Add $\overrightarrow{AB}$ and $\left(-\overrightarrow{AC}\right)$ to get $\overrightarrow{CB}$.

Quantity Product: Also known as dot product, it is used to calculate the angle between two vectors and the projection between them. For example:

$$\boldsymbol{a} \cdot \boldsymbol{b} = |\boldsymbol{a}|\,|\boldsymbol{b}| \cos\theta = a_x b_x + a_y b_y \tag{1.52}$$

where $\boldsymbol{a}$ and $\boldsymbol{b}$ are vectors, $|\boldsymbol{a}|$ and $|\boldsymbol{b}|$ are the lengths of $\boldsymbol{a}$ and $\boldsymbol{b}$, respectively. θ denotes the angle between $\boldsymbol{a}$ and $\boldsymbol{b}$; The value of $\boldsymbol{a} \cdot \boldsymbol{b}$ is equal to the sum of the multiplication of the components of $\boldsymbol{a}$ and $\boldsymbol{b}$ in the same direction, that is, the projections of $\boldsymbol{a}$ and $\boldsymbol{b}$ are multiplied and then summed; a_i and b_i represent the projections of vectors $\boldsymbol{a}$ and $\boldsymbol{b}$ on the i axis ($i = x, y, z$) in the spatial rectangular coordinate system, respectively.

Cross Product: Also known as a vector product, the result is a vector quantity. Used to calculate the vector between two vectors perpendicular to the plane in which they are located. For example:

$$\boldsymbol{a} \times \boldsymbol{b} = |\boldsymbol{a}|\,|\boldsymbol{b}| \sin\theta \vec{n} = \left(a_y b_z - a_z b_y, a_z b_x - a_x b_z, a_x b_y - a_y b_x\right) \tag{1.53}$$

where $\boldsymbol{a}$ and $\boldsymbol{b}$ are vectors, $|\boldsymbol{a}|$ and $|\boldsymbol{b}|$ are the lengths of $\boldsymbol{a}$ and $\boldsymbol{b}$, respectively. θ denotes the angle between $\boldsymbol{a}$ and $\boldsymbol{b}$. $\vec{n}$ denotes the normal vector of the plane in which they are placed. The value of $\boldsymbol{a} \times \boldsymbol{b}$ is equal to the product of the area on the plane where $\boldsymbol{a}$ and $\boldsymbol{b}$ are located and the product of $\vec{n}$.

Mixed Product: The mixed product $(\boldsymbol{a}, \boldsymbol{b}, \boldsymbol{c})$ of the vector $\boldsymbol{a}$, $\boldsymbol{b}$ and $\boldsymbol{c}$ is a scalar quantity.

$$(\boldsymbol{a}, \boldsymbol{b}, \boldsymbol{c}) = \boldsymbol{a} \cdot (\boldsymbol{b} \times \boldsymbol{c}) = \begin{vmatrix} a_x & a_y & a_z \\ b_x & b_y & b_z \\ c_x & c_y & c_z \end{vmatrix} \tag{1.54}$$

The vector positions of the mixed product $(\boldsymbol{a}, \boldsymbol{b}, \boldsymbol{c})$ can be rotated:

$$\boldsymbol{a} \cdot (\boldsymbol{b} \times \boldsymbol{c}) = \boldsymbol{b} \cdot (\boldsymbol{c} \times \boldsymbol{a}) = \boldsymbol{c} \cdot (\boldsymbol{a} \times \boldsymbol{b}) \tag{1.55}$$

The mixed product $(\boldsymbol{a}, \boldsymbol{b}, \boldsymbol{c})$ represents the volume of a parallelepiped composed of $\boldsymbol{a}$, $\boldsymbol{b}$ and $\boldsymbol{c}$.

1.3.2 Operations of Tensors

1. **The Addition of Tensors**
   When two tensors of the same order are added, their corresponding components are added, and the basis vector remains unchanged, and the tensor of the same order is obtained:

$$\boldsymbol{A} + \boldsymbol{B} = \boldsymbol{T} \tag{1.56}$$

For second-order tensors and fourth-order tensors:

$$A_{ij} + B_{ij} = T_{ij}$$
$$A_{ijkl} + B_{ijkl} = T_{ijkl} \tag{1.57}$$

2. **The Juxtaposition of Tensors**
   Definition: The juxtaposition of two tensors $\boldsymbol{A}$ and $\boldsymbol{B}$ is the juxtaposition of the basis vectors of the two tensors, and their corresponding components are multiplied to obtain a higher-order tensor whose order is equal to the sum of the orders $\boldsymbol{A}$ and $\boldsymbol{B}$
   For example, the m-order tensor $\boldsymbol{A}$ is the juxtaposition of the nth-order tensor

$$\boldsymbol{AB} = \boldsymbol{T} \tag{1.58}$$

where $\boldsymbol{T}$ is a tensor of order $m + n$.
Nature 1: If $\boldsymbol{A}$ and $\boldsymbol{B}$ are second-order tensors, then $\boldsymbol{T}$ is a fourth-order tensor, i.e. there is

$$A_{ij}B_{kl}\boldsymbol{e_i}\boldsymbol{e_j}\boldsymbol{e_k}\boldsymbol{e_l} = T_{ijkl}\boldsymbol{e_i}\boldsymbol{e_j}\boldsymbol{e_k}\boldsymbol{e_l} \tag{1.59}$$
$$\Rightarrow A_{ij}B_{kl} = T_{ijkl}$$

Nature 2: The scalar k is multiplied by the second-order tensor $\boldsymbol{B}$, and the resulting tensor $\boldsymbol{T}$ is still a second-order tensor:

$$k\mathbf{B} = \boldsymbol{T} \tag{1.60}$$
$$\Rightarrow kB_{ij} = T_{ij}$$

3. **The Dot Product of Tensors**
   Definition: The dot product of two tensors is to multiply two tensors together, and then dot product the arbitrarily specified base vectors of the two tensors. If not specified in advance, the dot product of the two basis vectors closest to the two tensors. At the same time, the tensor basis vector after the dot product is reduced by two.
   For example, the dot product of a m-order tensor $\boldsymbol{A}$ and n-order tensor $\boldsymbol{B}$:

$$\boldsymbol{A} \times \boldsymbol{B} = \boldsymbol{T} \tag{1.61}$$

where $\boldsymbol{T}$ is a tensor of order $m + n - 2$. Let $\boldsymbol{A}$ and $\boldsymbol{B}$ be second-order tensors, then $\boldsymbol{A} \cdot \boldsymbol{B} = \boldsymbol{T}$ is also a second-order tensor:

$$\boldsymbol{A} \cdot \boldsymbol{B} = \left(A_{ij}\boldsymbol{e_i}\boldsymbol{e_j}\right) \cdot (B_{kl}\boldsymbol{e_k}\boldsymbol{e_l}) = A_{ij}\delta_{jk}B_{kl}\boldsymbol{e_i}\boldsymbol{e_l} = A_{ik}B_{kl}\boldsymbol{e_i}\boldsymbol{e_l} = T_{il}\boldsymbol{e_i}\boldsymbol{e_l} = \boldsymbol{T}$$

namely:

$$A_{ik}B_{kl} = T_{il} \tag{1.62}$$

The double dot product is the product of two tensors after multiplying them together, and then dotting their base vectors twice in order. For example, if $\boldsymbol{A}$ and $\boldsymbol{B}$ are second-order tensors, then $\boldsymbol{AB} = k$ is a zero-order tensor (i.e. a scalar):

$$\boldsymbol{A} : \boldsymbol{B} = \left(A_{ij}e_ie_j\right) : (B_{kl}e_ke_l) = A_{ij}\delta_{ik}\delta_{jl}B_{kl} = A_{kj}B_{kj} = k \tag{1.63}$$

The dot product of the vector $\boldsymbol{U}$ and $\boldsymbol{V}$ is

$$\boldsymbol{U} \cdot \boldsymbol{V} = (U_ie_i) \cdot \left(V_je_j\right) = U_iV_i = k \tag{1.64}$$

Geometric meaning of scalar k: $k = |\boldsymbol{U}||\boldsymbol{V}| \cos\theta$. where θ is the angle between the vector $\boldsymbol{U}$ and $\boldsymbol{V}$, $|\boldsymbol{U}|$and $|\boldsymbol{V}|$, respectively, are the absolute values of the two vectors. When $\boldsymbol{V}$ is the unit vector, k represents the projection of the vector $\boldsymbol{U}$ in the $\boldsymbol{V}$ direction; When $\boldsymbol{U}$ and $\boldsymbol{V}$ are nonzero vectors, and $\boldsymbol{U} \cdot \boldsymbol{V} = 0$, then $\boldsymbol{U}$ is orthogonal to $\boldsymbol{V}$ and vice versa.

4. **Tensor Condensation**
   In the union vector notation of tensors, if a dot product is performed on one of the two basis vectors, the original tensor will be reduced by two orders, and this process is called tensor condensation, such as:

$$\boldsymbol{a} \cdot \boldsymbol{bcd} = (\boldsymbol{a} \cdot \boldsymbol{b})\boldsymbol{cd} \tag{1.65}$$

$$\boldsymbol{ab} \cdot \boldsymbol{cd} = \boldsymbol{a}(\boldsymbol{b} \cdot \boldsymbol{c})\boldsymbol{d} \tag{1.66}$$

$$\boldsymbol{abc} \cdot \boldsymbol{d} = \boldsymbol{ab}(\boldsymbol{c} \cdot \boldsymbol{d}) \tag{1.67}$$

   and other forms. The dot product in the right parentheses of the above is a number, so they are reduced to second-order parallel vectors. The second-order union is condensed and merged into a number, and the number can be regarded as the zero-order union.

5. **The Quotient of Tensors**
   If a physical quantity $\boldsymbol{T}$ is dotted with any tensor $\boldsymbol{B}$ to give a tensor $\boldsymbol{A}$, then the physical quantity $\boldsymbol{T}$ must be a tensor, and the tensor $\boldsymbol{T}$ is called the quotient of tensor $\boldsymbol{A}$ relative to tensor $\boldsymbol{B}$. For example, if the dot product of a physical quantity $\boldsymbol{T}$ and an arbitrary vector $\boldsymbol{V}$ gives the n-1 tensor $\boldsymbol{A}$, then the physical quantity $\boldsymbol{T}$ must be an n-order tensor.
   In the same way, if the product of a physical quantity $\boldsymbol{T}$ and any second-order tensor $\boldsymbol{B}$ gives the n-second order tensor $\boldsymbol{A}$, then the physical quantity $\boldsymbol{T}$ must be an nth-order tensor

6. **Transpose of Tensors**
   Definition: The transpose $\boldsymbol{T}^T$ of the second-order tensor $\boldsymbol{T} = T_{ij}\boldsymbol{e}_i\boldsymbol{e}_j$ is defined as:

$$\boldsymbol{T}^T = T_{ij}\boldsymbol{e}_i\boldsymbol{e}_j \tag{1.68}$$

   Nature 1:

$$\left(\boldsymbol{T}^T\right)^T = \boldsymbol{T} \tag{1.69}$$

   Nature 2: if $\boldsymbol{T}^T = \boldsymbol{T}$, or $T_{ij} = T_{ji}$, then $\boldsymbol{T}$ is said to be a symmetric tensor.

Nature 3: Let $\boldsymbol{A}$ and $\boldsymbol{B}$ be second-order tensors, and $\boldsymbol{U}$ be vectors

$$(\boldsymbol{A}\cdot\boldsymbol{B})^T = \boldsymbol{B}^T\cdot\boldsymbol{A}^T \tag{1.70}$$

$$\boldsymbol{A}\cdot\boldsymbol{U} = \boldsymbol{U}\cdot\boldsymbol{A}^T \tag{1.71}$$

1.3.3 Isotropic Tensors

Definition: If each component of a quantity remains unchanged in different coordinate systems, the tensor is said to be isotropic tensor.

Nature 1: The scalar is a zero-order tensor, which has nothing to do with the choice of coordinates, so any scalar is an isotropic tensor. The product of any scalar k and the unit tensor δ' is an isotropic tensor.

The second-order tensor $\delta = \delta_{ij}\boldsymbol{e}_i\boldsymbol{e}_j$ with δ as the coordinate component is called the unit tensor. In different coordinate systems, $\delta' = \delta'_{\boldsymbol{ij}}\boldsymbol{e}'_i\boldsymbol{e}'_j$, and $\delta'_{ij} = \boldsymbol{e}'_{\boldsymbol{i}}\cdot\boldsymbol{e}'_{\boldsymbol{j}} = \delta_{ij}$. Therefore, the product of the arbitrary scalar k and the unit tensor δ is an isotropic tensor.

Nature 2: Vectors are first-order tensors with no isotropic tensors.

Nature 3: A second-order isotropic tensor must be expressed as a form of $k\delta$ or $k\delta_{ij}$.

Proof that if T is a second-order isotropic tensor, then there is

$$T_{ij} = \beta_{ir'}\beta_{js'}T'_{rs} = T'_{ij} \tag{1.72}$$

Rotate the coordinate system 90° around the x_3 axis, and the coordinate conversion matrix is

$$[\beta_{ij'}] = \begin{bmatrix} 0 & -1 & 0 \\ 1 & 0 & 0 \\ 0 & 0 & 1 \end{bmatrix} \tag{1.73}$$

Substitution (1.72), there is $T_{11} = T'_{22} = T'_{11}, T_{22} = T'_{11} = T'_{22}$, so $T_{11} = T_{22}$. Rotate the coordinate system 90°around the x_2 axis to give $T_{11} = T_{33}$, so that $T_{11} = T_{22} = T_{33} = k$; In the same way, if the coordinate system is rotated 180° about the x_3 axis, there is $T_{13} = -T'_{13} = T'_{13}$, so $T_{13} = 0$, and similarly, when $i \neq j, T_{ij} = 0$, i.e. $T_{ij} = k\delta_{ij}$.

Definition: The third-order tensor $\varepsilon = e_{ijl}\boldsymbol{e_i}\,\boldsymbol{e_j}\boldsymbol{e_l}$ with the permutation symbol e_{ijl} as the coordinate component is called the transformation tensor.

Nature 4: A third-order isotropic tensor must be expressed as something like $k\varepsilon$ or ke_{ijl}, where k is an arbitrary scalar.

There are three basic forms of fourth-order isotropic tensors:

$$\boldsymbol{I}^{(0)} = \delta_{ij}\delta_{kl}\boldsymbol{e_i}\boldsymbol{e_j}\boldsymbol{e_k}\boldsymbol{e_l},\ \boldsymbol{I}^{(1)} = \delta_{ik}\delta_{jl}\boldsymbol{e_i}\boldsymbol{e_j}\boldsymbol{e_k}\boldsymbol{e_l},\ \boldsymbol{I}^{(2)} = \delta_{il}\delta_{jk}\boldsymbol{e_i}\boldsymbol{e_j}\boldsymbol{e_k}\boldsymbol{e_l} \tag{1.74}$$

The fourth-order isotropic tensor $\boldsymbol{I}^{(i)}$ must be expressed in the form of $a\boldsymbol{I}^{(0)} + b\boldsymbol{I}^{(1)} + c\boldsymbol{I}^{(2)}$, or

$$I^{(i)}_{ijkl} = a\delta_{ij}\delta_{kl} + b\delta_{ik}\delta_{jl} + c\delta_{il}\delta_{jk} \tag{1.75}$$

where a, b, and c are arbitrary scalars.

1.3.4 Tensor Functions and Calculus Operations

1. **A Function of Tensors**

In algebra, matrices can serve as elements of certain functions. For example, the k-power of the nth-order square, i.e. the matrix itself multiplied by $k-1$ times, is still an nth-order square:

$$[T]^2 = [T]\,[T] \tag{1.76}$$

$$[T]^k = \underbrace{[T]\,[T]\cdots[T]}_{k\times[T]} \tag{1.77}$$

Constructing the kth polynomial of the matrix $\boldsymbol{T}$, which is still an n-order square matrix, denoted $\boldsymbol{H}$, defines the matrix $\boldsymbol{H}$ as a function of the matrix $\boldsymbol{T}$:

$$[\boldsymbol{H}] = \boldsymbol{f}([\boldsymbol{T}]) = c_0\,[1] + c_1\,[\boldsymbol{T}] + c_2\,[\boldsymbol{T}]^2 + \cdots + c_k\,[\boldsymbol{T}]^k \tag{1.78}$$

It is also defined that the second-order tensor $\boldsymbol{H}$ is a function of the second-order tensor $\boldsymbol{T}$ as an independent variable:

$$\boldsymbol{H} = \boldsymbol{f}(\boldsymbol{T}) = c_0\boldsymbol{G} + c_1\boldsymbol{T} + c_2\boldsymbol{T}^2 + \cdots + c_k\boldsymbol{T}^k \tag{1.79}$$

Equation (1.79) actually reflects the functional relationship between the elements of the matrix $\boldsymbol{H}$ and the matrix $\boldsymbol{T}$, and the component of this equation:

$$H_{ij} = c_0\delta_{ij} + c_1T_{ij} + c_2T_{il}T_{li} + \cdots + c_kT_{il_1}T_{l_1l_2}\cdots T_{l_{k-1}j} \tag{1.80}$$

The functions listed above are polynomials. In general, if a tensor $\boldsymbol{H}$ (scalar, vector, tensor) depends on n tensors $\boldsymbol{T}_1, \boldsymbol{T}_2, \ldots, \boldsymbol{T}_n$ (vector, tensor), that is, when $\boldsymbol{T}_1, \boldsymbol{T}_2, \ldots, \boldsymbol{T}_n$ are given, $\boldsymbol{H}$ can be determined correspondingly, then $\boldsymbol{H}$ is said to be the tensor function of tensors $\boldsymbol{T}_1, \boldsymbol{T}_2, \ldots, \boldsymbol{T}_n$, denoted as $\boldsymbol{H}$ is the tensor function of tensors $\boldsymbol{T}_1, \boldsymbol{T}_2, \ldots, \boldsymbol{T}_n$.

2. **Differentiation and Derivatives of Tensors**

The following research methods are applicable to tensor field functions of any order (n order) in three-dimensional space, namely:

$$\boldsymbol{T} = \boldsymbol{T}(\boldsymbol{r}) \tag{1.81}$$

Its parallel vector expression is

$$\boldsymbol{T}(\boldsymbol{r}) = T_{ij\ldots kl}\boldsymbol{e}_i\boldsymbol{e}_j \ldots \boldsymbol{e}_k\boldsymbol{e}_l \tag{1.82}$$

where both the component and the base vector are functions of the vector diameter $\boldsymbol{r}$. In order to study its variation with a point, it is necessary to study the derivative of the nth tensor $\boldsymbol{T}$ to the sagittal $\boldsymbol{r}$. In the following studies, it is assumed that the partial derivative of the field function $\boldsymbol{T}$ to the coordinate x^l exists and is continuous.

Definition: The tensor field function $\boldsymbol{T}(\boldsymbol{r})$ is a finite differential of the delta $\boldsymbol{u}$ of $\boldsymbol{r}$

$$\boldsymbol{T}'(\boldsymbol{r};\boldsymbol{u}) = \lim_{h\to 0}\frac{1}{h}\,[\boldsymbol{T}(\boldsymbol{r}+h\boldsymbol{u}) - \boldsymbol{T}(\boldsymbol{r})] \tag{1.83}$$

It can be proved that the finite differential is linear for the increment $\boldsymbol{u}$ of the sagittal diameter, so it is denoted as:

$$\boldsymbol{T}'(\boldsymbol{r};\boldsymbol{u}) = \boldsymbol{T}'(\boldsymbol{r})\cdot\boldsymbol{u} \tag{1.84}$$

where $T'(r) = dT/dr$ is called the derivative of the nth-order tensor field function $T(r)$ to the sagittal r, which is $n + 1$ tensor according to the quotient of tensors.

It should be noted that for the tensor field function described in this chapter, the independent variable is the sagittal diameter r, and r is a function of the curvilinear coordinate x^l (1 = 1, 2,3):

$$r = r(x^l) \tag{1.85}$$

Thus, the tensor field function (the component of the tensor and the base vector) is also a function of coordinates, but the coordinate x^l is not a component of the vector r.

In any curve coordinate system r, the delta u of the vector diameter r at that point can be decomposed into

$$u = u_i e_i \tag{1.86}$$

In order to further obtain the components of the derivative $T(r)$ of the field function to the vector diameter in the coordinate system, the finite differentiation of the field function $T(r)$ when the increment is the basis vector e_i is first studied:

$$T'(r; u) = \lim_{h\to 0} \frac{1}{h} [T(r + hu) - T(r)] \tag{1.87}$$

As shown in Figure 1.2, the sagittal diameter r has an incremental hg_i, which is equivalent to the point where the field function is defined is moved from P' (sagittal r) to the point P' (sagittal r') on the tangent of the coordinate line x^i:

$$r' = r + hg_i \tag{1.88}$$

Its corresponding field function $T(r)$ actually becomes a composite function of the coordinates r^l $(l = 1, 2, 3)$ by Eq. (1.85):

$$T = T\left(r\left(x^l\right)\right) \tag{1.89}$$

Analogy with the definition of partial derivatives of multivariate functions, there are

$$\frac{\partial T}{\partial x^i} = \lim_{h\to 0} \frac{1}{h} \left[T\left(r\left(x^l + h\delta_i^l\right) - T\left(r\left(x^l\right)\right)\right)\right] \ (i = 1, 2, 3) \tag{1.90}$$

In Eq. (1.90), l is neither a free nor a dumb indicator, but only indicates that T is a function of x_1, x_2, x_3, i is the indicator of the determined coordinate for which the partial derivative is obtained, and $i = 1, 2$, and 3 correspond to three formulas. The change of the sagittal diameter $r\left(x^l\right)$ to $r'' = r\left(x^l h\delta_i^l\right)$, is equivalent to the point of the defined function in Figure 1.2 moving from P along the coordinate line x^i to the point P''. Comparing

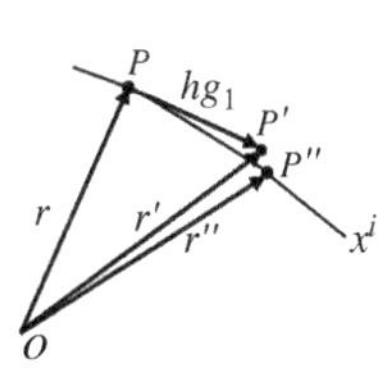

Figure 1.2 The change in the vector diameter in the tensor field function definition domain.

Eq. (1.87) with Eq. (1.90), it is noted that since the sagittal diameter at the P'' point can be expanded into a Taylor series:

$$\boldsymbol{r}\left(\boldsymbol{x}^l + \boldsymbol{h}\delta_i^l\right) = \boldsymbol{r}(\boldsymbol{x}^l) + \frac{\partial \boldsymbol{r}}{\partial \boldsymbol{x}^i}\boldsymbol{h} + \mathbf{o}(\boldsymbol{h}) = \boldsymbol{r}(\boldsymbol{x}^l) + \boldsymbol{h}\boldsymbol{g}_i + \mathbf{o}(\boldsymbol{h}) \tag{1.91}$$

Therefore the difference between the sagittal diameter $\boldsymbol{r}$ and $\boldsymbol{r}''$ is only a higher-order infinitesimal quantity, and therefore the difference between the values of the function at these two points is also a higher-order infinitesimal quantity, then the right-end limit of Eq. (1.87) and Eq. (1.90) must be equal, i.e.:

$$\boldsymbol{T}'(\boldsymbol{r}; \boldsymbol{g}_i) = \frac{\partial \boldsymbol{T}}{\partial \boldsymbol{x}^i} \tag{1.92}$$

According to the finite differentiation of the linear nature of the increment and Eq. (1.86) and Eq. (1.92), it can be seen:

$$\boldsymbol{T}'(\boldsymbol{r}\,; \boldsymbol{u}) = \boldsymbol{T}'(\boldsymbol{r}\,; \boldsymbol{u}^i\boldsymbol{g}_i) = \boldsymbol{u}^i\frac{\partial \boldsymbol{T}}{\partial \boldsymbol{x}^i} = \left(\frac{\partial \boldsymbol{T}}{\partial \boldsymbol{x}^i}\boldsymbol{g}^i\right)\boldsymbol{u} \tag{1.93}$$

Compare Eq. (1.93) with Eq. (1.84), since the increment is arbitrarily given:

$$\boldsymbol{T}'(\boldsymbol{r}) = \frac{d\boldsymbol{T}}{d\boldsymbol{r}} = \frac{\partial \boldsymbol{T}}{\partial \boldsymbol{x}^i}\boldsymbol{g}^i \tag{1.94}$$

It can be seen from the definition of finite differentiation and derivatives:

$$\boldsymbol{T}(\boldsymbol{r} + \boldsymbol{h}\boldsymbol{u}) - \boldsymbol{T}(\boldsymbol{r}) = \boldsymbol{T}'(\boldsymbol{r}) + \boldsymbol{h}\boldsymbol{u} + \mathbf{o}(\boldsymbol{h}) \tag{1.95}$$

Let

$$d\boldsymbol{r} = d\boldsymbol{u} \tag{1.96}$$

then the main part of Eq. (1.95) is called the differentiation of the tensor field function, denoted as $d\boldsymbol{T}$, which satisfies the derivative:

$$d\boldsymbol{T} = \boldsymbol{T}(\boldsymbol{r}) \cdot d\boldsymbol{r} \tag{1.97}$$

1.3.5 Gradient, Divergence, and Curl

1. **Gradient**

1. Gradient of Scalar Functions

Let $f(\boldsymbol{r})$ be a scalar function of the sagittal diameter $\boldsymbol{r}$, then the differentiation of $f(\boldsymbol{r})$:

$$df = f(\boldsymbol{r} + d\boldsymbol{r}) - f(\boldsymbol{r}) = \frac{df}{d\boldsymbol{r}} \cdot d\boldsymbol{r} \tag{1.98}$$

According to the quotient rule of tensors, $\partial f/\partial \boldsymbol{r}$ is a vector and can be expressed as

$$\frac{\partial f}{d\boldsymbol{r}} = \frac{\partial f}{\partial\,(r_i)}\boldsymbol{e}_i \tag{1.99}$$

A gradient called the function $f(\boldsymbol{r})$ and is usually denoted by ∇f or grad f. ∇ is a vector operator

$$\nabla = \frac{\partial}{\partial \boldsymbol{r}_i}\boldsymbol{e}_i \tag{1.100}$$

Let $f_{,i} = \partial f / \partial \boldsymbol{r}$, where the subscript i indicates that the partial derivative of the coordinate component r_i is obtained, and therefore there is

$$\nabla f = f_{,i}\boldsymbol{e}_i \tag{1.101}$$
$$df = \nabla f \cdot d\boldsymbol{x} = (f_{,i}\boldsymbol{e}_i) \cdot (dx_j\boldsymbol{e}_j) = f_{,i}dx_i \tag{1.102}$$

The geometric meaning of the gradient ∇f of the scalar function $f(\boldsymbol{r})$: $f(\boldsymbol{r}) = C$, which represents a surface in three-dimensional space, is called an equipotential surface, $f(\boldsymbol{r})$ is called a potential function, and the constant C takes different values and represents a different equipotential surface. Considering the neighbor point $\boldsymbol{r} + d\boldsymbol{r}$ at $\boldsymbol{r}$ on the equipotential surface, it is obvious that $f(\boldsymbol{r}) = C$, $f(\boldsymbol{r} + d\boldsymbol{r}) = C$, and the two equations are subtracted:

$$df = f(\boldsymbol{r} + d\boldsymbol{r}) - f(\boldsymbol{r}) = \nabla f \cdot d\boldsymbol{r} = 0 \tag{1.103}$$

Equation (1.103) shows that ∇f is orthogonal to $d\boldsymbol{r}$, i.e. the gradient ∇f of the scalar potential function $f(\boldsymbol{r})$, whose direction is orthogonal to the section of the equipotential surface, pointing to the direction in which the equipotential surface expands, which is the direction in which the potential function $f(\boldsymbol{r})$ changes the fastest.

Let $f(\boldsymbol{T})$ be a scalar function of the second-order tensor $\boldsymbol{T}$, then the differentiation of the function $f(\boldsymbol{T})$:

$$df = f(\boldsymbol{T} + d\boldsymbol{T}) - f(\boldsymbol{T}) = \frac{\partial f}{\partial \boldsymbol{T}} : d\boldsymbol{T} \tag{1.104}$$

From the quotient rule of tensors, $\frac{\partial f}{\partial \boldsymbol{T}}$ is a second-order tensor, which can be expressed as:

$$\frac{\partial f}{\partial \boldsymbol{T}} = \frac{\partial f}{\partial \boldsymbol{T}_{ij}}\boldsymbol{e}_i\boldsymbol{e}_j \tag{1.105}$$
$$df = \frac{\partial f}{\partial \boldsymbol{T}} : d\boldsymbol{T} = \frac{\partial f}{\partial \boldsymbol{T}_{ij}} d\boldsymbol{T}_{ij} \tag{1.106}$$

In a nine-dimensional Euclidean space consisting of the nine component tensors of the second-order tensor T_{ij}, $f(\boldsymbol{T}) = \mathrm{C}$ represents an equipotential surface, and the direction of the gradient $\partial f / \partial \boldsymbol{T}$ of the scalar potential function $f(\boldsymbol{T})$ is orthogonal to the "slice" of this equipotential surface, which is the direction in which the potential function $f(\boldsymbol{T})$ changes the fastest.

2. Gradient of Vector Functions

Let $\boldsymbol{V}(\boldsymbol{r})$ be a vector function of the vector diameter $\boldsymbol{r}$, then the differentiation of the function $\boldsymbol{V}(\boldsymbol{r})$ can be denoted as

$$d\boldsymbol{V} = \boldsymbol{V}(\boldsymbol{r} + d\boldsymbol{r}) - \boldsymbol{V}(\boldsymbol{r}) = \frac{\partial \boldsymbol{V}}{\partial \boldsymbol{r}} \cdot d\boldsymbol{r} \tag{1.107}$$

According to the quotient rule of tensors, $\partial V / \partial \boldsymbol{r}$ is a second-order tensor, called the gradient of the vector function $V(\boldsymbol{r})$, which can be represented by ∇V, and ∇ is still a vector operator:

$$\nabla \boldsymbol{V} = \frac{\partial \boldsymbol{V}}{\partial \boldsymbol{r}} = \frac{\partial \boldsymbol{V}}{\partial \boldsymbol{r}_j}\boldsymbol{e}_j = \frac{\partial V_i}{\partial \boldsymbol{r}_j}\boldsymbol{e}_i\boldsymbol{e}_j = V_{ij}\boldsymbol{e}_i\boldsymbol{e}_j \tag{1.108}$$

$$dV = \nabla V \cdot dx = V_{ij} x_j e_i \tag{1.109}$$

3. Gradient of Tensor Function
   Defined by the gradient of a vector function, it can be generalized to a gradient of a tensor function. Let $\boldsymbol{T}(\boldsymbol{r})$ be the nth-order tensor function of the vector $\boldsymbol{r}$, then the gradient ∇T of $\boldsymbol{T}(\boldsymbol{r})$ is an $n+1$ tensor

$$\nabla \boldsymbol{T} = \frac{\partial \boldsymbol{T}}{\partial \boldsymbol{r}} = \frac{\partial \boldsymbol{T}}{\partial \boldsymbol{r}_j} \boldsymbol{e}_j \tag{1.110}$$

Differentiation of the function $\boldsymbol{T}(\boldsymbol{r})$:

$$d\boldsymbol{T} = \boldsymbol{T}(r + d\boldsymbol{r}) - \boldsymbol{T}(\boldsymbol{r}) = \frac{\partial \boldsymbol{T}}{\partial \boldsymbol{r}} \cdot d\boldsymbol{r} \tag{1.111}$$

2. **Divergence**
   For example, the divergence of the gradient ∇f of the scalar function $f(\boldsymbol{x})$:

$$\nabla \cdot \nabla f = \nabla^2 f = \frac{\partial \nabla f}{\partial \boldsymbol{r}_j} \cdot \boldsymbol{e}_j = f_{ij} \boldsymbol{e}_i \cdot \boldsymbol{e}_j = f_{ij} \tag{1.112}$$

Divergence of vector $\boldsymbol{V}$:

$$\nabla \cdot \boldsymbol{V} = \frac{\partial \boldsymbol{V}}{\partial \boldsymbol{r}_i} \cdot \boldsymbol{e}_j = \frac{\partial V_i}{\partial \boldsymbol{r}_i} \boldsymbol{e}_i \cdot \boldsymbol{e}_j = V_{i,j}. \tag{1.113}$$

The divergence of the tensor $\boldsymbol{T}$ is defined as $\nabla \cdot \boldsymbol{T}$:

$$\nabla \cdot \boldsymbol{T} = \frac{\partial \boldsymbol{T}}{\partial \boldsymbol{r}_j} \cdot \boldsymbol{e}_j \tag{1.114}$$

3. **Curl**
   For example, the curl of the gradient ∇f of the scalar function: $f(\boldsymbol{r})$:

$$\nabla \times \nabla f = \frac{\partial \nabla f}{\partial \boldsymbol{r}_j} \times \boldsymbol{e}_j = f_{ij} \boldsymbol{e}_i \times \boldsymbol{e}_j = f_{ij} \boldsymbol{e}_{ijk} \boldsymbol{e}_k = 0 \tag{1.115}$$

Curl of vector $\boldsymbol{V}$:

$$\nabla \times \boldsymbol{V} = \frac{\partial \boldsymbol{V}}{\partial \boldsymbol{r}_j} \boldsymbol{e}_i \times \boldsymbol{e}_j = \boldsymbol{V}_{ij} \boldsymbol{e}_{ijk} \boldsymbol{e}_k \tag{1.116}$$

The curl of the tensor $\boldsymbol{T}$ is defined as $\nabla \times \boldsymbol{T}$:

$$\nabla \times \boldsymbol{T} = \frac{\partial \boldsymbol{T}}{\partial \boldsymbol{r}_i} \times \boldsymbol{e}_j \tag{1.117}$$

1.3.6 Green's Theorem and Stokes' Theorem

1. **Green's Theorem**
   The Green transformation formula gives the transformation formula of the volume integral of the tensor function and the area integral of the closed domain. If an nth-order tensor field function T (n is any positive integer or zero) is defined on the volume

domain V in three-dimensional space, and if dv is the micro-unit volume on the volume domain and dS is the micro-unit surface area vector, then there is:

$$\int_V \nabla\cdot\boldsymbol{T}d\boldsymbol{V} = \int_V \boldsymbol{T}\cdot\boldsymbol{n}dS \tag{1.118}$$

where the unit vector $\boldsymbol{n}$ represents the outer normal of dS. For example, if T is a second-order tensor, then there is:

$$\int_V T_{ij,j}e_i dV = \int_V T_{ij}n_j e_i dS \quad \text{or} \quad \int_V T_{ij,j}n_j dS = \int_V T_{ij}n_j dS \tag{1.119}$$

2. **Stokes' Theorem**
As shown in Figure 1.3, if the closed boundary of the open surface a is l, and the direction of l is right-handed with the direction n of the outer normal of da, then there is

$$\int_a d\boldsymbol{a}\cdot(\nabla\times\boldsymbol{v}) = \int_l ds\cdot\boldsymbol{v} \tag{1.120}$$

This is the Stokes formula. Generalizes $\boldsymbol{v}$ to tensors of any order φ:

$$\begin{cases} \int_a d\boldsymbol{a}\cdot(\nabla\times\varphi) = \int_l d\boldsymbol{s}\cdot\varphi \\ \int_a (\varphi\times\nabla)\cdot d\boldsymbol{a} = -\int_l \varphi\cdot d\boldsymbol{s} \end{cases} \tag{1.121}$$

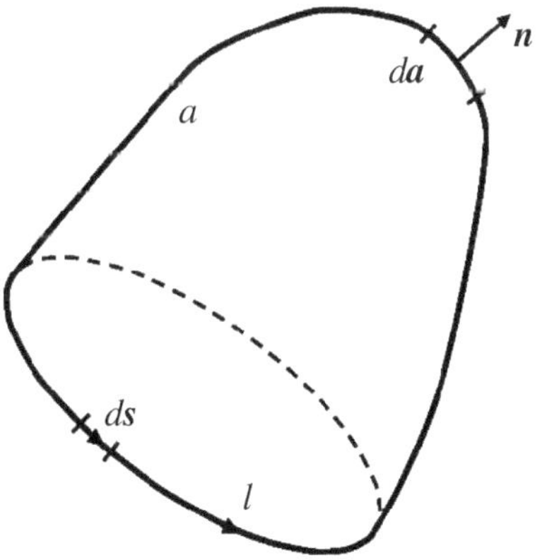

Figure 1.3 Open surface a and closed curve l.

1.4 Application of Tensors in Mechanics

Tensors, serving as both mathematical constructs and representations of physical quantities, play a foundational role in continuum mechanics. They provide the essential framework for expressing and interpreting physical laws in a manner that is independent of coordinate systems, thereby ensuring objectivity and consistency across different frames of reference. Within theoretical domains such as elasticity, plasticity, damage mechanics, and micromechanics, tensors are not merely symbolic tools; they are indispensable in the formulation of constitutive laws, the characterization of material anisotropy, and the description of mechanical and structural evolution.

In the theory of elasticity, second-order tensors are used to describe the internal state of stress and strain in materials, and to establish constitutive relations through stiffness or compliance tensors. In plasticity theory, it is critical to distinguish among the total strain tensor, the elastic strain tensor, and the plastic strain tensor, which together define the incremental response of materials. Yield criteria are typically constructed using tensor invariants, and consistent tangent operators—expressed as fourth-order tensors—are employed in the numerical integration of elastoplastic constitutive models.

In damage mechanics, tensor-valued internal variables are introduced to describe the degradation of material stiffness. These damage tensors modify the stress-strain relationship, leading to the concept of effective stress, which underpins the formulation of damage-coupled constitutive laws.

At the microscale, tensors are also used to characterize crystal orientation, lattice distortion, and slip systems in polycrystalline materials. In micromechanics, the deformation gradient tensor serves as a key mathematical tool for establishing multiscale relations between microstructural behavior and macroscopic responses.

Table 1.1 summarizes representative tensors frequently used in mechanics and indicates their respective tensor orders, reflecting their structural and physical roles across various scales and formulations.

Table 1.1 Some commonly used tensors in mechanics.

Tensor type	Order	Application
Stress/strain tensor	2	Continuum mechanics, structural mechanics
Inertia tensor	2	Rigid body dynamics, rotor mechanics
Flexibility/stiffness tensor	4	Constitutive relations, material modeling
Damage tensor	2	Fracture mechanics, material degradation, fatigue
Deformation gradient tensor	2	Large deformation theory, nonlinear mechanics

1.5 Fundamental Laws of Continuum Mechanics

The subject of continuum mechanics is the mechanical behavior of continuous media—systems or materials that satisfy the laws of Newtonian mechanics in three-dimensional Euclidean space under the assumption of continuous time. Specifically, it addresses the macroscopic mechanical properties of matter treated as a continuum. As a discipline, continuum mechanics is broadly categorized into solid mechanics, fluid mechanics, and rheology. Solid mechanics can further be subdivided into elastoplastic mechanics and viscoelastoplastic mechanics based on material response characteristics.

Rational mechanics, also known as mathematical mechanics, provides the theoretical foundation for continuum mechanics. It emphasizes the use of rigorous logical reasoning and abstract mathematical formulations to study mechanical systems in a unified and systematic way. D'Alembert, in his 1743 treatise *Dynamics*, asserted that rational mechanics, like geometry, must be founded upon axioms that are self-evident and universally valid [11]. A core objective of rational mechanics is to establish a set of axioms and general principles that are applicable to all types of continua.

In the context of scientific methodology, axioms are fundamental truths distilled from extensive empirical experience. They are considered self-evident, irreducible, and beyond the need for proof. Axioms serve as starting points for deriving more complex theories. Euclid's seminal work *Elements* exemplifies the axiomatic approach: all its theorems are logically deduced from a finite set of axioms. This method has since become a hallmark of rigorous scientific reasoning and has influenced the development of other knowledge systems.

The deformation and motion of a continuum must obey fundamental physical laws, which are commonly expressed in the form of conservation equations. During deformation, changes in physical quantities arise from internal sources and external influences. Typically, internal sources are represented by volume integrals, while external influences are modeled via surface integrals. The fundamental conservation laws of continuum mechanics include:

- **a.** Conservation of mass,
- **b.** Conservation of linear momentum,
- **c.** Conservation of angular momentum, and
- **d.** Conservation of energy.

When thermodynamic effects are considered, the Clausius–Duhem inequality, representing the second law of thermodynamics, must also be satisfied. Collectively, these laws constitute the field equations of continuum mechanics and form the core theoretical framework for modeling physical behavior in solids and fluids.

1.5.1 Law of Conservation of Mass

In the process of deformation and movement, the total mass of the continuum remains unchanged, and the total mass of the same continuum remains unchanged for any selected configuration. Mass is a metric that every substance possesses, and is both nonnegative and additive. If the mass is absolutely continuous with respect to spatial coordinates, there is a mass density $\rho\ (\boldsymbol{x},\ t)$. If the volume of the continuum is υ, its mass is:

$$m = \int_{\upsilon} \rho\ (\boldsymbol{x},\ t)\ \mathrm{d}\upsilon \tag{1.122}$$

The law of conservation of mass can be expressed as:

$$\int_{\upsilon} \rho\ (\boldsymbol{x},\ t)\ \mathrm{d}\upsilon = \int_{V_0} \rho_0\ (\boldsymbol{x},\ t)\ \mathrm{d}V \tag{1.123}$$

where V_0 represents the volume occupied by the continuum in the reference configuration; ρ_0 is the mass density in the reference configuration.

Since the volume change of the two configurations is equal to the value of the deformation gradient determinant, Eq. (1.123) can be expressed as:

$$\int_{V_0} (\rho_0 - \rho J)\ \mathrm{d}V = 0 \tag{1.124}$$

Equation (1.124) is the continuity equation in the form of Lagrange integral, which requires that it is valid for any volume element, then:

$$\rho_0 = \rho J \tag{1.125}$$

Since ρ_0 is a constant, both sides of Eq. (1.125) take the derivative of time at the same time, there is:

$$\frac{\mathrm{d}}{\mathrm{d}t}(\rho J) = J\frac{\mathrm{d}\rho}{\mathrm{d}t} + \rho\frac{\mathrm{d}J}{\mathrm{d}t} = J\frac{\mathrm{d}\rho}{\mathrm{d}t} + \rho J \mathrm{div}\boldsymbol{\upsilon} = 0 \tag{1.126}$$

thus obtain:

$$\frac{\mathrm{d}\rho}{\mathrm{d}t} + \rho \mathrm{div}\boldsymbol{\upsilon} = 0 \tag{1.127}$$

Equation (1.127) is the differential form of the continuity equation, also known as the Euler continuity equation.

1.5.2 Law of Conservation of Momentum

At a certain time t, the volume of the continuum is υ, and the outer surface area is s. The physical force of the continuum is $\boldsymbol{b}$, the surface force is $\boldsymbol{t}_n$, $\boldsymbol{n}$, which is the unit vector of the microelement's out-of-plane normal, and there is a velocity field $\boldsymbol{\upsilon}$ in the region. According to the definition of momentum, there is

$$\boldsymbol{M} = \int_{\upsilon} \rho\boldsymbol{\upsilon}\mathrm{d}\upsilon \tag{1.128}$$

If the resultant force acting on the volume υ is recorded as $\boldsymbol{F}$, it can be obtained according to the force acting on the continuum.

$$\boldsymbol{F} = \oint_{s} \boldsymbol{t}_n \mathrm{d}s + \int_{\upsilon} \rho\boldsymbol{b}\mathrm{d}\upsilon \tag{1.129}$$

Since the rate of change of momentum with respect to time is equal to the net force acting on the object, we can get from Eq. (1.128)

$$\frac{\mathrm{d}\boldsymbol{M}}{\mathrm{d}t} = \boldsymbol{F} = \frac{\mathrm{d}}{\mathrm{d}t}\int_{v} \rho\boldsymbol{v}\mathrm{d}v = \oint_{s}\boldsymbol{t}_{n}\mathrm{d}s + \int_{v} \rho\boldsymbol{b}\mathrm{d}v \tag{1.130}$$

Combining Eq. (1.122), there is

$$\frac{\mathrm{d}}{\mathrm{d}t}\int_{v} \rho\boldsymbol{v}\mathrm{d}v = \int_{v} \rho\frac{\mathrm{d}\boldsymbol{v}}{\mathrm{d}t}\mathrm{d}v = \oint_{s}\boldsymbol{\sigma}\boldsymbol{n}\mathrm{d}s + \int_{v} \rho\boldsymbol{b}\mathrm{d}v \tag{1.131}$$

According to Gauss's theorem, the area integral of Eq. (1.131) is converted into a volume integral:

$$\int_{v} \rho\frac{\mathrm{d}\boldsymbol{v}}{\mathrm{d}t}\mathrm{d}v = \int_{v} \mathrm{div}\boldsymbol{\sigma}\mathrm{d}v + \int_{v} \rho\boldsymbol{b}\mathrm{d}v \tag{1.132}$$

Eq. (1.132) is the law of conservation of momentum in integral form. Since the acceleration $\boldsymbol{a}(\boldsymbol{x}, t) = \mathrm{d}\boldsymbol{v}/\mathrm{d}t$, there is

$$\mathrm{div}\boldsymbol{\sigma} + \rho\boldsymbol{b} = \rho\boldsymbol{a} \tag{1.133}$$

where div is the divergence operator in the current configuration.

Eq. (1.133) is the Cauchy momentum equation, also known as the equation of motion.

1.5.3 Law of Conservation of Moment of Momentum

The moment of momentum, also known as angular momentum, is the moment of momentum with respect to a point O. If the volume of the element is v and $\boldsymbol{x}$ is the position vector of the element in the coordinate system, then the moment of momentum of the element with respect to the coordinate origin O is

$$\boldsymbol{L} = \int_{v} (\boldsymbol{x} \times \boldsymbol{v})\,\rho\mathrm{d}v \tag{1.134}$$

Or

$$L_i = \int_{v} e_{ijk}x_j v_k \rho\mathrm{d}v \tag{1.135}$$

The rate of change with time of the moment of momentum of any part of the continuum with respect to a point is equal to the combined moment of the forces acting on the continuum, including physical and surface forces. Therefore, the integral form of the law of conservation of moment of momentum is

$$\int_{s} e_{ijk}x_j t_k^{(n)}\mathrm{d}s + \int_{v} e_{ijk}x_j b_k \rho\mathrm{d}v = \frac{\mathrm{d}}{\mathrm{d}t}\int_{v} e_{ijk}x_j v_k \rho\mathrm{d}v \tag{1.136}$$

1.5.4 Law of Conservation of Energy

At any time t, the kinetic energy of a substance of volume v in the current configuration is

$$K = \frac{1}{2}\int_{v} \rho\boldsymbol{v}\cdot\boldsymbol{v}\mathrm{d}v \tag{1.137}$$

Take the first derivative of time on both sides of Eq. (1.137):

$$\dot{K} = \frac{1}{2}\int_{v}\frac{\mathrm{d}}{\mathrm{d}t}(\boldsymbol{v}\cdot\boldsymbol{v})\,\rho\mathrm{d}v = \int_{v}\boldsymbol{v}\cdot\rho\boldsymbol{a}\mathrm{d}v \tag{1.138}$$

By replacing the motion equation of Eq. (1.133) into Eq. (1.138) and using the formula$(\boldsymbol{v}\boldsymbol{\sigma})$ div $= \boldsymbol{v}\,(\boldsymbol{\sigma}\mathrm{div}) + (\boldsymbol{v}\otimes\nabla) : \boldsymbol{\sigma}$, Eq. (1.138) can finally be written in the following integral form:

$$\dot{K} + \int_{v}\boldsymbol{\sigma} : \boldsymbol{D}\mathrm{d}v = \oint_{s}\boldsymbol{v}\cdot\boldsymbol{t}_{(n)}\mathrm{d}s + \int_{v}\rho\boldsymbol{v}\cdot\boldsymbol{b}\mathrm{d}v \tag{1.139}$$

The second term on the left of Eq. (1.139) represents the rate of change of strain rate with respect to time, denoted as dU/dt. The right side of the equation is the power of the surface force and the physical force.

If only mechanical energy and heat energy are considered, and other energies, such as chemical energy and electromagnetic energy, are not considered, the law of conservation of energy is the first law of thermodynamics. Therefore, the first law of thermodynamics can be understood as a narrow form of conservation of energy.

If the internal energy per unit mass is denoted as u, the total internal energy of the substance

$$U = \int_{v}\rho u\mathrm{d}v \tag{1.140}$$

The total energy P of matter is the sum of kinetic energy and internal energy, that is, $P = K + U$.

According to the first law of thermodynamics, the first derivative of total energy with respect to time is equal to the power of the external force acting on the volume and the heat applied externally to the volume per unit time. So, the first derivative of the total energy with respect to time has

$$\dot{U} = \int_{v}\rho\dot{u}\mathrm{d}v = \oint_{s}\boldsymbol{v}\cdot\boldsymbol{t}_{(n)}\mathrm{d}s + \int_{v}\rho\boldsymbol{v}\cdot\boldsymbol{b}\mathrm{d}v - \int_{s}\boldsymbol{q}\cdot\boldsymbol{n}\mathrm{d}s + \int_{v}\rho\lambda\mathrm{d}v \tag{1.141}$$

where $\boldsymbol{q}$ is the heat flow vector, heat flux, per unit time per unit area; λ is the radiation heat obtained per unit time per unit mass; $\boldsymbol{n}$ is the unit external normal vector of micro-plane ds. The heat flow is specified to be positive in and negative out.

According to the kinetic energy theorem Eq. (1.139), it can be obtained that the rate of internal energy with respect to time is

$$\dot{U} = \int_{v}\rho\dot{u}\mathrm{d}v = -\int_{s}\boldsymbol{q}\cdot\boldsymbol{n}\mathrm{d}s + \int_{v}\rho\lambda\mathrm{d}v + \int_{v}\boldsymbol{\sigma} : \boldsymbol{D}\mathrm{d}v \tag{1.142}$$

Equation (1.142) is the first law of thermodynamics in integral form.

According to Gauss's theorem, converting the area integral of Eq. (1.142) into a volume integral gives the first law of thermodynamics in differential form:

$$\rho\dot{u} = -div\boldsymbol{q} + \rho\lambda + \boldsymbol{\sigma} : \boldsymbol{D} \tag{1.143}$$

Equation (1.143) is the energy conservation equation of the current configuration, also known as the local energy equation, which shows that the rate of change of internal energy with time is equal to the sum of the stress power and the heat supplied to the medium.

1.5.5 Second Law of Thermodynamics

The first law of thermodynamics establishes the conservation and conversion relationship between mechanical and thermal energy but does not account for the irreversibility observed in many real-world processes. In nature, most physical and mechanical transformations are inherently irreversible, necessitating a more restrictive principle to describe them.

The second law of thermodynamics addresses this limitation by introducing the concept of entropy and providing criteria for determining whether a process is reversible. In the context of continuum mechanics, the second law serves as a fundamental constraint to assess the thermodynamic admissibility of deformation and heat transfer processes. It can be expressed in various equivalent forms, one of which states that: it is impossible for heat to spontaneously transfer from a colder body to a hotter body without the involvement of external work or additional effects. To establish a mathematical description of this law, two state functions, thermodynamic temperature T and entropy S of the system, are used. The total entropy of the system is equal to the sum of the entropy of each part. In continuum mechanics, specific entropy, entropy of per unit mass, or entropy density s is commonly used, so total entropy $S = \int_v \rho s \mathrm{d}v$. Changes in system entropy are affected by changes in the surrounding environment and within the system

$$\mathrm{d}s = \mathrm{d}s^e + \mathrm{d}s^i \tag{1.144}$$

where $\mathrm{d}s$ is the amount of change in specific entropy; $\mathrm{d}s^e$ is the amount of change caused by external action; $\mathrm{d}s^i$ is the increment caused by internal change. $\mathrm{d}s^i$ is not negative, for reversible processes, it equals to zero; for irreversible processes, $\mathrm{d}s^i$ is positive.

According to the second law of thermodynamics, in a continuum of volume v, the rate of change of total entropy with time is no less than the sum of the entropy flowing through the surface of the continuum and the entropy generated by the internal volume sources. This principle is expressed in integral form as Clausius inequality:

$$\dot{S} = \frac{\mathrm{d}}{\mathrm{d}t}\int_v \rho s \mathrm{d}v \geq \int_v \frac{\lambda}{T}\rho \mathrm{d}v - \int_s \frac{1}{T}\boldsymbol{q}\cdot\boldsymbol{n}\mathrm{d}s \tag{1.145}$$

For reversible process, Eq. (1.145) takes equal sign; For irreversible processes, take a greater than sign.

The difference between the two ends of the inequality is entropy generation Eq. (1.145) rate Γ in the volume v, which represents the entropy generation rate per unit mass by γ, that is

$$\Gamma = \int_v \rho\gamma \mathrm{d}v, \gamma \geq 0 \tag{1.146}$$

From Eq. (1.145) and Eq. (1.146) written as the entropy balance equation in integral form:

$$\int_v \rho\dot{s}\mathrm{d}v = \int_v \frac{\lambda}{T}\rho\mathrm{d}v - \int_s \frac{1}{T}\boldsymbol{q}\cdot\boldsymbol{n}\mathrm{d}s + \int_v \rho\gamma\mathrm{d}v \tag{1.147}$$

By using Gauss theorem, the aspect integral of Eq. (1.148) is transformed into volume integral, and the entropy balance equation in differential form can be obtained as follows:

$$\rho \dot{s} = \rho \left(\frac{\lambda}{T} + \gamma \right) - \operatorname{div} \left(\frac{\boldsymbol{q}}{T} \right) \tag{1.148}$$

1.6 Constitutive Model Construction Principle

The constitutive relation in continuum mechanics refers to the fundamental relationship between stress and strain within a given material. While the foundational principles of continuum mechanics—namely the conservation of mass, momentum, and energy, along with the second law of thermodynamics (principle of entropy increase)—govern the general behavior of all continua, they are insufficient to uniquely determine the mechanical response of specific materials. Due to variations in material type, microstructure, and internal mechanisms, different substances may exhibit significantly different responses under identical external stimuli. To capture these unique behaviors, it is essential to develop constitutive equations tailored to specific materials. These equations enable the differentiation of materials within the continuum framework and provide a precise description of deformation and motion for particular substances.

The formal establishment of constitutive axioms began in the 1950s. In 1950, Oldroyd emphasized that rheological constitutive relations must possess proper invariance properties [12]. Subsequently, Noll introduced the three constitutive principles for continuum mechanics: the principle of determinacy (stress certainty), the local action principle, and the principle of objectivity (or frame-indifference) [13]. Later, researchers such as Wang and Truesdell and Eringen expanded this foundation [14, 15]. Eringen ultimately proposed eight constitutive axioms:

a. Axiom of Causality
b. Axiom of Determinacy (Certainty)
c. Axiom of Equal Existence
d. Axiom of Objectivity (Frame-Indifference)
e. Axiom of Material Invariance
f. Axiom of Domain
g. Axiom of Memory
h. Axiom of Compatibility

Each of these axioms plays a critical role in constraining and formulating constitutive models that are physically meaningful and mathematically consistent. A detailed explanation of each axiom is provided below.

1. **Causality Axiom**
   In continuum thermomechanics, the motion, temperature, and electric charge of a material point are typically treated as independent constitutive variables, while quantities such as internal energy density, entropy density, stress tensor, and heat flux vector are regarded as dependent constitutive variables. These dependent variables are determined by the evolution and interaction of the independent variables and represent the material's thermomechanical response.

When analyzing thermodynamic processes in a continuum without considering the coupling effects of mechanical deformation, electromagnetic fields, or chemical reactions, the set of independent constitutive variables is simplified to include only the motion $\mathbf{x}$ and temperature T. These variables are functions of both the material position vector $\boldsymbol{X}$ and time t, and can be expressed as:

$$\boldsymbol{x} = \boldsymbol{x}(\boldsymbol{X}, t), \quad T = T(\boldsymbol{X}, t) \tag{1.149}$$

When the independent variable is determined, other dependent variables are also determined, such as velocity vector, deformation gradient, etc., which can be obtained from the continuity equation.

2. **Axiom of Certainty**
   The thermodynamic constitutive functional and stress state of a point of matter at a given time are determined only by the motion and temperature history of all points in the matter, and have nothing to do with future motions. This axiom excludes the influence of any other factors outside the object and in the future on the properties of the material point X. As long as all the past motions and temperature history of the object are known, the present behavior of the object is completely determined. From the axiom of certainty, we know that the stress $\boldsymbol{\sigma}$ at the point X of a substance is completely determined by the entire history of the motion of the object up to the present time.
3. **Axiom of Equal Existence**
   Initially, all constitutive functionals should be expressed in terms of the same set of independent constitutive variables, and no variable should be arbitrarily excluded. The exclusion of any variable must be justified by contradictions that arise from its inclusion, either theoretically or experimentally.

   Eringen emphasized that this axiom requires treating all variables on an equal footing—no class of variable may be disregarded or rejected without sufficient justification [15]. Only when restrictions are imposed by fundamental principles—such as the axiom of objectivity, the axiom of compatibility, or material symmetry considerations—can the admissibility of certain variables be limited.
4. **Axiom of Objectivity**
   The mechanical behavior of materials must be independent of the choice of reference frame. In other words, a constitutive model should remain invariant under changes in the coordinate system or reference configuration. This requirement is formalized in the axiom of objectivity (or frame-indifference), which states that the constitutive relations must be invariant under superposed rigid body motions.

   Consequently, all physical quantities appearing in constitutive equations must be objective quantities—their values and physical meaning must not change under transformations such as rotation or translation of the reference frame. For two physically identical processes observed from different frames, the constitutive functional must remain the same.

   When formulated using tensor notation, which inherently respects transformation laws and objectivity, this axiom is naturally satisfied. Thus, the adoption of tensorial descriptions is essential for ensuring the physical consistency and frame-independence of constitutive models in continuum mechanics.

5. **Axiom of Invariance of Matter**
   The constitutive equation must faithfully reflect the material symmetry of the substance. The mechanical response of materials is significantly influenced by their microstructural characteristics, which often give rise to specific symmetry properties. For instance, the crystallographic orientation in crystalline solids can lead to anisotropic behavior, where material responses vary depending on the direction of loading.

   These inherent symmetries impose constraints on permissible forms of the constitutive relations by restricting the functional dependence of material response on deformation or motion. Accordingly, the constitutive equations must be formulated in a manner that is consistent with the symmetry group of the material.

   It is important to note that different physical properties of the same material may exhibit different symmetry characteristics. For example, a material may be mechanically isotropic—exhibiting the same mechanical response in all directions—yet electrically anisotropic due to directional variations in its conductive properties. Therefore, when constructing constitutive models for coupled physical fields (e.g. thermo-electro-mechanical systems), one must carefully account for the specific symmetry properties relevant to each physical phenomenon.
6. **Axiom of Domain**
   The motion of a material point within a body is generally influenced by the motion of surrounding points. Among these, the influence is strongest from those within its local neighborhood, often referred to as the material point's field of influence. As the distance increases, the effect of other points on the motion of the given point typically decays rapidly and becomes negligible.

   From the perspective of microscopic or molecular mechanics, the deformation behavior of a material point is ultimately governed by interatomic and intermolecular interactions. However, such interactions decay exponentially with distance, often becoming insignificant beyond approximately 10 interatomic spacings. This physical observation provides the basis for the local action axiom in continuum mechanics, which states that the stress at a material point is determined solely by the state of neighboring points within a finite, localized region, and is independent of the state of points that lie beyond this region.

   Consequently, constitutive models are commonly formulated under the assumption of locality, wherein stress is a function of local strain and its derivatives, excluding the need for long-range interactions in standard continuum formulations.
7. **Axiom of Memory**
   The influence of past constitutive variables on the present state diminishes with increasing temporal distance. This axiom, often referred to as the memory fading principle, reflects the intrinsic history-dependence of material behavior: the current state of a material (e.g. stress, strain, internal energy) is affected by its deformation history. However, the influence of past states decays over time, and values of constitutive variables from the distant past eventually become negligible.

   In essence, this axiom captures the temporal nonlocality in materials with memory (e.g. viscoelastic materials), while asserting that the weighting of historical effects decreases as one move further away from the present. Mathematically, this often leads to formulations where historical effects are incorporated through convolution integrals with decaying kernel functions, ensuring that long-past events exert progressively less influence on the current constitutive response.

8. **Consistency Axiom**
   The constitutive relation must be consistent with the fundamental framework of continuum mechanics, including the geometrical relations, equilibrium equations, and the governing laws of thermodynamics. Specifically, a valid constitutive model must satisfy all the essential conservation laws—namely, the conservation of mass, linear momentum, angular momentum, and energy—as well as the Clausius–Duhem inequality, which expresses the second law of thermodynamics in differential form.

   This requirement ensures that the constitutive formulation does not contradict the foundational physical principles governing the behavior of continuous media. Accordingly, this principle is referred to as the axiom of consistency, underscoring the necessity for constitutive equations to align with both mechanical and thermodynamic admissibility.

References

1 Huang, K., Xue, M., and Lu, M. (2003). *Tensor Analysis*. Beijing: Tsinghua University Press.
2 Huang, K. (1989). *Nonlinear Continuum Mechanics*. Beijing: Tsinghua University Press.
3 Lai, W. M., Rubin D., and Krempl E. (2010). *Introduction to Continuum Mechanics*. Burlington: Elsevier.
4 Allix, O., and Hild, F. (2002). Continuum damage mechanics of materials and structures. Amsterdam: Elsevier.
5 Zhao, Y. (2016). *Modern Continuum Mechanics*. Beijing: Science Press.
6 Wang, Z. (2000). *Foundations of Rational Mechanics*. Beijing: Science Press.
7 Shedov. J. I. I. (2007). *Continuum Mechanics*. Translated by Li Zhi. Beijing: Higher Education Press.
8 Gao, W. (1999). *Fundamentals of Solid Mechanics*. Beijing: China Railway Publishing House.
9 Li, Y. (2012). *Introduction to Tensor Principles and Modern Continuum Mechanics*. Hefei, Anhui: University of Science and Technology of China Press.
10 Mikhail, I. (2007). *Tensor Algebra and Tensor Analysis for Engineers*. Leipzig: LE-TEX. Jelonek, Schmidt & Vöckler GbR.
11 D'Alembert J. L. R. (1743). *Traité de Dynamique*. Paris: David l'aîné.
12 Oldroyd, J. G. (1950). On the formulation of rheological equations of state. *Proceedings of the Royal Society of London Series A-Mathematical and Physical Sciences*. 200 (1063): 523–541.
13 Noll, W. A. (1958). Mathematical theory of the mechanical behavior of continuous media. *Archive for Rational Mechanics and Analysis* 2 (1): 197–226.
14 Wang, C., and Truesdell, C. (1973). *Introduction to rational elasticity*. New York: Springer Science & Business Media.
15 Eringen, A. C. (2002). *Nonlocal Continuum Field Theories*. New York: Springer.

2

Fundamentals of Elasticity Theory

2.1 Fundamental Assumptions of Elasticity Theory

In problems of elasticity theory, the geometry of the object (i.e. the shape and size of its boundary), the elastic constants of the material, the body forces acting on the object, and the boundary constraints or prescribed surface forces are typically known. The unknowns to be determined include the stress components, strain components, and displacement components.

To solve for these unknowns, elasticity theory approaches the problem from three fundamental perspectives: statics, geometry, and physics. Accordingly, three corresponding sets of equations are established within the elastic domain. Specifically, the equilibrium differential equations are derived from the force balance conditions of an infinitesimal element; the geometric (kinematic) equations arise from the differential relationships between strain and displacement; and the physical (constitutive) equations describe the material specific relationship between stress and strain. Additionally, appropriate boundary conditions must be applied on the surface of the elastic body. For surfaces with specified external forces, traction boundary conditions are imposed based on the equilibrium of differential surface elements; for surfaces with prescribed constraints, displacement boundary conditions are formulated according to the compatibility between the imposed constraints and the displacement field.

Solving an elasticity problem, therefore, entails determining the stress, strain, and displacement fields that simultaneously satisfy the equilibrium equations, geometric relations, and constitutive laws under the prescribed boundary conditions.

In any scientific or engineering investigation, it is neither practical nor necessary to consider all possible influencing factors, as doing so would make the problem overly complex or analytically intractable. Instead, a systematic simplification is performed: the problem is first analyzed to identify the dominant factors, while effects deemed negligible are omitted. This process of abstraction leads to the formulation of a simplified physical model, which captures the essential behavior of the system and serves as the focus of theoretical or computational analysis. The results obtained from such a model are valid for any real system that reasonably satisfies the underlying assumptions of the model.

Constitutive Models of Solid Materials: From Mechanical Principles to Engineering Applications, First Edition.

In the context of elasticity theory, the modeling process involves reducing the complexity of real materials and structures to a set of fundamental assumptions. As a starting point, four basic assumptions are typically made regarding the material properties of the object [1].

2.1.1 Continuity Assumption

It is initially assumed that the object is a continuum; that is, the material is distributed uniformly throughout the entire volume, with no voids or discontinuities present. This assumption allows physical quantities such as stress, strain, and displacement to be represented as continuous and differentiable functions of spatial coordinates, thereby permitting the application of differential calculus to analyze their spatial variations within the body.

In reality, all materials exhibit a discrete microstructure composed of atoms or molecules, which fundamentally contradicts the continuum assumption. Nevertheless, this idealization remains a valid and widely adopted approximation in continuum mechanics, provided that the length scale of analysis significantly exceeds the characteristic microscopic dimensions of the material. However, if the characteristic length scales associated with the microstructure—such as particle size and interatomic spacing—are several orders of magnitude smaller than the dimensions of the body under consideration, the continuum assumption remains a valid and practical idealization. In such cases, the errors introduced by treating the material as continuous are negligible, and the model can still provide accurate predictions of macroscopic mechanical behavior.

2.1.2 Perfect Elasticity Assumption

The object is assumed to be perfectly elastic. Perfect elasticity is defined as the complete recovery of original geometry upon removal of externally applied deformation-causing forces, leaving no residual deformation. Under this assumption, the instantaneous deformation of the object is exclusively determined by the contemporaneous external forces, exhibiting no dependence on previous loading history.

According to material mechanics principles, ductile materials demonstrate near-perfect elastic behavior when stresses remain below the yield limit, while brittle materials exhibit quasi-perfect elasticity within the proportional limit. In classical elasticity theory, the perfect elasticity assumption further implies a proportional relationship between deformation and applied stress, establishing a fundamental linear constitutive relationship. Consequently, stress and deformation in such linearly elastic bodies conform to Hooke's law, with material-specific elastic constants that maintain invariance across all stress and strain magnitudes.

2.1.3 Homogeneity Assumption

It is assumed that the object is homogeneous; that is, the material is assumed to be homogeneous, implying uniform composition throughout the entire body. This assumption guarantees identical elastic properties in all regions, resulting in spatial invariance of elastic behavior. For composite materials (e.g. concrete), homogeneity can still be maintained if:

1. constituent material phases are sufficiently small relative to macroscopic dimensions, and
2. their distribution is statistically uniform at the scale of observation.

2.1.4 Isotropy Assumption

The material is assumed to be isotropic, meaning its elastic properties are directionally invariant. This assumption implies that all elastic constants are orientation-independent. While natural materials like wood and bamboo exhibit pronounced anisotropy, engineered materials like structural steel demonstrate effective isotropy at macroscopic scales. Although steel comprises anisotropic crystalline grains, their random spatial orientation and small size relative to macroscopic dimensions yield statistically isotropic behavior when considered as a continuum polycrystalline aggregate.

Any material satisfying these four fundamental assumptions (continuity, homogeneity, perfect elasticity, and isotropy) constitutes an ideal elastic body. Additionally, the fifth hypothesis, the small deformation assumption, is introduced as follows, which limits the kinematic description of deformation to infinitesimal strains.

2.1.5 Small Deformation Assumption

This assumption implies that, upon loading, the displacement of every point within the object is considerably smaller than the original dimensions of the object, and both strains and rotation angles are significantly smaller than unity. Therefore, when establishing the equilibrium equations after deformation, it is acceptable and convenient to use the original dimensions instead of the deformed ones without introducing significant errors. Moreover, in analyzing the relationship between strains and displacements, second-order and higher-order terms or products involving rotation angles and strains can be neglected relative to the first-order terms. For example, for a small rotation angle α:

$$\begin{aligned}
&\lim_{\alpha \to 0} \sin \alpha = \alpha - \frac{1}{3!}\alpha^3 + \cdots \approx \alpha \\
&\lim_{\alpha \to 0} \cos \alpha = 1 - \frac{1}{2!}\alpha^2 + \cdots \approx 1 \\
&\lim_{\alpha \to 0} \tan \alpha = \alpha + \frac{1}{3}\alpha^3 + \cdots \approx \alpha
\end{aligned} \tag{2.1}$$

For a small normal strain ε_x:

$$\frac{1}{1+\varepsilon} = 1 - \varepsilon_x + \varepsilon_x^2 - \varepsilon_x^3 + \cdots \approx 1 - \varepsilon_x \tag{2.2}$$

Under these assumptions, the geometric relations of elasticity reduce to linear formulations. This linearization renders the elasticity problem mathematically tractable, permitting direct application of superposition principles.

2.2 Three Fundamental Equations

To determine the unknown quantities from the known quantities in elasticity problems, it is necessary to establish relationships between these known and unknown quantities, as well as relationships among the various unknown quantities, thereby deriving a solvable set of equations. When deriving these equations, analysis can be approached from three perspectives. First, from the perspective of statics, equilibrium equations relating stresses, body forces, and surface forces are established. Second, from the perspective of geometry, geometric equations relating strains, displacements, and displacement gradients are constructed. Lastly, from the perspective of material physics, physical equations relating strains and stresses are established [1].

2.2.1 Equilibrium Equations

Consider an infinitesimal rectangular parallelepiped PACB around point P, its dimensions in the x-and y-directions are dx and dy, respectively, and its dimension in the z-direction is a unit length of "1." This infinitesimal parallelepiped is then used for stress analysis, as illustrated in the following Figure 2.1.

Generally, stress components are functions of the spatial coordinates x and y. Therefore, stresses acting on opposite faces (left-right or top-bottom) are not identical, but differ slightly. To simplify the analysis, the stress vectors acting on the inclined faces AC and BC are expanded using a Taylor series about a reference point, and only the first-order terms are retained, while higher-order contributions are omitted due to their negligible influence at infinitesimal scales. Thus, the magnitudes of the stresses shown in Figure 2.1 can be obtained directly (either explicitly or by calculating differentials).

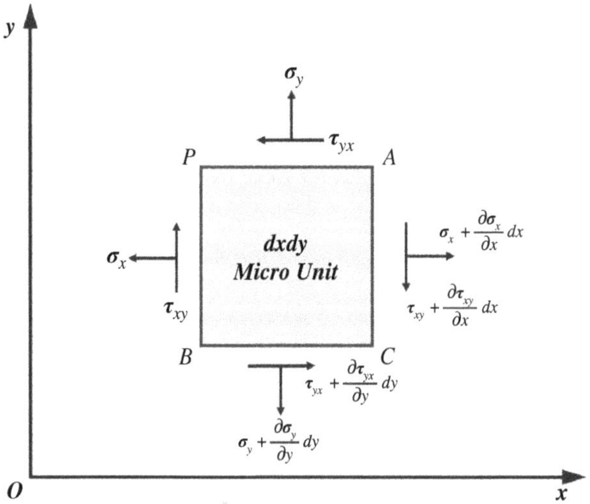

Figure 2.1 Stress components acting on an infinitesimal element.

Next, the equilibrium equations for forces and moments are established through force analysis.

Firstly, moment equilibrium is considered:

$$\tau_{xy} + \frac{1}{2}\frac{\partial \tau_{xy}}{\partial x}dx = \tau_{yx} + \frac{1}{2}\frac{\partial \tau_{yx}}{\partial y}dy \tag{2.3}$$

From the requirement of moment equilibrium, the principle of symmetry of shear stresses is obtained, namely:

$$\tau_{xy} = \tau_{yx} \tag{2.4}$$

Based on force analysis, equilibrium in the x-direction gives:

$$\begin{aligned}&\left(\sigma_x + \frac{\partial \sigma_x}{\partial x}dx\right) dy \times 1 - \sigma_x dy \times 1\\&+\left(\tau_{yx} + \frac{1}{2}\frac{\partial \tau_{yx}}{\partial y}dy\right) dx \times 1 - \tau_{yx} dx \times 1\\&+X dx dy \times 1 = 0\end{aligned} \tag{2.5}$$

Simplifying yields:

$$\frac{\partial \sigma_x}{\partial x} + \frac{\partial \tau_{yx}}{\partial y} + X = 0 \tag{2.6}$$

Similarly, equilibrium in the y-direction yields:

$$\frac{\partial \sigma_y}{\partial y} + \frac{\partial \tau_{xy}}{\partial x} + Y = 0 \tag{2.7}$$

Thus, the equilibrium differential equations for plane problems are established as follows:

$$\begin{cases}\dfrac{\partial \sigma_x}{\partial x} + \dfrac{\partial \tau_{yx}}{\partial y} + X = 0\\[2mm]\dfrac{\partial \sigma_y}{\partial y} + \dfrac{\partial \tau_{xy}}{\partial x} + Y = 0\end{cases} \tag{2.8}$$

The equilibrium equations do not contain $\boldsymbol{E}$ or $\boldsymbol{\mu}$; therefore, these equations are independent of material properties and are satisfied throughout the entire elastic body, including the boundary.

2.2.2 Geometric Equations

In plane problems, discussing the geometric equations involves examining the relationship between strains and displacements.

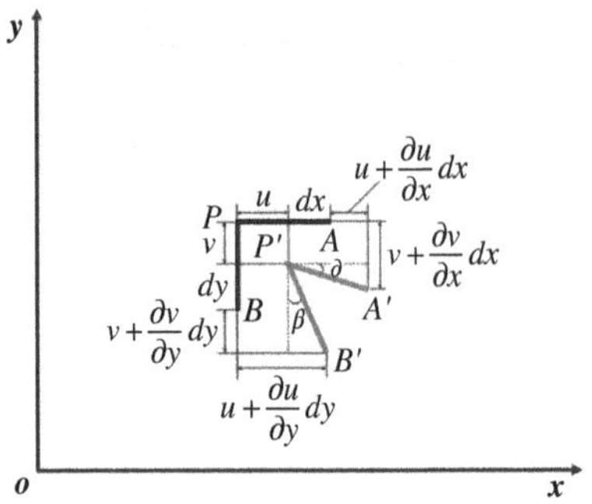

Figure 2.2 Displacement state at spatial point P.

As shown in Figure 2.2, consider the displacement state at the spatial point P. Firstly, examine the displacement of line segments near the point P:

$$
\begin{aligned}
&P \to P' \begin{cases} u \\ v \end{cases} \\
&A \to A' \begin{cases} u + \dfrac{\partial u}{\partial x}dx \\ v + \dfrac{\partial v}{\partial x}dx \end{cases} \\
&B \to B' \begin{cases} u + \dfrac{\partial u}{\partial y}dy \\ v + \dfrac{\partial v}{\partial y}dy \end{cases}
\end{aligned}
\tag{2.9}
$$

The above expressions, of course, use Taylor expansions, omitting second-order and higher-order terms. Additionally, line segments are defined as $PA = dx$, $PB = dy$.

Before deriving the equations, it is necessary to explain several concepts clearly: normal strain and shear strain. Normal strain, to be intuitively understood, is the ratio between the change in length and the original length. Shear strain refers to the change in the angle between two initially perpendicular line segments at a given point.

Thus, normal strains can be expressed as follows:

$$
\begin{aligned}
\varepsilon_x &= \frac{u + \frac{\partial u}{\partial x}dx - u}{dx} = \frac{\partial u}{\partial x} \\
\varepsilon_y &= \frac{v + \frac{\partial v}{\partial y}dy - v}{dy} = \frac{\partial v}{\partial y}
\end{aligned}
\tag{2.10}
$$

Regarding shear strain, similarly, it can be obtained with slightly more complicated expressions. From geometric relationships:

$$\gamma_{xy} = \alpha + \beta$$
$$\tan\alpha = \frac{v + \frac{\partial v}{\partial x}dx - v}{dx} = \frac{\partial v}{\partial x} \approx \alpha \tag{2.11}$$
$$\tan\beta = \frac{u + \frac{\partial u}{\partial y}dy - u}{dy} = \frac{\partial u}{\partial y} \approx \alpha$$

Therefore, the shear strain simplifies to:

$$\gamma_{xy} = \frac{\partial v}{\partial x} + \frac{\partial u}{\partial y} \tag{2.12}$$

Finally, the geometric equations governing plane elasticity problems are formulated by consolidating the preceding results:

$$\begin{cases} \varepsilon_x = \dfrac{\partial u}{\partial x} \\ \varepsilon_y = \dfrac{\partial v}{\partial y} \\ \gamma_{xy} = \dfrac{\partial v}{\partial x} + \dfrac{\partial u}{\partial y} \end{cases} \tag{2.13}$$

2.2.3 Physical Equations

The physical equations are derived from generalized Hooke's law and Poisson's ratio. The derivation process is included in most standard materials mechanics textbooks. These equations are:

$$\begin{cases} \varepsilon_x = \dfrac{1}{E}\left[\sigma_x - \mu\left(\sigma_y + \sigma_z\right)\right] \\ \varepsilon_y = \dfrac{1}{E}\left[\sigma_y - \mu\left(\sigma_x + \sigma_z\right)\right] \\ \varepsilon_z = \dfrac{1}{E}\left[\sigma_z - \mu\left(\sigma_y + \sigma_x\right)\right] \end{cases} \tag{2.14}$$

$$\begin{cases} \gamma_{xy} = \dfrac{1}{G}\tau_{xy} \\ \gamma_{yz} = \dfrac{1}{G}\tau_{yz} \\ \gamma_{zx} = \dfrac{1}{G}\tau_{zx} \end{cases} \tag{2.15}$$

The above are the three-dimensional physical equations. In two-dimensional problems, only three of them are needed:

$$\begin{cases} \varepsilon_x = \dfrac{\sigma_x - \mu\sigma_y}{E} \\ \varepsilon_y = \dfrac{\sigma_y - \mu\sigma_x}{E} \end{cases} \tag{2.16}$$

$$\gamma_{xy} = \frac{2(1+\mu)}{E}\tau_{xy} \tag{2.17}$$

In summary, the fundamental unknowns in elasticity are three displacement components, six stress components, and six strain components, totaling fifteen unknowns. Correspondingly, the fundamental equations include: three equilibrium differential equations, six geometric equations, and six physical equations—fifteen equations in total. Therefore, these equations can be simultaneously solved to determine all unknowns.

2.3 Deformation and Strain

2.3.1 Reference Configuration and Transient Configuration

In continuum mechanics, matter is seen as a collection of material points [2]. The matter point here is an idealized mechanical model, and is not the microscopic particles in physics, such as atoms, molecules, etc. Such idealized "points of matter" must be small enough at the macro level that they can be treated as points in geometry, thus establishing the field of continuum mechanics and using the relevant tools of field theory. At the same time, it should be large enough to contain enough microscopic particles to avoid significant changes in various physical quantities of material points caused by the movement of a small number of microscopic particles, so that the classical field theory of continuum mechanics can be established. Therefore, the material point in continuum mechanics can be considered as a macroscopically small and microscopically large mass of matter.

Both matter and material points occupy or exist at specific locations in space. In order to describe changes in the positions of material points and relative positions between material points in the process of deformation and movement, and to relate material points to their spatial positions, it is necessary to establish a coordinate system to correspond the material points to their positions in space. In this manner, material points can be replaced by spatial points. The set of position points corresponding to all material points is the configuration of the object. From this point of view, the movement and deformation of matter is the movement and deformation of configuration. Unless otherwise specified, Cartesian coordinate systems are used throughout the book. The configuration at the initial time ($t = 0$) is called the initial configuration, which can be used as a reference, called the reference configuration, and is denoted as B. Any other instantaneous t configuration is an instantaneous configuration, and the instantaneous configuration at the current moment can also be called the current configuration. The position of the reference configuration material point is represented by X_J ($J = 1, 2, 3$), and the current configuration is represented by X_i ($i = 1, 2, 3$), then the relation between the reference configuration midpoint **X** and its position $\boldsymbol{x}$ in the instantaneous configuration can be expressed as:

$$x_i = \psi_i(X_1, X_2, X_3, t) = \psi_i(\boldsymbol{X}_J, t), \quad (i, J = 1, 2, 3) \tag{2.18}$$

or

$$\boldsymbol{x} = \boldsymbol{\psi}(\boldsymbol{X}, t) \tag{2.19}$$

The deformation function $\boldsymbol{\psi}(\boldsymbol{X}, t)$ is a vector function. Eqs. (2.18) and (2.19) are component and vector expressions in the rectangular coordinate system, respectively. The two equations are Lagrangian (L's) descriptions of the law of point motion.

If the deformation function $\boldsymbol{\psi}(\boldsymbol{X}, t)$ is smooth, continuous, and one-to-one, then Eq. (2.20) has a single-valued, continuous, and smooth inverse function:

$$\boldsymbol{X} = \boldsymbol{\psi}^{-1}(\boldsymbol{x}, t) \tag{2.20}$$

By giving the position of the point in the current configuration at time t, the position of the point in the initial configuration can be traced from Eq. (2.20), which is called the Euler method (E's) description.

By taking the partial derivative of Eq. (2.19) with respect to X_J, a second-order tensor is obtained:

$$\frac{\partial \boldsymbol{x}}{\partial \boldsymbol{X}} = \frac{\partial x_i}{\partial X_J}\boldsymbol{e}_i \otimes \boldsymbol{e}_J = F_{iJ}\boldsymbol{e}_i \otimes \boldsymbol{e}_J = \boldsymbol{F} \tag{2.21}$$

In Eq. (2.21), F is called the deformation gradient tensor, which is also called the material deformation gradient because it is obtained under the Lagrange description. The two points in the current configuration and the reference configuration in the F connection are therefore also called "two-point tensors," whose components are expressed as:

$$\boldsymbol{F} = \begin{bmatrix} x_1 \\ x_2 \\ x_3 \end{bmatrix} \begin{bmatrix} \dfrac{\partial}{\partial X_1} & \dfrac{\partial}{\partial X_2} & \dfrac{\partial}{\partial X_3} \end{bmatrix} = \begin{bmatrix} \dfrac{\partial x_1}{\partial X_1} & \dfrac{\partial x_1}{\partial X_2} & \dfrac{\partial x_1}{\partial X_3} \\ \dfrac{\partial x_2}{\partial X_1} & \dfrac{\partial x_2}{\partial X_2} & \dfrac{\partial x_2}{\partial X_3} \\ \dfrac{\partial x_3}{\partial X_1} & \dfrac{\partial x_3}{\partial X_2} & \dfrac{\partial x_3}{\partial X_3} \end{bmatrix} \tag{2.22}$$

By differentiating Eq. (2.20) partially with respect to x, a second-order tensor is obtained.

$$\frac{\partial \boldsymbol{X}}{\partial \boldsymbol{x}} = \frac{\partial X_J}{\partial x_i}\boldsymbol{e}_J \otimes \boldsymbol{e}_i = F_{Ji}^{-1}\boldsymbol{e}_J \otimes \boldsymbol{e}_i = \boldsymbol{F}^{-1} \tag{2.23}$$

Or it can be obtained directly from Eq. (2.21)

$$d\boldsymbol{X} = \boldsymbol{F}^{-1} d\boldsymbol{x} \tag{2.24}$$

Obviously, $\boldsymbol{F}^{-1}$ is the inverse of the deformation gradient tensor. $\boldsymbol{F}$ is also called the spatial deformation gradient, and its components are of the form

$$\boldsymbol{F} = \begin{bmatrix} X_1 \\ X_2 \\ X_3 \end{bmatrix} \begin{bmatrix} \frac{\partial}{\partial x_1} & \frac{\partial}{\partial x_2} & \frac{\partial}{\partial x_3} \end{bmatrix} = \begin{bmatrix} \frac{\partial X_1}{\partial x_1} & \frac{\partial X_1}{\partial x_2} & \frac{\partial X_1}{\partial x_3} \\ \frac{\partial X_2}{\partial x_1} & \frac{\partial X_2}{\partial x_2} & \frac{\partial X_2}{\partial x_3} \\ \frac{\partial X_3}{\partial x_1} & \frac{\partial X_3}{\partial x_2} & \frac{\partial X_3}{\partial x_3} \end{bmatrix} \tag{2.25}$$

According to the implicit function theorem, in order to have a continuously smooth inverse function, the determinant of the deformation gradient $\boldsymbol{F}$ must be satisfied

$$J = \det \boldsymbol{F} = \left| \frac{\partial x_i}{\partial X_J} \right| \neq 0 \tag{2.26}$$

The determinant $\boldsymbol{J}$ of $\boldsymbol{F}$ gives the volume ratio of the current configuration to the reference configuration,

$$\frac{dv}{dV} = \det \boldsymbol{F} = J \tag{2.27}$$

where $\boldsymbol{v}$ and $\boldsymbol{V}$ represent the volume of the body element in the current configuration and the reference configuration, respectively.

2.3.2 Strain Tensor

The Green deformation tensor is

$$\boldsymbol{C} = \boldsymbol{F}^T \boldsymbol{F} \tag{2.28}$$

where $\boldsymbol{F}^T$ is the transpose of the deformation gradient tensor $\boldsymbol{F}$. It can be obtained from Eq. (2.21):

$$d\boldsymbol{x} = \boldsymbol{F} d\boldsymbol{X} = d\boldsymbol{X} \boldsymbol{F}^T \tag{2.29}$$

So, the square variance of the distance between two adjacent points is:

$$\begin{aligned} |d\boldsymbol{x}|^2 - |d\boldsymbol{X}|^2 &= (\boldsymbol{F}d\boldsymbol{X}) \cdot (\boldsymbol{F}d\boldsymbol{X}) - d\boldsymbol{X} \cdot d\boldsymbol{X} = \left(d\boldsymbol{X}\boldsymbol{F}^T\right) \cdot (\boldsymbol{F}d\boldsymbol{X}) - d\boldsymbol{X} \cdot d\boldsymbol{X} \\ &= d\boldsymbol{X} \cdot \left(\boldsymbol{F}^T\boldsymbol{F} - \boldsymbol{I}\right) d\boldsymbol{X} = d\boldsymbol{X} \cdot 2\boldsymbol{E} \cdot d\boldsymbol{X} \end{aligned} \tag{2.30}$$

In Eq. (2.30), the second-order tensor E is a Green's (or Lagrange's) finite strain tensor. The tensor is a symmetric tensor:

$$\boldsymbol{E} = \frac{1}{2}\left(\boldsymbol{F}^T\boldsymbol{F} - \boldsymbol{I}\right) = \frac{1}{2}(\boldsymbol{C} - \boldsymbol{I}) \tag{2.31}$$

According to Eq. (2.24),

$$\begin{aligned} |d\boldsymbol{x}|^2 - |d\boldsymbol{X}|^2 &= d\boldsymbol{x} \cdot d\boldsymbol{x} - \left(\boldsymbol{F}^{-1}d\boldsymbol{X}\right) \cdot \left(\boldsymbol{F}^{-1}d\boldsymbol{X}\right) \\ &= d\boldsymbol{x} \cdot d\boldsymbol{x} - \left(d\boldsymbol{X}\boldsymbol{F}^{-T}\right) \cdot \left(\boldsymbol{F}^{-1}d\boldsymbol{X}\right) \\ &= d\boldsymbol{x} \cdot \left(\boldsymbol{I} - \boldsymbol{F}^{-T}\boldsymbol{F}^{-1}\right) d\boldsymbol{x} = d\boldsymbol{x} \cdot 2\boldsymbol{e} \cdot d\boldsymbol{x} \end{aligned} \tag{2.32}$$

In Eq. (2.32), the second-order tensor $\boldsymbol{e}$ is the Almansi (or Euler) finite strain tensor.

$$\boldsymbol{e} = \frac{1}{2}\left(\boldsymbol{I} - \boldsymbol{F}^{-T}\boldsymbol{F}^{-1}\right) \tag{2.33}$$

The Green strain tensor and the Almansi strain tensor can be expressed, respectively, in the Cartesian coordinate system as:

$$E_{IJ} = \frac{1}{2}\left(\frac{\partial u_I}{\partial X_J} + \frac{\partial u_J}{\partial X_I} + \frac{\partial u_K}{\partial X_I}\frac{\partial u_K}{\partial X_J}\right) = \frac{1}{2}\left(u_{I,J} + u_{J,I} + u_{K,I}u_{K,J}\right) \tag{2.34}$$

$$e_{ij} = \frac{1}{2}\left(\frac{\partial u_i}{\partial x_j} + \frac{\partial u_j}{\partial x_i} + \frac{\partial u_k}{\partial x_i}\frac{\partial u_k}{\partial x_j}\right) = \frac{1}{2}\left(u_{i,j} + u_{j,i} + u_{k,i}u_{k,j}\right) \tag{2.35}$$

It is generally considered that the elastic and plastic deformation is small when the deformation is less than 0.1%, and each displacement gradient component $\partial u_I/\partial X_J$ or $\partial u_i/\partial X_j$ is much less than 1, so their product terms can be omitted and become from Eqs. (2.34) and (2.35),

$$E_{IJ} = \frac{1}{2}\left(\frac{\partial u_I}{\partial X_J} + \frac{\partial u_J}{\partial X_I}\right) = \frac{1}{2}\left(u_{I,J} + u_{J,I}\right) \tag{2.36}$$

$$e_{ij} = \frac{1}{2}\left(\frac{\partial u_i}{\partial x_j} + \frac{\partial u_j}{\partial x_i}\right) = \frac{1}{2}\left(u_{i,j} + u_{j,i}\right) \tag{2.37}$$

If the displacement itself is also a small amount under small deformation, then the difference between the L-coordinate and E-coordinate of the material point is very small, and the material displacement gradient $\partial u_I/\partial X_J$ component and the spatial displacement gradient $\partial u_i/\partial X_j$ component are very close. Therefore, the Green strain tensor and the Almansi strain tensor under small deformation can be approximately considered to be equal,

$$E_{ij} \approx e_{ij} \approx \varepsilon_{ij} = \frac{1}{2}\left(u_{i,j} + u_{j,i}\right) \tag{2.38}$$

ε is the Cauchy strain tensor, is the second-order real symmetric tensor, its component form is:

$$\varepsilon_{ij} = \begin{bmatrix} \varepsilon_{11} & \varepsilon_{12} & \varepsilon_{13} \\ \varepsilon_{21} & \varepsilon_{22} & \varepsilon_{23} \\ \varepsilon_{31} & \varepsilon_{32} & \varepsilon_{33} \end{bmatrix} \tag{2.39}$$

2.3.3 Principal Strain and Volumetric Strain

At a point in the body, there must be three mutually perpendicular directions, and the three directions remain mutually perpendicular after deformation, that is, the shear strain is equal to zero. These three directions are called the principal direction of strain or the principal axis of strain, referred to as the principal direction of strain. The positive strain of these three strains is called the principal strain. Then the principal strain and principal direction of strain are solved under the assumption of small deformation.

If the strain tensor at any point in the body is $\boldsymbol{\varepsilon}$, then the positive strain along the unit vector $\boldsymbol{n}$ is:

$$\varepsilon_i^{(n)} = \varepsilon_{ij} n_j \tag{2.40}$$

If $\boldsymbol{n}$ is one of the principal directions of strain at this point, the principal strain is ε, then the shear strain in this direction is equal to zero, so:

$$\varepsilon_i^{(n)} = \varepsilon n_i = \varepsilon \delta_{ij} n_j \tag{2.41}$$

Let the left side of Eqs. (2.40) and (2.41) be equal, then there is:

$$\left(\varepsilon_{ij} - \delta_{ij}\varepsilon\right) n_j = 0 \tag{2.42}$$

Let the unit vector $\boldsymbol{n}$ be $(l,\ m,\ n)$, where $l,\ m,\ n$ is the cosine of the Angle between the vector and the three coordinate axes, then Eq. (2.42) can be written in the following form:

$$\begin{cases} l\left(\varepsilon_{11} - \varepsilon\right) + m\varepsilon_{12} + n\varepsilon_{13} = 0 \\ l\varepsilon_{21} + n\left(\varepsilon_{22} - \varepsilon\right) + n\varepsilon_{23} = 0 \\ l\varepsilon_{31} + m\varepsilon_{32} + n\left(\varepsilon_{33} - \varepsilon\right) = 0 \end{cases} \tag{2.43}$$

The direction cosine l, m, n in Eq. (2.43) is regarded as an unknown quantity. If the system of equations has a nonzero solution, its corresponding coefficient determinant is equal to zero, that is:

$$\left|\varepsilon_{ij} - \delta_{ij}\varepsilon\right| = 0 \tag{2.44}$$

or

$$\begin{vmatrix} \varepsilon_{11} - \varepsilon & \varepsilon_{12} & \varepsilon_{13} \\ \varepsilon_{21} & \varepsilon_{22} - \varepsilon & \varepsilon_{23} \\ \varepsilon_{31} & \varepsilon_{32} & \varepsilon_{33} - \varepsilon \end{vmatrix} = 0 \tag{2.45}$$

Expanding the above determinant, the cubic equation about ε is obtained:

$$\left|\varepsilon_{ij} - \delta_{ij}\varepsilon\right| = -\varepsilon^3 + I_1\varepsilon^2 + I_2\varepsilon + I_3 = 0 \tag{2.46}$$

In Eq. (2.46), the coefficients $I_1,\ I_2$, and I_3 do not change with the change of the coordinate system and are called the first, second, and third invariants of the strain, respectively. The sum of the three real roots of Eq. (2.46) corresponds to the three principal strains. Their relation to the strain component and principal strain is:

$$I_1 = \varepsilon_{ii} = \varepsilon_1 + \varepsilon_2 + \varepsilon_3 = \varepsilon_{11} + \varepsilon_{22} + \varepsilon_{33} \tag{2.47}$$

$$I_2 = -\frac{1}{2}\left(\varepsilon_{ii}\varepsilon_{ii} - \varepsilon_{ij}\varepsilon_{ji}\right) = -\left(\varepsilon_1\varepsilon_2 + \varepsilon_1\varepsilon_3 + \varepsilon_2\varepsilon_3\right)$$
$$= -\begin{vmatrix} \varepsilon_{11} & \varepsilon_{12} \\ \varepsilon_{21} & \varepsilon_{22} \end{vmatrix} - \begin{vmatrix} \varepsilon_{22} & \varepsilon_{23} \\ \varepsilon_{32} & \varepsilon_{33} \end{vmatrix} - \begin{vmatrix} \varepsilon_{11} & \varepsilon_{13} \\ \varepsilon_{31} & \varepsilon_{33} \end{vmatrix} \tag{2.48}$$

$$I_3 = e_{ijk}\varepsilon_{i1}\varepsilon_{j2}\varepsilon_{k3} = \varepsilon_1\varepsilon_2\varepsilon_3 = \begin{vmatrix} \varepsilon_{11} & \varepsilon_{12} & \varepsilon_{13} \\ \varepsilon_{21} & \varepsilon_{22} & \varepsilon_{23} \\ \varepsilon_{31} & \varepsilon_{32} & \varepsilon_{33} \end{vmatrix} \tag{2.49}$$

After the three principal strains are obtained from Eq. (2.46) and then substituted into Eq. (2.42) or Eq. (2.43), respectively, the corresponding principal directions of the three principal strains can be obtained.

If the three principal strains are not equal, that is $\varepsilon_1 \neq \varepsilon_2 \neq \varepsilon_3$, the corresponding principal directions of the three strains are orthogonal in pairs. When two of these principal strains are equal, for example $\varepsilon_1 = \varepsilon_2 \neq \varepsilon_3$, in this case ε_3 any direction in the plane perpendicular to the corresponding principal direction is ε_1 and ε_2 the corresponding principal direction, so there are infinitely many groups of principal directions. When the three principal strains are equal, $\varepsilon_1 = \varepsilon_2 = \varepsilon_3$, any direction in the space is the principal strain direction. The corresponding deformation in this case is uniform spherical expansion (or compression).

Next, the physical significance of the first strain invariant is discussed. The orthogonal coordinate system is established with three principal strain directions as axis directions. The sides that intersect before deformation in the object and are parallel to the three coordinate axes are respectively dX_1, dX_2, dX_3 Cube element. At this time, the shear strain is zero, and the element remains as a cube after deformation. The length of each side after deformation is $(1+\varepsilon_1)dX_1$, $(1+\varepsilon_2)dX_2$, and $(1+\varepsilon_3)dX_3$.

The volume strain is introduced in a similar way to define line strain θ

$$\theta = \frac{\Delta V}{V_0} = \frac{V_1 - V_0}{V_0} = \frac{(1+\varepsilon_1)(1+\varepsilon_2)(1+\varepsilon_3)\, dX_1dX_2dX_3 - dX_1dX_2dX_3}{dX_1dX_2dX_3} \tag{2.50}$$

In Eq. (2.50), V_0 Initial unit volume, V_1 is the volume after deformation, ΔV is Volume change. Under the assumption of small deformation, ignoring small quantities, Eq. (2.31) can be approximated as

$$\theta \approx \varepsilon_1 + \varepsilon_2 + \varepsilon_3 = \varepsilon_{11} + \varepsilon_{22} + \varepsilon_{33} = I_1 \tag{2.51}$$

It can be seen that in the case of small deformation, the first strain invariant is approximately equal to the volume strain.

The Cauchy strain tensor can be divided into two parts: the strain sphere tensor $\boldsymbol{\varepsilon}^m$ and the strain partial tensor $\boldsymbol{\varepsilon}'$

$$\boldsymbol{\varepsilon}^m = \frac{1}{3}\mathrm{Tr}\,(\boldsymbol{\varepsilon})\,\boldsymbol{I} = \frac{1}{3}\theta\boldsymbol{I} \tag{2.52}$$

$$\boldsymbol{\varepsilon}' = \boldsymbol{\varepsilon} - \boldsymbol{\varepsilon}^m \tag{2.53}$$

There is obviously

$$\begin{bmatrix} \varepsilon_{11} & \varepsilon_{12} & \varepsilon_{13} \\ \varepsilon_{21} & \varepsilon_{22} & \varepsilon_{23} \\ \varepsilon_{31} & \varepsilon_{32} & \varepsilon_{33} \end{bmatrix} = \frac{1}{3}\begin{bmatrix} \theta & 0 & 0 \\ 0 & \theta & 0 \\ 0 & 0 & \theta \end{bmatrix} + \begin{bmatrix} \varepsilon_{11} - \frac{1}{3}\theta & \varepsilon_{12} & \varepsilon_{13} \\ \varepsilon_{21} & \varepsilon_{22} - \frac{1}{3}\theta & \varepsilon_{23} \\ \varepsilon_{31} & \varepsilon_{32} & \varepsilon_{33} - \frac{1}{3}\theta \end{bmatrix} \tag{2.54}$$

The strain ball tensor represents isotropic expansion or contraction of volume. The strain partial tensor is related to shear deformation. Obviously, the first invariant of the strain partial tensor is zero.

2.3.4 Strain Coordination Equation

As long as a continuously differentiable displacement function is given, the Cauchy strain tensor can be directly obtained by Eq. (2.38). Conversely, if each component of the strain tensor is given, it is not necessarily possible to obtain the corresponding displacement component through Eq. (2.38). To obtain three continuously differentiable displacement components through six differential equations, additional constraints are required. From the perspective of continuum deformation, in the deformation process, the strain of each point in the deformation body must be coordinated with each other, and should not be due to different points or micro-elements due to strain uncoordinated pores, or overlap. In order to meet the requirements of mathematical and physical coordination, the components of strain must satisfy the coordination equation (compatibility equation).

For ease of acceptance, the relationship between each component of Eq. (2.38) and displacement is written in the following form:

$$\begin{cases} \varepsilon_{11} = \dfrac{\partial u_1}{\partial X_1} \\ \varepsilon_{22} = \dfrac{\partial u_2}{\partial X_2} \\ \varepsilon_{33} = \dfrac{\partial u_3}{\partial X_3} \\ \varepsilon_{12} = \dfrac{1}{2}\left(\dfrac{\partial u_1}{\partial X_2} + \dfrac{\partial u_2}{\partial X_1}\right) \\ \varepsilon_{23} = \dfrac{1}{2}\left(\dfrac{\partial u_2}{\partial X_3} + \dfrac{\partial u_3}{\partial X_2}\right) \\ \varepsilon_{31} = \dfrac{1}{2}\left(\dfrac{\partial u_3}{\partial X_1} + \dfrac{\partial u_1}{\partial X_3}\right) \end{cases} \tag{2.55}$$

Take the second derivative of the first two equations of Eq. (2.55) with respect to X_2 and X_1, respectively, and add them together to get:

$$\frac{\partial^2 \varepsilon_{11}}{\partial X_2^2} + \frac{\partial^2 \varepsilon_{22}}{\partial X_1^2} = \frac{\partial^2}{\partial X_1 \partial X_2}\left(\frac{\partial u_1}{\partial X_2} + \frac{\partial u_2}{\partial X_1}\right) = 2\frac{\partial^2 \varepsilon_{12}}{\partial X_1 \partial X_2} \tag{2.56}$$

Equation (2.56) is ε_{11}, ε_{22} and ε_{12} the coordination relationship should be satisfied. Using the same method for the first three equations in Eq. (2.55), two other coordination relationships similar to that in Eq. (2.56) can be obtained. Take the derivative of the last three equations in Eq. (2.55) with respect to X_1, X_2, and X_3, respectively:

$$\frac{\partial}{\partial X_1}\left(-\frac{\partial \varepsilon_{23}}{\partial X_1}+\frac{\partial \varepsilon_{31}}{\partial X_2}+\frac{\partial \varepsilon_{12}}{\partial X_3}\right)=\frac{\partial^2 \varepsilon_{11}}{\partial X_2 \partial X_3} \tag{2.57}$$

Equation (2.57) indicates ε_{11} the coordination relationship that should be satisfied with the three shear strains. A similar coordination relationship between ε_{22} and ε_{33} with shear strain can be obtained by using the same method. There are six coordination equations as follows:

$$\begin{cases}\dfrac{\partial^2 \varepsilon_{11}}{\partial X_2^2}+\dfrac{\partial^2 \varepsilon_{22}}{\partial X_1^2}=2\dfrac{\partial^2 \varepsilon_{12}}{\partial X_1 \partial X_2} \\ \dfrac{\partial^2 \varepsilon_{22}}{\partial X_3^2}+\dfrac{\partial^2 \varepsilon_{33}}{\partial X_2^2}=2\dfrac{\partial^2 \varepsilon_{23}}{\partial X_2 \partial X_3} \\ \dfrac{\partial^2 \varepsilon_{33}}{\partial X_1^2}+\dfrac{\partial^2 \varepsilon_{11}}{\partial X_3^2}=2\dfrac{\partial^2 \varepsilon_{31}}{\partial X_1 \partial X_3} \\ \dfrac{\partial}{\partial X_1}\left(-\dfrac{\partial \varepsilon_{23}}{\partial X_1}+\dfrac{\partial \varepsilon_{31}}{\partial X_2}+\dfrac{\partial \varepsilon_{12}}{\partial X_3}\right)=\dfrac{\partial^2 \varepsilon_{11}}{\partial X_2 \partial X_3} \\ \dfrac{\partial}{\partial X_2}\left(-\dfrac{\partial \varepsilon_{31}}{\partial X_2}+\dfrac{\partial \varepsilon_{12}}{\partial X_3}+\dfrac{\partial \varepsilon_{23}}{\partial X_1}\right)=\dfrac{\partial^2 \varepsilon_{22}}{\partial X_1 \partial X_3} \\ \dfrac{\partial}{\partial X_3}\left(-\dfrac{\partial \varepsilon_{12}}{\partial X_3}+\dfrac{\partial \varepsilon_{23}}{\partial X_1}+\dfrac{\partial \varepsilon_{31}}{\partial X_2}\right)=\dfrac{\partial^2 \varepsilon_{33}}{\partial X_1 \partial X_2}\end{cases} \tag{2.58}$$

The coordination equation in Eq. (2.58) can also be written in the following form

$$\varepsilon_{ij,kl}+\varepsilon_{kl,ij}-\varepsilon_{ik,jl}-\varepsilon_{jl,ik}=0 \tag{2.59}$$

Equation (2.59) contains a total of 81 equations, but it can be proved by the symmetry of the strain tensor that there are only 6 independent equations, that is, deformation coordination equations in Eq. (2.58).

2.4 Stress Analysis

2.4.1 Cauchy Principle of Stress

The word “Stress” was first used in Rankine’s 1856 paper [3]. In 1822, Cauchy announced the “birth certificate of modern continuum mechanics,” also known as the “Cauchy principle of stress.” This principle is considered to be the basis of rational continuum mechanics [4] and paved the way for the theory of continuum mechanics for solids and fluids [5]. The principle of stress can be expressed in various forms, and it is expressed in one sentence as “the stress vector of a section with normal n at a certain point inside an object is independent of the shape of the section” [6].

Many literatures and books on the definition of stress have been introduced in detail and will not be repeated here. The Cauchy stress tensor is a second-order symmetric tensor defined on the current configuration and is often $\boldsymbol{\sigma}$ expressed. As shown in Figure 2.3, if the unit external normal direction $\boldsymbol{n}$ of a section in the object is, the surface force on the section is

$$\boldsymbol{t}_n = \boldsymbol{\sigma}\boldsymbol{n} \text{ or } t_{n_i} = \sigma_{ij} n_j \tag{2.60}$$

This is the Cauchy stress formula. The surface force vector can be decomposed into normal stress perpendicular to the section σ and tangent stress vector parallel to the section $\boldsymbol{\tau}$:

$$\boldsymbol{\sigma} = \boldsymbol{t}_n \cdot \boldsymbol{n} \tag{2.61}$$

$$\boldsymbol{\tau} = \boldsymbol{t}_n - \boldsymbol{\sigma}\boldsymbol{n} = \boldsymbol{t}_n(\boldsymbol{I} - \boldsymbol{n} \times \boldsymbol{n}) \tag{2.62}$$

Take the cube element in the field of point $\boldsymbol{P}$ in the object and establish the rectangular coordinate system as shown in Figure 2.4. Note the positive unit vector of each coordinate

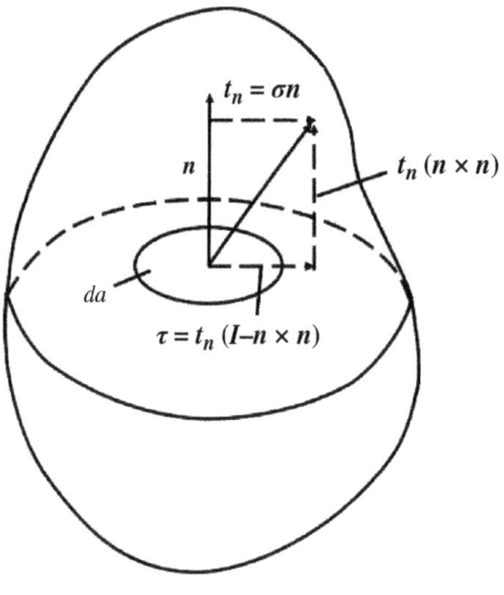

Figure 2.3 Schematic diagram of the Cauchy surface force.

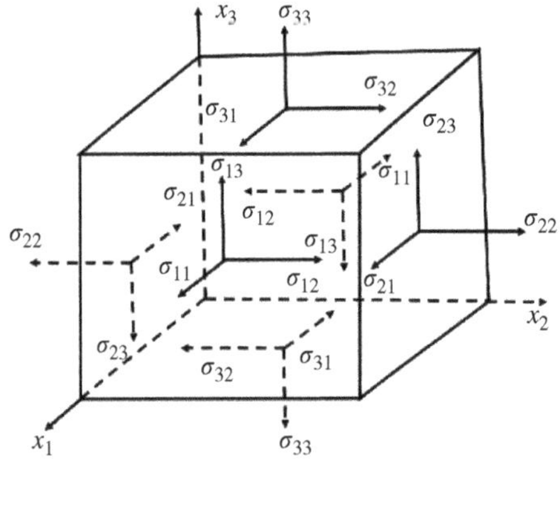

Figure 2.4 Components of the stress tensor.

axis as $\boldsymbol{e}_1$,$\boldsymbol{e}_2$,$\boldsymbol{e}_3$. $\boldsymbol{n}$ is taken as the three unit direction vectors respectively, and the surface force on each surface of the element can be decomposed in the direction of the three coordinate axes to obtain one normal stress and two shear stresses parallel to the three coordinate axes, respectively, and the value of each stress component is $\sigma_i = \boldsymbol{t}_n \boldsymbol{e}_i$.The subscript of the surface force vector $\boldsymbol{t}_n$ is defined as the direction of the normal outside the surface of the surface force, denoted as, that is $\boldsymbol{t}_i$, That is, the surface force acting on a surface with the external normal pointing in the direction of $\boldsymbol{e}_i$. Then the stress values on each face of the cube element can be obtained by the following equation:

$$\sigma_{ij} = t_i e_j \tag{2.63}$$

Note that Eq. (2.63) is only one equation for the calculation of stress components on each surface and is not used as a definition of the stress tensor. It can be shown that the second-order stress tensor $\boldsymbol{\sigma}$, whose component is determined by Eq. (2.63), uniquely determines the stress state at the point $\boldsymbol{P}$. The first subscript of the stress tensor component σ_{ij} indicates the outer normal direction of the surface of the stress, that is, the surface of the force. The second subscript indicates that it points to.

2.4.2 Cauchy Stress Formula

Next, the Cauchy stress formula of Eq. (2.60) is proved. In order to establish the relationship between the stress tensor $\boldsymbol{\sigma}$ and the stress vector $\boldsymbol{t}_n$ at point P in the object, the tetrahedral element PABC with the edges along the three coordinate axes is taken in the field of point P, as shown in Figure 2.5. If the area of plane ABC is dS, and the area of ΔPBC is dS_1, ΔPAC is dS_2, and ΔPAB is dS_3, then there is

$$dS_i = dS \cdot \cos(\boldsymbol{n}, \boldsymbol{e}_i) = dS \cdot (\boldsymbol{n} \cdot \boldsymbol{e}_i) = dSn_i \tag{2.64}$$

where n is the unit external normal of the ABC plane. n_i is the cosine of the Angle between n and the axis. Note that the surface force vectors on each face are $\boldsymbol{t}_{(n)}$ and $\boldsymbol{t}(\boldsymbol{e}_i)$ respectively, b_i is physical strength.

According to the equilibrium condition, the vector sum of the surface force and the physical force on the element is equal to zero:

$$\boldsymbol{t}_{(n)} dS - \boldsymbol{t}(\boldsymbol{e}_1)\, dS_1 - \boldsymbol{t}(\boldsymbol{e}_2)\, dS_2 - \boldsymbol{t}(\boldsymbol{e}_3)\, dS_3 + \rho \boldsymbol{b} dV = 0 \tag{2.65}$$

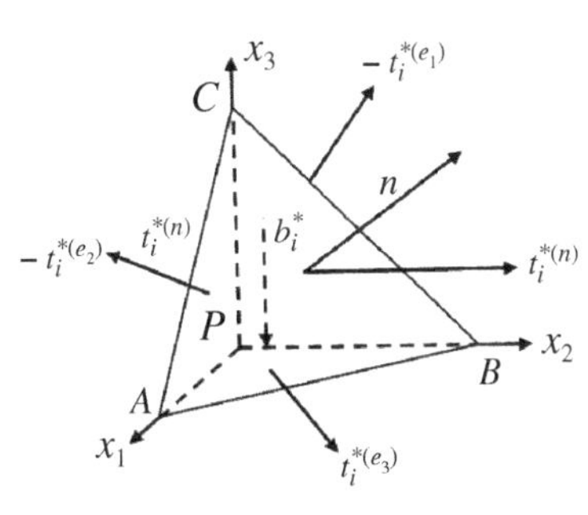

Figure 2.5 Force diagram of a tetrahedral infinitesimal element.

In the above expressions, ρ denotes the material density of the infinitesimal element. When the lengths of the edges of the tetrahedron approach zero, its volume becomes a higher-order infinitesimal, and thus the body force term can be neglected. Accordingly, Eq. (2.64) yields:

$$\boldsymbol{t}_{(n)}dS = \boldsymbol{t}((\boldsymbol{e}_1)\, n_1 dS + \boldsymbol{t}(\boldsymbol{e}_2)\, n_2 dS + \boldsymbol{t}(\boldsymbol{e}_3)\, n_3 dS \tag{2.66}$$

Neglecting dS on both sides and based on $t_j(\boldsymbol{e}_i) = \sigma_{ij}$:

$$\boldsymbol{t}_{(n)} = \boldsymbol{\sigma n} \tag{2.67}$$

Equation (2.67) can be expressed in the following form:

$$\begin{cases} t_{(n_1)} = n_1\sigma_{11} + n_2\sigma_{12} + n_3\sigma_{13} \\ t_{(n_2)} = n_1\sigma_{21} + n_2\sigma_{22} + n_3\sigma_{23} \\ t_{(n_2)} = n_1\sigma_{31} + n_2\sigma_{32} + n_3\sigma_{33} \end{cases} \tag{2.68}$$

where $t_{(n_i)}$ represents the components of the surface force vector on face ABC projected onto the three coordinate axes. It is evident that once the stress state at point P is determined, the stress state on any plane passing through that point can also be determined accordingly. When face ABC lies on the boundary of the object, $\boldsymbol{t}_{(n)}$ becomes the surface force acting on the boundary. Thus, Eq. (2.67) or Eq. (2.68) can be used to represent the stress boundary condition:

$$\bar{\boldsymbol{t}}_{(n)} = (\boldsymbol{\sigma})_s\, \boldsymbol{n} \tag{2.69}$$

or

$$\begin{cases} \bar{t}_{(n_1)} = n_1\,(\sigma_{11})_s + n_2\,(\sigma_{12})_s + n_3\,(\sigma_{13})_s \\ \bar{t}_{(n_2)} = n_1\,(\sigma_{21})_s + n_2(\sigma_{22})_s + n_3\,(\sigma_{23})_s \\ \bar{t}_{(n_2)} = n_1\,(\sigma_{31})_s + n_2\,(\sigma_{32})_s + n_3\,(\sigma_{33})_s \end{cases} \tag{2.70}$$

2.4.3 Equilibrium Different Equations and Stress Tensor Symmetry

A microelement of volume V is taken at any time in the continuous body, which is balanced under the surface force $t_{(n)}$ and the physical force b (including the inertial force). According to the force balance condition, the sum of surface force and physical force is

$$\int_S \boldsymbol{t}_{(n)} dS + \int_V \rho \mathbf{b} dV = 0 \tag{2.71}$$

According to Gauss's theorem, the area integral is transformed into a volume integral, and the Cauchy stress formula is applied. Equation (2.71) becomes:

$$\int_V \left(\sigma_{ji,j} + \rho b_i\right) dV = 0 \tag{2.72}$$

Obviously, the integrand function in Eq. (2.72) should be equal to zero, so:

$$\sigma_{ji,j} + \rho b_i = 0 \text{ or } \nabla\boldsymbol{\sigma} + \rho\boldsymbol{b} = 0 \tag{2.73}$$

This is the stress differential equilibrium equation.

Take the moment from the origin of the coordinate system according to the torque balance:

$$\int_S e_{ijk}x_j t_k dS + \int_V e_{ijk}x_j \rho b_k dV = 0 \tag{2.74}$$

Using Gauss's theorem and Eq. (2.73), Eq. (2.74) can be transformed into:

$$\int_V e_{ijk}\sigma_{jk} dV = 0 \tag{2.75}$$

Likewise, to meet the above conditions, the requirements:

$$e_{ijk}\sigma_{jk} = 0 \tag{2.76}$$

Equation (2.76) is written in component form, resulting in the following equations: $\sigma_{23} = \sigma_{32}$, $\sigma_{13} = \sigma_{31}$, $\sigma_{12} = \sigma_{21}$. It is proven that the shear stress is equal to each other, and the stress tensor changes from nine components to six independent components.

2.4.4 Principal Stress and Stress Invariants

If the stress tensor of a point P in a continuum body is known, the stress vector $\boldsymbol{t}_n$ across any plane of that point can be obtained according to Cauchy's stress formula. When the obtained stress vector of a given plane is parallel to the normal vector $\boldsymbol{n}$ of the plane, the shear stress obtained by Eq. (2.62) is zero, and only the normal stress perpendicular to the plane σ. At this time, the normal stress is also called the principal stress, and the normal direction of the plane is called the direction of the principal stress, referred to as the principal direction of stress. At this time,

$$\boldsymbol{t}_{(n)} = \sigma\boldsymbol{n} \tag{2.77}$$

According to Cauchy's stress formula and $n_i = \delta_{ij}n_j$, it can be obtained from Eq. (2.77):

$$\left(\sigma_{ij} - \delta_{ij}\sigma\right) n_j = 0 \tag{2.78}$$

Equation (2.78) is written as a component equation in the form:

$$\begin{bmatrix} \sigma_{11} - \sigma & \sigma_{12} & \sigma_{13} \\ \sigma_{21} & \sigma_{22} - \sigma & \sigma_{23} \\ \sigma_{31} & \sigma_{32} & \sigma_{33} - \sigma \end{bmatrix} \begin{bmatrix} n_1 \\ n_2 \\ n_3 \end{bmatrix} = 0 \tag{2.79}$$

Taking n_1, n_2, and n_3 as unknowns, Eq. (2.79) can have a nonzero solution if the coefficient determinant is equal to zero:

$$\left|\sigma_{ij} - \delta_{ij}\sigma\right| = 0 \tag{2.80}$$

or

$$\begin{vmatrix} \sigma_{11}-\sigma & \sigma_{12} & \sigma_{13} \\ \sigma_{21} & \sigma_{22}-\sigma & \sigma_{23} \\ \sigma_{31} & \sigma_{32} & \sigma_{33}-\sigma \end{vmatrix} = 0 \tag{2.81}$$

Expand Eq. (2.81) to obtain a cubic equation about σ

$$\begin{aligned} &\sigma^3 - (\sigma_{11}+\sigma_{22}+\sigma_{33})\,\sigma^2 + \left(\sigma_{11}\sigma_{22}+\sigma_{22}\sigma_{33}+\sigma_{11}\sigma_{33}-\sigma_{23}^2-\sigma_{13}^2-\sigma_{12}^2\right)\sigma- \\ &\left(\sigma_{11}\sigma_{22}\sigma_{33}-\sigma_{11}\sigma_{23}^2-\sigma_{22}\sigma_{13}^2-\sigma_{33}\sigma_{12}^2+2\sigma_{23}\sigma_{12}\sigma_{13}\right)=0 \end{aligned} \tag{2.82}$$

To solve Eq. (2.82), if three real roots σ_1, σ_2 and σ_2, then they are the three principal stresses of the P point are obtained. Each principal stress corresponds to a principal direction, so if there are three different principal stresses, there are three principal directions. By substituting each principal stress σ^k into Eq. (2.78) or Eq. (2.79), the corresponding direction cosine $\boldsymbol{n}^k$ is obtained, that is, the direction of the principal stress corresponding to the principal stress. It can be proved that the principal directions corresponding to the three different principal stresses are orthogonal to each other.

When the stress state of point P is determined, its principal stress is also determined, and does not change with the change of the coordinate system. If the coordinate axis is established in the three principal directions of the stress, then the stress state at point P can be expressed as:

$$\sigma_{ij} = \begin{bmatrix} \sigma_1 & 0 & 0 \\ 0 & \sigma_2 & 0 \\ 0 & 0 & \sigma_3 \end{bmatrix} \tag{2.83}$$

By substituting the stress tensor of Eq. (2.83) into Eq. (2.78) and solving the principal stress in the same way, the cubic equation for σ can be obtained:

$$\sigma^3 - (\sigma_1+\sigma_2+\sigma_3)\,\sigma^2 + (\sigma_1\sigma_2+\sigma_2\sigma_3+\sigma_1\sigma_3)\,\sigma - (\sigma_{11}\sigma_{22}\sigma_{33}) = 0 \tag{2.84}$$

Since the principal stress does not change with the change of the coordinate system, Eqs. (2.82) and (2.84) have the same solution. Therefore, the coefficients corresponding to the two equations are equal, and the values of these coefficients do not change with the change of the coordinate system:

$$J_1 = \sigma_1+\sigma_2+\sigma_3 = \sigma_{11}+\sigma_{22}+\sigma_{33} = \mathrm{tr}\boldsymbol{\sigma} \tag{2.85}$$

$$\begin{aligned} J_2 &= \sigma_1\sigma_2+\sigma_2\sigma_3+\sigma_1\sigma_3 = \sigma_{11}\sigma_{22}+\sigma_{22}\sigma_{33}+\sigma_{11}\sigma_{33}-\sigma_{23}^2-\sigma_{13}^2-\sigma_{12}^2 \\ &= \frac{1}{2}\left(\sigma_{ii}\sigma_{jj}-\sigma_{ij}\sigma_{ji}\right) \end{aligned} \tag{2.86}$$

$$J_3 = \sigma_{11}\sigma_{22}\sigma_{33} = \sigma_{11}\sigma_{22}\sigma_{33} - \sigma_{11}\sigma_{23}^2 - \sigma_{22}\sigma_{13}^2 - \sigma_{33}\sigma_{12}^2 + 2\sigma_{23}\sigma_{12}\sigma_{13} = \det \boldsymbol{\sigma} \tag{2.87}$$

In Eq. (2.87), J_1, J_2 and J_3 are called the first, second, and third stress invariants, respectively.

As can be seen from Eq. (2.85), the sum of normal stresses on any three perpendicular surfaces in a continuous body is a constant, equal to the sum of the three principal stresses at that point, which is called the volume stress. In online elastic bodies, there is a linear relationship between volume stress and volume strain:

$$J_1 = \frac{E}{1-2\upsilon} I_1 = KI_1 \tag{2.88}$$

where K is called the bulk modulus.

2.5 Foundation of the Elastic Constitutive

2.5.1 Elastic Constitutive Theory

When studying unidirectional tension and compression, if the known material does not yield and the stress is less than the yield stress σ_0, the relationship between stress σ and strain ε is linear, satisfying Hooke's law:

$$\sigma = E\varepsilon \tag{2.89}$$

where E is the elastic modulus (Young's modulus). In addition, axial deformation also causes a change in the transverse dimension. The transverse strain can be calculated according to Poisson's ratio ε':

$$\varepsilon' = -\mu\varepsilon = -\frac{\sigma}{E} \tag{2.90}$$

Under the stress condition of pure shear, when the shear stress does not exceed the shear proportion limit, the relationship between the shear stress and the shear strain is also subject to Hooke's law, that is

$$\tau - G\gamma \; or \; \gamma - \frac{\tau}{G} \tag{2.91}$$

2.5.2 Generalized Hooke's Law

In a three-dimensional stress state, it is usually necessary to describe the stress state of a point of an object, and the corresponding strain component also needs nine [7]. But according to the law of mutual equivalence of shear stress, $\tau_{xy} = \tau_{yx}$, $\tau_{xz} = \tau_{zx}$, $\tau_{yz} = \tau_{zy}$, only six of the nine stress components and nine strain components are independent of each other. For isotropic materials with small deformation within the on-line elastic range, the linear strain is only related to the normal stress component and has no relation to the shear stress, while the shear strain is only related to the shear stress and has nothing to do with the normal stress. The shear stress only causes the change of the corresponding shear strain component, and does not affect the shear strain in other directions. Therefore, for an ideal elastomer, the above relation has the following form:

$$\begin{pmatrix} \sigma_x \\ \sigma_y \\ \sigma_z \\ \tau_{xy} \\ \tau_{yz} \\ \tau_{zx} \end{pmatrix} = \begin{pmatrix} c_{11} & c_{12} & c_{13} & c_{14} & c_{15} & c_{16} \\ c_{21} & c_{22} & c_{23} & c_{24} & c_{25} & c_{26} \\ c_{31} & c_{32} & c_{33} & c_{34} & c_{35} & c_{36} \\ c_{41} & c_{42} & c_{43} & c_{44} & c_{45} & c_{46} \\ c_{51} & c_{52} & c_{53} & c_{54} & c_{55} & c_{56} \\ c_{61} & c_{62} & c_{63} & c_{64} & c_{65} & c_{66} \end{pmatrix} \begin{pmatrix} \varepsilon_x \\ \varepsilon_y \\ \varepsilon_z \\ \gamma_{xy} \\ \gamma_{yz} \\ \gamma_{zx} \end{pmatrix} \tag{2.92}$$

where c_{ij} is the elastic constant. From the homogeneity of the material, the constant c_{ij} is independent of the coordinates x, y, and z. If tensor notation is used, Eq. (2.92) can be abbreviated as

$$\sigma_{ij} = c_{ijkl}\varepsilon_{kl}\ (i, j, k, l = 1, 2, 3, 4, 5, 6) \tag{2.93}$$

where c_{ijkl} is the elastic constant.

Equation (2.92) establishes the relationship between stress and strain, called generalized Hooke's law or elastic constitutive equation. In Eq. (2.92), there are 36 elastic constants. These 36 constants are not independent; for isotropic materials, there are only two independent elastic constants.

First, it is proven that the principal stress direction coincides with the principal strain direction in the elastic state. Therefore, if x, y, and z are the main stress directions, then the shear strain components γ_{xy}, γ_{yz}, and γ_{zx} should be equal to zero. Therefore, according to Eq. (2.92), there is

$$\tau_{xy} = c_{41}\varepsilon_x + c_{42}\varepsilon_y + c_{43}\varepsilon_z \tag{2.94}$$

Now the coordinate system $Ox'y'z'$ is introduced, and the original coordinate system $Oxyz$ can be coincident with it after rotating 180° about the Y-axis. The direction cosine between the old and new axes is shown in Table 2.1.

Table 2.1 Direction cosine between old and new coordinate axes.

Coordinate Axes	x	y	z
x'	l_{11}	l_{12}	l_{13}
y'	l_{21}	l_{22}	l_{23}
z'	l_{31}	l_{32}	l_{33}

Therefore, there are:

$$l_{11} = l_{33} = \cos 180° = -1$$

$$l_{22} = \cos 0° = 1$$

$$l_{21} = l_{31} = l_{12} = l_{32} = l_{13} = l_{23} = \cos 90° = 0$$

For isotropic materials, the elastic constant should be independent of direction. So for the new coordinate system, there is:

$$\tau_{x'y'} = c_{41}\varepsilon_{x'} + c_{42}\varepsilon_{y'} + c_{43}\varepsilon_{z'} \tag{2.95}$$

It is obtained by the coordinate transformation equation of the stress component:

$$\begin{cases} \tau_{x'y'} = l_{11}l_{22}\tau_{xy} = -\tau_{xy} \\ \varepsilon_{x'} = l_{11}{}^2\varepsilon_x = \varepsilon_x \\ \varepsilon_{y'} = l_{22}{}^2\varepsilon_y = \varepsilon_y \\ \varepsilon_{z'} = l_{33}{}^2\varepsilon_z = \varepsilon_z \end{cases} \tag{2.96}$$

It can be obtained from Eqs. (2.95) and (2.96):

$$-\tau_{xy} = c_{41}\varepsilon_x + c_{42}\varepsilon_y + c_{43}\varepsilon_z \tag{2.97}$$

By comparing Eqs. (2.94) and (2.97), there is $\tau_{xy} = -\tau_{xy}$, therefore:

$$\tau_{xy} = 0$$

The same can be obtained

$$\tau_{yz} = \tau_{zx} = 0$$

It can be seen that for an isotropic elastomer, if the x, y, and z axes are the main strain directions, then the main stress direction must be the same; that is, the strain principal axis coincides with the stress principal axis.

Then consider the number of independent elastic constants of the isotropic material. First, coordinate axes Ox, Oy, and Oz with the direction of the principal stress. Therefore, the relationship between stress and principal strain can be obtained from Eq. (2.92) as follows:

$$\begin{pmatrix} \sigma_x \\ \sigma_y \\ \sigma_z \end{pmatrix} = \begin{pmatrix} c_{11} & c_{12} & c_{13} \\ c_{21} & c_{22} & c_{23} \\ c_{31} & c_{32} & c_{33} \end{pmatrix} \begin{pmatrix} \varepsilon_x \\ \varepsilon_y \\ \varepsilon_z \end{pmatrix} \tag{2.98}$$

In an isotropic medium, the influence of σ_x on ε_x is the same as that of σ_y on ε_y and σ_z on ε_z, that is, there should be $c_{11} = c_{22} = c_{33}$. Similarly, ε_x and ε_y have the same influence on σ_x, that is, $c_{12} = c_{13}$, and the similar ones are $c_{21} = c_{23}$, $c_{31} = c_{32}$ and so on

$$\begin{cases} c_{11} = c_{22} = c_{33} = a \\ c_{12} = c_{21} = c_{13} = c_{31} = c_{23} = c_{32} = b \end{cases} \tag{2.99}$$

It follows that for the principal axes of strain (denoted by 1, 2, and 3), the elastic constant has only two a's and b's. Substitute Eq. (2.100) into Eq. (2.98), and let $a - b = 2\mu$, $b = \lambda$, $\theta = \varepsilon_1 + \varepsilon_2 + \varepsilon_3$, the following elastic constitutive relation can be obtained:

$$\begin{cases} \sigma_1 = \lambda\theta + 2\mu\varepsilon_1 \\ \sigma_2 = \lambda\theta + 2\mu\varepsilon_2 \\ \sigma_3 = \lambda\theta + 2\mu\varepsilon_3 \end{cases} \tag{2.100}$$

where λ and μ are called Lame constants.

After the coordinate transformation, the constitutive relation in any coordinate system O-xyz can be obtained as

$$\begin{cases} \sigma_x = \lambda\theta + 2\mu\varepsilon_x, \tau_{xy} = \mu\gamma_{xy} \\ \sigma_y = \lambda\theta + 2\mu\varepsilon_y, \tau_{yz} = \mu\gamma_{yz} \\ \sigma_z = \lambda\theta + 2\mu\varepsilon_z, \tau_{zx} = \mu\gamma_{zx} \end{cases} \tag{2.101}$$

or abbreviated to

$$\sigma_{ij} = \lambda\delta_{ij}\theta + 2\mu\varepsilon_{ij} \tag{2.102}$$

The above proves that the elastic constants of isotropic homogeneous elastomers are only 2.

Some common engineering materials, such as reinforced concrete members with different bidirectional reinforcement, wood, etc., have obvious asymmetric elasticity. The elastic properties of these materials can often be considered symmetric with respect to the planes x = 0, y = 0, and z = 0 in a suitably chosen coordinate system. Because these three planes are orthogonal to each other, they are called orthogonal anisotropic materials.

The constitutive relation of orthotropic elastic materials can be obtained from the generalized Hooke's law Eq. (2.92) according to the requirement that the elastic constant c_{ij} remains unchanged when any coordinate axis is reversed:

$$\begin{cases} \sigma_x = c_{11}\varepsilon_x + c_{12}\varepsilon_y + c_{13}\varepsilon_z \\ \sigma_y = c_{21}\varepsilon_x + c_{22}\varepsilon_y + c_{23}\varepsilon_z \\ \sigma_z = c_{31}\varepsilon_x + c_{32}\varepsilon_y + c_{33}\varepsilon_z \\ \tau_{xy} = c_{44}\gamma_{xy} \\ \tau_{yz} = c_{55}\gamma_{yz} \\ \tau_{zx} = c_{66}\gamma_{zx} \end{cases} \tag{2.103}$$

In Eq. (2.103), there are nine elastic constants c_{ij}.

After ε_{ij} in Eq. (2.101) is solved, the strain component σ_{ij} can be represented by the stress component ε_{ij}, and the expression is:

$$\varepsilon_x = \frac{\lambda + \mu}{\mu\,(3\lambda + 2\mu)}\sigma_x - \frac{\lambda}{2\mu\,(3\lambda + 2\mu)}\left(\sigma_y + \sigma_z\right) \tag{2.104}$$

Equation (2.104) is slightly transformed and $\sigma = \sigma_{ii}$ can be abbreviated as:

$$\varepsilon_{ij} = \frac{\lambda + \mu}{\mu\,(3\lambda + 2\mu)}\sigma_{ij} - \frac{\lambda\delta_{ij}\sigma}{2\mu\,(3\lambda + 2\mu)} \tag{2.105}$$

Now consider the special case where the sides of the object are parallel to the coordinate axis, and derive the elastic constant and generalized Hooke's law commonly used in engineering. When the two pairs of edges where the normal direction of the boundary of the object is reconnected with the X-axis are used uniformly σ_x, and the other edges are free, it is known as the mechanics of materials:

$$\varepsilon_x = \frac{\sigma_x}{E} \tag{2.106}$$

$$\varepsilon_y = \varepsilon_z = -\upsilon\varepsilon_x = -\upsilon\frac{\sigma_x}{E} \tag{2.107}$$

where E and υ are the elastic modulus and Poisson's ratio, respectively.

By comparing Eq. (2.104) with Eqs. (2.106) and (2.107), it can be obtained:

$$\begin{cases} E = \dfrac{\mu\,(3\lambda + 2\mu)}{\lambda + \mu} \\ \upsilon = \dfrac{\lambda}{2\,(\lambda + \mu)} \end{cases} \tag{2.108}$$

In engineering, generalized Hooke's law is often expressed as E and υ, so Eq. (2.104) can be expressed as:

$$\begin{aligned} \varepsilon_x &= \frac{1}{E}\left(\sigma_x - \upsilon(\sigma_y + \sigma_z)\right), \gamma_{xy} = \frac{\tau_{xy}}{G} \\ \varepsilon_y &= \frac{1}{E}\left(\sigma_y - \upsilon(\sigma_x + \sigma_z)\right), \gamma_{yz} = \frac{\tau_{yz}}{G} \\ \varepsilon_z &= \frac{1}{E}\left(\sigma_z - \upsilon(\sigma_y + \sigma_x)\right), \gamma_{zx} = \frac{\tau_{zx}}{G} \end{aligned} \tag{2.109}$$

where

$$G = \frac{E}{2(1 + \upsilon)} \tag{2.110}$$

Is the shear modulus of an isotropic body. It can be seen from the expression of G that G is not an independent elastic constant. For isotropic elastomers, there are only two independent elastic constants: λ and μ, or E and υ. After a slight transformation of Eq. (2.109), it can be abbreviated as

$$\varepsilon_{ij} = \frac{1}{E}\sigma_{ij} - \frac{\upsilon\delta_{ij}\sigma}{E} \tag{2.111}$$

where $\sigma = \sigma_{ii}$. If the stress is relieved σ_{ij}, then Eq. (2.111) is converted to

$$\sigma_{ij} = \frac{E}{1+\upsilon}\varepsilon_{ij} - \frac{\upsilon E\delta_{ij}\theta}{(1+\upsilon)(1-2\upsilon)} \tag{2.112}$$

If ream

$$\begin{cases} \sigma_m = \dfrac{1}{3}\left(\sigma_x + \sigma_y + \sigma_z\right) \\ \varepsilon_m = \dfrac{1}{3}\left(\varepsilon_x + \varepsilon_y + \varepsilon_z\right) \end{cases} \tag{2.113}$$

Then generalized Hooke's law can be written as

$$\begin{cases} \sigma_m = 3K\varepsilon_m \\ S_{ij} = 2Ge_{ij} \end{cases} \tag{2.114}$$

In Eq. (2.114), S_{ij} and e_{ij} are the stress and strain partial tensors, respectively; $K = \frac{E}{3(1-2\upsilon)}$.

In the case of plane stress, since $\sigma_z = \tau_{yz} = \tau_{zx} = 0$, Eq. (2.109) can be converted to:

$$\begin{cases} \varepsilon_x = \dfrac{1}{E}\left(\sigma_x - \upsilon\sigma_y\right) \\ \varepsilon_y = \dfrac{1}{E}\left(\sigma_y - \upsilon\sigma_x\right) \\ \varepsilon_z = \dfrac{\upsilon}{E}\left(\sigma_y + \sigma_x\right) \end{cases} \tag{2.115}$$

If the strain component is used to represent the stress component, it can be obtained from Eq. (2.112):

$$\begin{cases} \sigma_x = \dfrac{E}{(1-\upsilon^2)}\left(\varepsilon_x + \upsilon\varepsilon_y\right) \\ \sigma_y = \dfrac{E}{(1-\upsilon^2)}\left(\varepsilon_y + \upsilon\varepsilon_x\right) \\ \tau_{xy} = G\gamma_{xy} \end{cases} \tag{2.116}$$

For a plane strain problem, since $\varepsilon_z = \gamma_{yz} = \gamma_{zx} = 0$, can be obtained from Eq. (2.109)

$$\begin{cases} \varepsilon_x = \dfrac{1+\upsilon}{E}\left((1-\upsilon)\,\sigma_x - \upsilon\sigma_y\right) \\ \sigma_y = \dfrac{1+\upsilon}{E}\left((1-\upsilon)\,\sigma_y - \upsilon\sigma_x\right) \\ \gamma_{xy} = \dfrac{\tau_{xy}}{G} \end{cases} \tag{2.117}$$

The stress is obtained by solving the equation as:

$$\begin{cases} \sigma_x = \dfrac{E}{(1+\upsilon)(1-2\upsilon)}\left((1-\upsilon)\,\varepsilon_x + \upsilon\varepsilon_y\right) \\ \sigma_y = \dfrac{E}{(1+\upsilon)(1-2\upsilon)}\left((1-\upsilon)\,\varepsilon_y + \upsilon\varepsilon_x\right) \\ \sigma_z = \dfrac{\upsilon E}{(1+\upsilon)(1-2\upsilon)}\left(\varepsilon_x + \varepsilon_y\right) \\ \tau_{xy} = G\gamma_{xy} \end{cases} \tag{2.118}$$

By comparing the generalized Hooke's law of the plane stress and plane strain problems above, it can be seen that, for example, E is replaced by E_1 and υ is replaced by υ_1 in the stress–strain relation Eq. (2.115) of the plane stress problem

$$E_1 = \frac{E}{1-\upsilon^2},\ \upsilon_1 = \frac{\upsilon}{1-\upsilon} \tag{2.119}$$

The stress–strain relation Eq. (2.117) for the plane strain problem can then be obtained.

As can be seen from Eq. (2.114), the deformation of the object can be divided into two parts: one is the relative volume deformation caused by the equal normal stress (hydrostatic pressure); The other part is the change of the geometry of the object caused by the action of the stress partial tensor. It can also be considered that the former deformation does not include a change in the shape of the object (that is, distortion), and the latter deformation does not include a change in volume, so that the deformation can be decomposed into two parts, which is very useful in plastic theory.

If the original volume of the tiny hexahedron element in the deformed body is V_0, then:

$$V_0 = dx \cdot dy \cdot dz \tag{2.120}$$

The volume after deformation is:

$$\begin{aligned} V_0 &= (1+\varepsilon_x)\,dx \cdot \left(1+\varepsilon_y\right) dy \cdot (1+\varepsilon_z)\,dz \\ &= dxdydz\left(\left(1+\varepsilon_x+\varepsilon_y+\varepsilon_z\right) + o\left(\varepsilon^2\right)\right) \end{aligned} \tag{2.121}$$

Omit the higher order trace, get:

$$V = V_0 + V_0\theta \tag{2.122}$$

where $\theta = 1 + \varepsilon_x + \varepsilon_y + \varepsilon_z$, or

$$\theta = \frac{\Delta V}{V_0} \tag{2.123}$$

It can be seen that θ is the relative volume change per unit volume before and after deformation, or called volume strain. It is derived from generalized Hooke's law:

$$\theta = \frac{1-2\upsilon}{E}\left(\sigma_x + \sigma_y + \sigma_z\right) \tag{2.124}$$

When $\sigma_x = \sigma_y = \sigma_z = \sigma_m$,

$$\theta = \frac{3(1-2\upsilon)}{E}\sigma_m \tag{2.125}$$

or

$$K = \frac{\sigma_m}{\theta} = \frac{E}{3\,(1-2\upsilon)} \tag{2.126}$$

Here, K is the bulk modulus of elasticity, also known as the volumetric modulus. Substituting $\theta = 3\sigma_m$ into Eq. (2.126), the first equation in Eq. (2.114) can be obtained.

References

1 Yang, G. (2018). *Elasticity Mechanics*. Beijing: Higher Education Press.
2 Zhao, Y. (2016). *Modern Continuum Mechanics*. Beijing: Science Press.
3 Rankine, W. J. M. (1856). On axes of elasticity and crystalline forms. *Philosophical Transactions of the Royal Society of London* 146: 261–285.
4 Truesdell, C. (1968). *Essays in the History of Mechanics*. New York: Springer-Verlag.
5 Eckart, C. (1940). The thermodynamics of irreversible processes. III. Relativistic theory of the simple fluid. *Physical Review* 58 (10): 919.
6 Brillouin, M. (1900). Théorie moléculaire des gaz. Diffusion du mouvement et de l'énergie. *Annals of Physics and Chemistry* 20: 440–485.
7 Chaves, E. W. V. (2013). *Notes on Continuum Mechanics*. Springer Nature.

3

Plastic Constitutive Theory

In most cases, constitutive models of materials are formulated based on a combination of elasticity and plasticity to describe their macroscopic mechanical behavior. In reality, however, the macroscopic mechanical performance of materials is closely related to structural deformation and plastic evolution at the mesoscopic and microscopic scales. Investigating the evolution mechanisms and fundamental characteristics of micro- and meso-scale structures under mechanical loading is essential for formulating a physically sound macroscopic constitutive model [1–3]. This chapter focuses exclusively on existing macroscopic phenomenological constitutive models and provides an introduction to plastic behavior and plastic constitutive theories of materials.

3.1 Plastic Behavior of Solid Materials

In structural engineering and solid mechanics, understanding the plastic behavior of materials is essential for evaluating their service performance, failure mechanisms, and processing or forming characteristics. Plastic deformation refers to the irreversible strain that occurs once the material exceeds its yield limit. Under the influence of microstructural characteristics, loading conditions, and environmental factors, different materials exhibit distinct plastic responses. Metals and concrete, as two representative classes of engineering materials, demonstrate fundamentally different mechanisms of plasticity, yet both significantly impact mechanical performance, structural stability, and service life assessment. A thorough investigation of the plastic evolution and constitutive characteristics of various solid materials is of great theoretical and practical significance for developing unified and predictive mechanical models.

The plasticity of metallic materials primarily originates from the slip and accumulation of dislocations within the crystal lattice, and the deformation mechanisms are closely tied to the crystal structure. Face-centered cubic (FCC) metals, such as copper and aluminum, possess abundant slip systems and thus exhibit excellent ductility. Body-centered cubic (BCC) metals, such as low-carbon steel, show sensitivity to temperature and strain rate; Figure 3.1 illustrates the tensile curve of low-carbon steel. In contrast, hexagonal close-packed (HCP) metals like titanium and magnesium have limited slip systems and tend to develop texture and twinning during deformation. From a macroscopic modeling perspective, metallic plasticity is often described using elastoplastic constitutive theories, including isotropic

Constitutive Models of Solid Materials: From Mechanical Principles to Engineering Applications, First Edition.

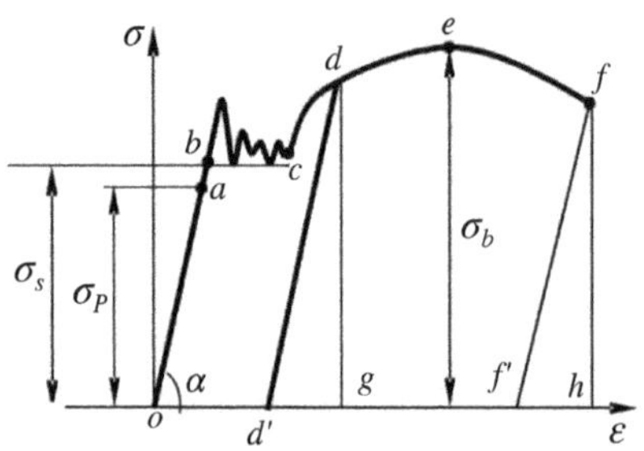

Figure 3.1 Tensile stress–strain curve of low-carbon steel.

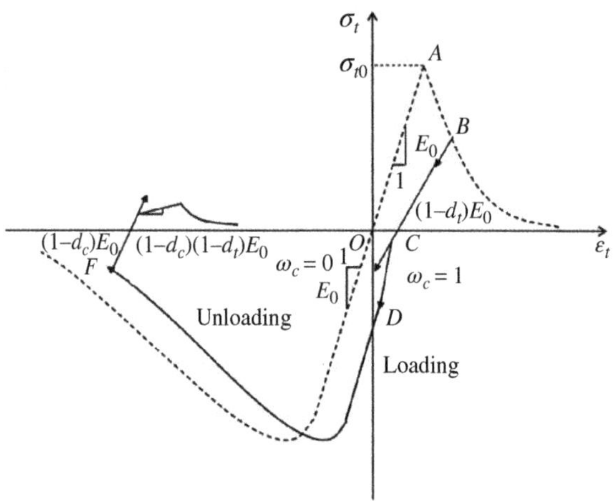

Figure 3.2 Uniaxial tensile/compressive stress–strain curve of concrete.

hardening, anisotropic yield criteria (e.g. Hill and Barlat models), and crystal plasticity theory [4–7]. The latter is particularly suited for predicting the microscale response, stress localization, and texture evolution of polycrystalline materials under complex loading. These models serve as foundational tools for conventional manufacturing processes and offer robust theoretical support for emerging technologies such as additive manufacturing, microforming, and high-temperature service applications.

Concrete, as a multiphase composite and heterogeneous material, exhibits plastic behavior mechanisms that differ significantly from those of metals. Figure 3.2 illustrates a schematic stress–strain curve for concrete under tension and compression. During loading, concrete often demonstrates coupled mechanisms such as crack initiation and interfacial failure between aggregates and mortar, which together shape its apparent plastic characteristics. Under uniaxial compression, concrete shows a certain degree of plastic flow and stress redistribution. However, under tension, energy is primarily released through crack propagation, resulting in limited observable macroscopic plasticity. To capture its complex damage-plasticity coupling behavior, researchers have proposed a variety of non-linear constitutive models, such as the concrete damage plasticity (CDP) model, continuum damage mechanics models, and mesoscale constitutive models. These models typically introduce internal variables—such as crack density or plastic strain tensors—to describe degradation characteristics and nonassociated flow rules. Such models are widely applied in areas including ultimate load analysis, prediction of impact and explosion responses, and performance evaluation under high-temperature or freeze-thaw conditions. A detailed discussion on damage modeling in solid materials will be presented in Chapter 4.

Overall, although metals and concrete differ in their microscopic mechanisms, constitutive formulations, and numerical treatment approaches, their plastic behaviors commonly exhibit strong nonlinearity, path dependence, and localization tendencies. The development of plastic constitutive models, together with multiscale mechanical analysis, provides effective tools for understanding the deformation and failure processes of complex structures. Advancements in experimental techniques and numerical simulations will further promote research on the plasticity of metals and concrete toward the exploration of microscale mechanisms, data-driven modeling, and multiphysics coupling. These efforts are expected to play an increasingly central role in the design and performance prediction of advanced structural materials.

3.2 Variable Definitions

3.2.1 Decomposition of the Strain Tensor

In continuum mechanics and elasticity theory, the strain tensor is the fundamental quantity used to characterize material deformation. To elucidate the contributions of different physical processes to the material response, the strain tensor is often decomposed. Depending on the specific research context, several standard decomposition approaches are employed.

1. Additive Decomposition of Elastic and Plastic Strain
   This is the most commonly used method in classical elastoplastic theory, particularly under small deformation conditions. Under this assumption, the total strain tensor ε is decomposed into the sum of its elastic and plastic components:

   $$\varepsilon = \varepsilon^e + \varepsilon^p \tag{3.1}$$

   Here, ε^e denotes the elastic strain, corresponding to the recoverable part of the deformation, while ε^p represents the plastic strain, corresponding to the permanent, nonrecoverable deformation induced by mechanisms such as dislocation slip. This form of decomposition is commonly used in classical metal constitutive modeling (e.g. J_2 plasticity theory).
2. Multiplicative Decomposition of the Deformation Gradient
   When finite deformation (large deformation) is involved, the deformation gradient tensor $\boldsymbol{F}$ is adopted and decomposed multiplicatively as:

   $$\boldsymbol{F} = \boldsymbol{F}^e \cdot \boldsymbol{F}^p \tag{3.2}$$

   Here, $\boldsymbol{F}^p$ denotes the plastic part of the deformation caused by mechanisms such as crystal slip. resulting in irreversible deformation; $\boldsymbol{F}^e$ reflects the elastic response that remains after $\boldsymbol{F}^p$, such as lattice stretching and rotation. This type of decomposition is widely applied in crystal plasticity theory and high-precision finite element analysis, enabling the clear separation of plastic slip and elastic deformation.

3. Decomposition of Mechanical and Thermal Strains
In thermo-mechanical coupling analyses, the strain tensor is also commonly decomposed into the mechanical strain and the thermal strain caused by thermal expansion:

$$\boldsymbol{\varepsilon} = \boldsymbol{\varepsilon}^{\mathbf{mech}} + \boldsymbol{\varepsilon}^{\mathrm{th}} \tag{3.3}$$

Here, $\boldsymbol{\varepsilon}^{\mathbf{mech}}$ represents the mechanical strain due to loading, and $\boldsymbol{\varepsilon}^{\mathrm{th}}$ is the strain induced by temperature changes. In general, the thermal expansion strain of a material is linearly related to the temperature variation. This decomposition is widely used in the structural analysis of high-temperature environments, such as in civil engineering structures, aerospace materials, and nuclear components.

3.2.2 Strain Rate and Strain Increment

As previously discussed, the stress–strain relationship in plastic deformation can only be determined by understanding the deformation path or loading history. To track the deformation path, it is necessary to introduce the concepts of strain rate and strain increment. When the medium is in motion, let the velocity of a material point be $v_i\ (x, y, z; t)$, Considering a specific moment during the deformation process as the reference point, if the material point undergoes an infinitesimal displacement $du_i = v_i dt$ after an infinitesimal time interval t, then

$$d\varepsilon_{ij} = \frac{1}{2}\left(du_{i,j} + du_{j,i}\right) = \frac{1}{2}\left(v_{i,j} + v_{j,i}\right) dt \tag{3.4}$$

Let

$$\dot{\varepsilon}_{ij} = \frac{1}{2}\left(v_{i,j} + v_{j,i}\right) \tag{3.5}$$

It is referred to as the strain rate tensor.

For slow deformation processes at room temperature, plastic deformation is independent of time. The time mentioned above is only a table showing the order of deformation or loading. To reflect the characteristic that is not affected by the time parameters, it is more appropriate to use the strain increment tensor $d\varepsilon_{ij}$ instead of the parameter $\dot{\varepsilon}_{ij}$. If the instantaneous displacement increment is expressed by $d\varepsilon_{ij}$, the strain increment tensor is

$$d\varepsilon_{ij} = \frac{1}{2}\left(du_{i,j} + du_{j,i}\right) \tag{3.6}$$

Similarly, the strain increment strength can be defined as

$$\boldsymbol{d\varepsilon_i} = \frac{\sqrt{2}}{3}\sqrt{\left(d\varepsilon_1 - d\varepsilon_2\right)^2 + \left(d\varepsilon_2 - d\varepsilon_3\right)^2 + \left(d\varepsilon_3 - d\varepsilon_1\right)^2} \tag{3.7}$$

Note that ε_{ij} cycles are calculated from the initial position, while $\dot{\varepsilon}_{ij}$ cycles are calculated from the instantaneous state, so in general $\dot{\varepsilon}_{ij} \neq d\varepsilon_{ij}/dt$ only when there is a small deformation. In addition, the spindles of $\dot{\varepsilon}_{ij}$ and ε_{ij} generally do not coincide.

In plastic mechanics, stress is not only related to strain, but also to the whole deformation history [3]. If the deformation history is still characterized by a set of internal variables $\xi_\beta\ (\beta = 1, 2, ..., n)$ deformation history, the stress can be written as

$$\sigma_{ij} = \sigma_{ij}\left(\varepsilon_{kl}, \xi_\beta\right) \tag{3.8}$$

When the stress is fixed, there is a single corresponding relationship between σ_{ij} and ε_{ij}, and the strain can also be expressed by stress

$$\varepsilon_{ij} = \varepsilon_{ij}\left(\sigma_{kl}, \xi_\beta\right) \tag{3.9}$$

Thus, the rate of change or increment of stress and strain can be expressed as follows

$$\dot{\sigma}_{ij} = \frac{\partial \sigma_{ij}\left(\varepsilon_{kl}, \xi_\beta\right)}{\partial \varepsilon_{kl}}\dot{\varepsilon}_{kl} + \frac{\partial \sigma_{ij}\left(\varepsilon_{kl}, \xi_\beta\right)}{\partial \xi_\beta}\dot{\xi}_\beta = \dot{\sigma}_{ij}^{\mathrm{e}} + \dot{\sigma}_{ij}^{\mathrm{p}} \tag{3.10}$$

$$d\sigma_{ij} = \frac{\partial \sigma_{ij}\left(\varepsilon_{kl}, \xi_\beta\right)}{\partial \varepsilon_{kl}}d\varepsilon_{kl} + \frac{\partial \sigma_{ij}\left(\varepsilon_{kl}, \xi_\beta\right)}{\partial \xi_\beta}d\xi_\beta \tag{3.11}$$

$$\dot{\varepsilon}_{ij} = \frac{\partial \varepsilon_{ij}\left(\sigma_{kl}, \xi_\beta\right)}{\partial \sigma_{kl}}\dot{\sigma}_{kl} + \frac{\partial \varepsilon_{ij}\left(\sigma_{kl}, \xi_\beta\right)}{\partial \xi_\beta}\dot{\xi}_\beta = \dot{\varepsilon}_{ij}^{\mathrm{e}} + \dot{\varepsilon}_{ij}^{\mathrm{p}} \tag{3.12}$$

$$d\varepsilon_{ij} = \frac{\partial \varepsilon_{ij}\left(\varepsilon_{kl}, \xi_\beta\right)}{\partial \sigma_{kl}}d\sigma_{kl} + \frac{\partial \varepsilon_{ij}\left(\varepsilon_{kl}, \xi_\beta\right)}{\partial \xi_\beta}d\xi_\beta \tag{3.13}$$

Let

$$L_{ijkl} = \frac{\partial \sigma_{ij}\left(\varepsilon_{kl}, \xi_\beta\right)}{\partial \varepsilon_{kl}}, M_{ijkl} = \frac{\partial \varepsilon_{ij}\left(\sigma_{kl}, \xi_\beta\right)}{\partial \sigma_{kl}} \tag{3.14}$$

where elastic stress rate: $\dot{\sigma}_{ij}^{\mathrm{e}} = L_{ijkl}\dot{\varepsilon}_{kl}$; plastic stress rate $\dot{\sigma}_{ij}^{\mathrm{p}} = \frac{\partial \sigma_{ij}}{\partial \dot{\xi}_\beta}\dot{\xi}_\beta$; elastic strain rate: $\dot{\varepsilon}_{ij}^{\mathrm{e}} = M_{ijkl}\dot{\sigma}_{kl}$; and plastic strain rate: $\dot{\varepsilon}_{ij}^{\mathrm{p}} = \frac{\partial \varepsilon_{ij}}{\partial \xi_\beta}\dot{\xi}_\beta$.

Here, L_{ijkl} and M_{ijkl} are fourth-order elastic tensors. In general, the former is not only related to strain, but also related to internal variables; while the latter is not only related to stress, but also related to internal variables. It shows that the elastic property depends on the plastic deformation, so the elastic deformation and the plastic deformation are coupled. To simplify the analysis, subsequent discussions will consider only the case without coupling between elastic and plastic deformation. Here

$$\sigma_{ij} = \sigma_{ij}^{\mathrm{e}} + \sigma_{ij}^{\mathrm{p}}; \varepsilon_{ij} = \varepsilon_{ij}^{\mathrm{e}} + \varepsilon_{ij}^{\mathrm{p}} \tag{3.15}$$

and

$$\sigma_{ij}^{\mathrm{e}} = L_{ijkl}\varepsilon_{kl}^{\mathrm{e}}; \varepsilon_{ij}^{\mathrm{e}} = M_{ijkl}\sigma_{kl}^{\mathrm{e}} \tag{3.16}$$

In the form, L_{ijkl} and M_{ijkl} are constants that depend only on the properties of the material. Plastic stress σ_{ij}^{p} and plastic strain $\varepsilon_{ij}^{\mathrm{p}}$ are only functions of internal variables, and only when the internal variables change, σ_{ij}^{p} and $\varepsilon_{ij}^{\mathrm{p}}$ will change accordingly.

When $\xi_\beta = 0$, it is the generalized Hooke's law

$$\dot{\sigma}_{ij} = C_{ijkl}\dot{\varepsilon}_{kl} \tag{3.17}$$

For isotropic materials, write them in the form of components

$$\left.\begin{aligned}\dot{\varepsilon}_{11}&=\frac{1}{E}\left[\dot{\sigma}_{11}-\nu\left(\dot{\sigma}_{22}+\dot{\sigma}_{33}\right)\right], \quad \dot{\varepsilon}_{23}=\left(\frac{1+\nu}{E}\right)\dot{\sigma}_{23}\\ \dot{\varepsilon}_{22}&=\frac{1}{E}\left[\dot{\sigma}_{22}-\nu\left(\dot{\sigma}_{11}+\dot{\sigma}_{33}\right)\right], \quad \dot{\varepsilon}_{31}=\left(\frac{1+\nu}{E}\right)\dot{\sigma}_{31}\\ \dot{\varepsilon}_{33}&=\frac{1}{E}\left[\dot{\sigma}_{33}-\nu\left(\dot{\sigma}_{11}+\dot{\sigma}_{22}\right)\right], \quad \dot{\varepsilon}_{12}=\left(\frac{1+\nu}{E}\right)\dot{\sigma}_{12}\end{aligned}\right\} \tag{3.18}$$

Equation (3.18) can be rewritten as the relationship between partial stress rate and partial strain rate:

$$\begin{aligned}\dot{e}_{ij}&=\frac{1+\nu}{E}\dot{S}_{ij}=\frac{1}{2\mu}\dot{S}_{ij}\\ \dot{\varepsilon}_{kk}&=\frac{1-2\nu}{E}\dot{\sigma}_{kk}=\frac{1}{3K}\dot{\sigma}_{kk}\end{aligned} \tag{3.19}$$

where E is Young's modulus; υ is Poisson's ratio; μ is the shear modulus; and $\frac{1}{2\mu}=(1+\nu)/\mathrm{E}$. Bulk modulus $\mathrm{K}=\mathrm{E}/3\,(1-2\nu)$. Equation (3.48) can also be expressed in incremental form:

$$\begin{aligned}de_{ij}&=\frac{1}{2\mu}dS_{ij}\\ d\varepsilon_{kk}&=\frac{1}{3K}d\sigma_{kk}\end{aligned} \tag{3.20}$$

The former is five independent equations (since $S_{ii}=0$), while the latter is a single independent equation, resulting in six independent equations.

To facilitate the extension to plastic situations and to be consistent with the form of plastic constitutive equations, the general Hooke's law relationship can also be expressed in full form as:

$$\begin{aligned}e_{ij}&=\frac{3\varepsilon_i}{2\sigma_i}S_{ij}\\ \sigma_i&=3\mu\varepsilon_i\end{aligned} \tag{3.21}$$

The above discussion indicates that once the elastic tensor is given, the establishment of the elastoplastic constitutive equation reduces to the problem of correctly expressing the relationships regarding plastic stress rates and plastic strain rates, or the expressions for plastic stress increments and plastic strain increments.

3.3 Drucker Postulate

Previously, the conditions for material strengthening during plastic deformation processes, as well as the criteria for loading, unloading, and neutral loading, were briefly introduced. Based on a large number of macroscopic experiments, the American mechanic (Drucker) proposed an important hypothesis regarding material strengthening during the loading process under general stress states, known as the Drucker strengthening postulate. According to this postulate, a fundamental and universal geometric property of the loading surface, namely its convexity, can be derived, along with the loading and unloading criteria. Moreover, this postulate provides the basis for formulating the laws of plastic deformation, that is, the plastic constitutive relations.

The tensile stress–strain curve of materials can take several forms, as shown in Figure 3.3.

For the material shown in Figure 3.3a, with the increment of loading stress $d\sigma > 0$, the corresponding strain increment $d\varepsilon > 0$ is produced, the stress–strain curve increases monotonously, and the material is strengthened. In this deformation process, $d\sigma \cdot d\varepsilon > 0$ deformation indicates that the additional stress $d\sigma$ does positive work on

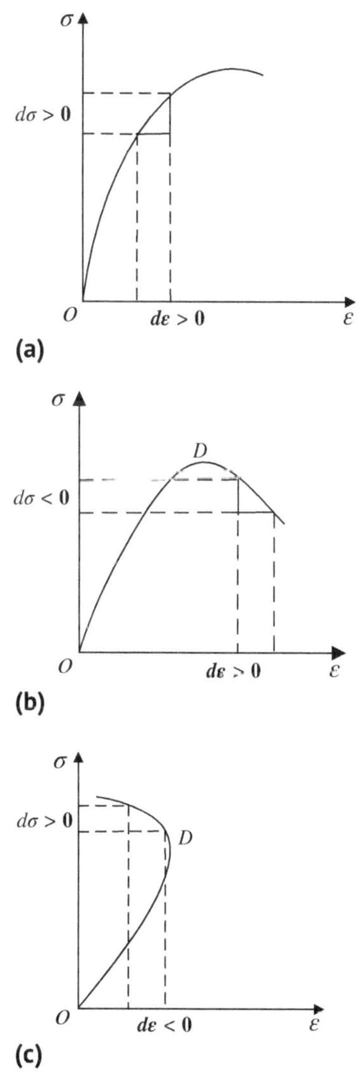

Figure 3.3 Tensile stress–strain curves: (a) stable material; (b) unstable material; and (c) physically impossible case.

the strain increment $d\varepsilon$, and the material with this property is called stable material or strengthening material. The material without strengthening effect is also a stable material $d\sigma = 0$, so $d\sigma \cdot d\varepsilon = 0$. For stable materials, it should generally be written as

$$d\sigma \cdot d\varepsilon \geq 0 \tag{3.22}$$

For the material shown in Figure 3.3b, the stress–strain curve decreases after point D, and the stress decreases ($d\varepsilon > 0$), as the strain increases ($d\sigma < 0$). At this time, although the total stress still does positive work, the stress increment does negative work, that is $d\sigma \cdot d\varepsilon < 0$. Such materials are called unstable materials or softened materials, and the descending part of the curve is called the softening stage.

For the material shown in Figure 3.3c, the strain of the stress–strain curve decreases with the increase of stress in the section after the point, which is equivalent to a suspender with a heavy object, which increases when the weight of the suspended object is increased, which violates the law of conservation of energy, so it is impossible.

Only stable materials, including reinforced materials and ideal elastoplastic materials, will be discussed.

As shown in Figure 3.4, it can be seen from the tensile curve that when the stress level σ^0 of the material under a certain loading history begins to slowly load to a certain stress σ after yield, an additional stress increment $d\sigma$ is added at this time. It will cause a corresponding plastic deformation increment $d\varepsilon^p$. Then, the stress is slowly reduced back to the original stress level σ^0. The load path is shown in the ABCDE in Figure 3.4. In this stress cycle $\sigma^0 \to \sigma \to \sigma + d\sigma \to \sigma^0$, the elastic strain produced in the loading stage can be recovered in the unloading stage, and the corresponding elastic strain energy can be completely released. The rest is the plastic work consumed by irrecoverable plastic deformation (the shaded part of Figure 3.4), which is irreversible and will always be greater than zero. This part of plastic work can be divided into two shaded areas of M_1, M_2, and two parts shown in Figure 3.4, from which the following two inequalities can be written

$$\begin{aligned} (\sigma - \sigma^0) \cdot d\varepsilon^p &> 0 \\ d\sigma \cdot d\varepsilon^p &\geq 0 \end{aligned} \tag{3.23}$$

The work done by stress on plastic strain is nonnegative, in which the equal sign of the second formula is suitable for ideal plastic materials.

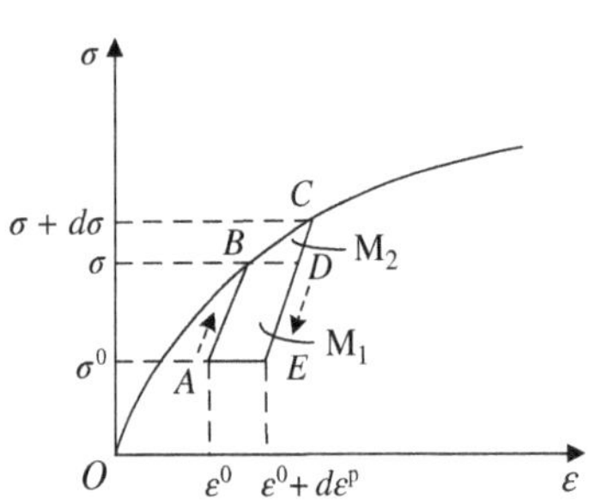

Figure 3.4 Irreversibility of plastic work.

In 1952, combined with the first law of thermodynamics, Drucker extended the above two formulas to the loading process of complex stress states, and put forward an axiom about the irreversible plastic function of stable materials, which is now called the Drucker postulate. It takes the form of:

$$\left(\sigma_{ij}-\sigma_{ij}^{0}\right)\cdot d\varepsilon_{ij}^{P}\geqslant 0,\ d\sigma_{ij}\cdot d\varepsilon_{ij}^{P}\geqslant 0 \tag{3.24}$$

The postulate is described as follows.

As shown in Figure 3.5, any point in the object is in equilibrium under a certain stress level σ_{ij}^{0} (point A in the figure) on the loading surface $f_1 = 0$ after a certain loading history. Now, through the loading path a, the stress state just enters the yield stress state (point B in the figure) corresponding to the loading surface $f_1 = 0$, and the stress level is σ_{ij}, which is an elastic deformation process. Then continue to apply a small load (path b), so that the stress state of this point reaches a neighboring loading surface $f_2 = 0$ (point C), and the corresponding stress increment is $d\sigma_{ij}$, then a new plastic strain increment will occur, which is set to $d\varepsilon_{ij}$. Then unload along a certain path c and return the stress level to σ_{ij}^{0}, which is also an elastic deformation process.

The Drucker postulate requires that in the above stress cycle, the work done by the additional stress is nonnegative, which is needed

$$\oint_{\sigma_{ij}^{0}}\left(\sigma_{ij}-\sigma_{ij}^{0}\right)d\varepsilon_{ij}\geqslant 0 \tag{3.25}$$

In Eq. (3.25), the integral symbol $\oint_{\sigma_{ij}}()$ denotes a cyclic integral that starts at σ_{ij}^{0} and returns to σ_{ij}^{0}. This is consistent with a large number of experimental observations. Because after each stress cycle, there must be residual deformation in the body, so that the strain in the stress cycle cannot return to the original value, this cycle is not a closed cycle in the sense of thermodynamics. The Drucker axiom is usually called the quasi-thermodynamic axiom.

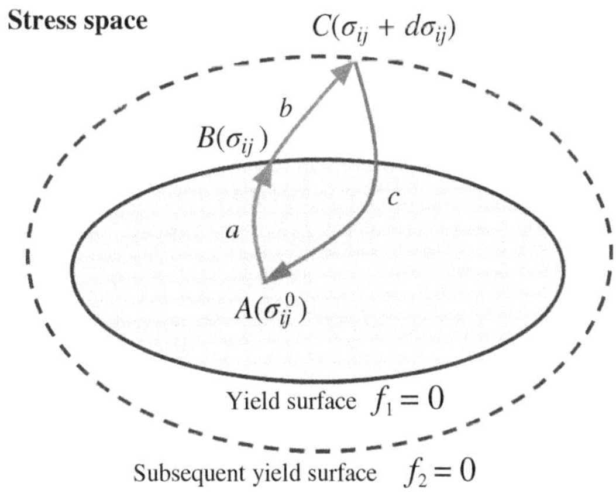

Figure 3.5 Description of the Drucker axiom.

Since the elastic strain is reversible, the sum of the work done by the stress on the elastic strain is zero in the above stress cycle, which requires:

$$\oint_{\sigma_{ij}^0} \left(\sigma_{ij} - \sigma_{ij}^0\right) d\varepsilon_{ij}^p \geq 0 \tag{3.26}$$

In the above stress cycle, plastic deformation occurs only in the loading process (path *b*). Therefore, if high-order small quantities are omitted, the Drucker axiom is required:

$$\left(\sigma_{ij} - \sigma_{ij}^0\right) \cdot d\varepsilon_{ij}^p + \frac{1}{2} d\sigma_{ij} \cdot d\varepsilon_{ij}^p \geqslant 0 \tag{3.27}$$

Two situations can be distinguished here:

1. If the initial state is in the interior of the loading surface, that is $\sigma_{ij} \neq \sigma_{ij}^0$. Since $d\sigma_{ij}$ is an arbitrary infinitesimal quantity, the higher-order small quantity compared with σ_{ij} inverse can be omitted; it can be concluded that

$$\left(\sigma_{ij} - \sigma_{ij}^0\right) \cdot d\varepsilon_{ij}^p \geq 0 \tag{3.28}$$

   Note that there is an equal sign on it, which is slightly different from that in uniaxial tension, considering that a neutral variable load is allowed under complex stress conditions, where $d\varepsilon_{ij}^p = 0$.
2. If $\sigma_{ij} = \sigma_{ij}^0$, when the starting state is above the loading surface, there is

$$d\sigma_{ij} \cdot d\varepsilon_{ij}^p \geqslant 0 \tag{3.29}$$

   The equal sign in the equation holds in two cases, that is, in the case of neutral load change $\left(d\varepsilon_{ij}^p = 0\right)$ and for perfectly plastic materials $\left(d\sigma_{ij} = 0\right)$. Equation (3.29) is called Drucker stability condition. According to this condition, it can be inferred that the material must be stable if the stress closed cycle can be constructed. It is also regarded as the mathematical definition of reinforcement, and it is called the uniqueness condition of reinforcement.

When the plastic strain increment is determined, the plastic work caused by the plastic strain increment caused by the stress on the yield surface is the largest among the possible stresses corresponding to the plastic solids (without violating the yield condition). Therefore, the above inequality is also known as the maximum plastic work principle (principle of maximum plastic work). It is equivalent to the Drucker axiom, and any material that satisfies these inequalities is a stable material. Elastic materials, ideal elastic-plastic materials, and strengthening materials are all stable materials.

3.3.1 Important corollary of the Drucker axioms

Corollary 1: The convexity of yield surfaces (including initial yield surfaces and subsequent yield surfaces)

Let σ_{ij}^0 denote a point on the yield surface, and σ_{ij} is the point on the yield surface. If the stress space coincides with the plastic strain space, and the origin of $d\varepsilon_{ij}^p$ is placed at the stress point of the yield surface, as shown in Figure 3.6a, σ_{ij}^0 is represented by vector $\overrightarrow{OA}$, σ_{ij}

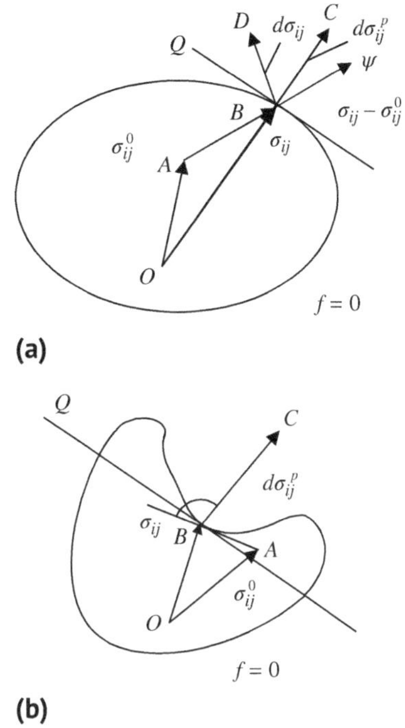

Figure 3.6 Convexity of yield surface: (a) convex yield surface and (b) non-convex yield surface.

is represented by vector $\overrightarrow{\mathrm{OB}}$, $d\varepsilon_{ij}^p$ is represented by vector $\overrightarrow{\mathrm{BC}}$, $d\sigma_{ij}$ is represented by vector $\overrightarrow{\mathrm{BD}}$, then it requires

$$\vec{\mathrm{AB}} \cdot \vec{\mathrm{BC}} \geqslant 0 \tag{3.30}$$

where

$$\left|\vec{\mathrm{AB}}\right| \cdot \left|\vec{\mathrm{BC}}\right| \cdot \cos\psi \geqslant 0 \tag{3.31}$$

This means that the angle ψ between the two vectors is the acute angle $(-\pi/2 \leqslant \psi \leqslant \pi/2)$. Make the tangent plane Q perpendicular to $d\varepsilon_{ij}^p$ through point B, that is, σ_{ij}^0 must be on the opposite side of $d\varepsilon_{ij}^p$. While σ_{ij}^0 is an arbitrary point on the yield plane, it can be seen that the whole yield plane is on the opposite side of the tangent plane to $d\varepsilon_{ij}^p$. Because σ_{ij} is an arbitrary point on the yield surface, it can be known that the yield surface is convex everywhere.

On the other hand, if the yield surface is not convex, then $\overrightarrow{\mathrm{AB}}$ is not always on the same side of Q, as shown in Figure 3.6b, that is, make $\overrightarrow{\mathrm{BC}} \perp \mathrm{Q}$, but a point A can always make $\overrightarrow{\mathrm{AB}}$ and $\overrightarrow{\mathrm{BC}}$ obtuse.

Corollary 2: The normality of the plastic strain increment vector along the yield surface (i.e. the orthogonal flow rule).

When the yield surface is smooth at the stress point σ_{ij}, the plane that passes through σ_{ij} and is tangent to the yield surface will be unique. If $d\varepsilon_{ij}^{p}$ and the yield surface are not orthogonal at σ_{ij}, as shown in Figure 3.7, then a plane perpendicular to $d\varepsilon_{ij}^{p}$ through σ_{ij} will be cut from the yield surface. In this way, there will be interior points of the yield surface on both sides of the tangent plane Q, so there must be a certain point σ_{ij}^{0} on the yield surface, which makes the inequality Eq. (3.25) or Eq. (3.26) not true. According to the Drucker axiom, this is impossible. Therefore, Q must be the tangent plane passing through the yield surface at σ_{ij}, and $d\varepsilon_{ij}^{p}$ must be along the outer normal direction of the yield surface. On the other hand, the direction of the outer normal of any point on the bending surface is consistent with its gradient direction, so

$$d_{ij}^{p} = d\lambda \cdot \frac{\partial f}{\partial \sigma_{ij}} \qquad d\lambda \geqslant 0 \tag{3.32}$$

where $d\lambda$ is a nonnegative proportional coefficient and a scalar. Eq. (3.32) is called the orthogonal flow rule (normality rule, orthogonality). It is proven that the magnitude of $d\varepsilon_{ij}^{p}$ is related to $d\sigma_{ij}$, but the direction has nothing to do with $d\sigma_{ij}$, because the direction is only determined by the yield surface, which is determined by σ_{ij} and has nothing to do with $d\sigma_{ij}$.

Equation (3.32) shows that if f is regarded as a plastic potential function and equals to the yield function, Eq. (3.32) can be called the plastic flow law associated with the yield bar. It connects the plastic strain increment with the yield condition, and once the yield condition is determined, the plastic strain increment can be determined according to this relation [3]. Therefore, it is of great significance to the study of physical relations in plasticity. In fact, as demonstrated in the following section, the flow law serves as a fundamental basis for deriving the incremental plastic constitutive relation and provides the theoretical foundation for plastic strain analysis. The flow rule is the third basic element to establish the plastic constitutive relation.

In summary, it has been proved that the inequality can be established only if the yield surface is convex at the same time and the plastic strain increment vector is along the normal direction of the yield surface. This is the geometric meaning of Drucker axiom.

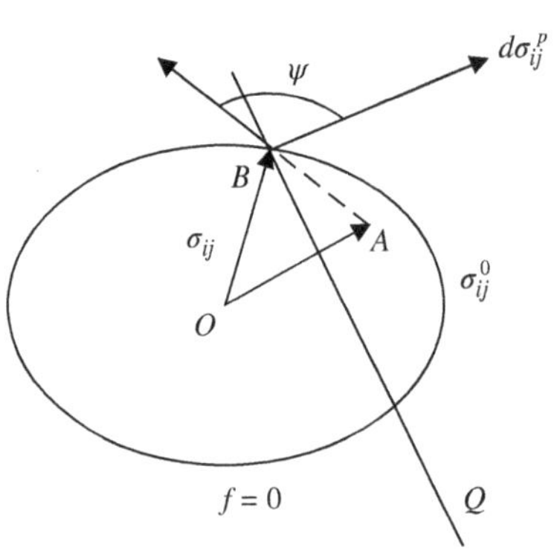

Figure 3.7 Normality of plastic strain vector.

3.4 Incremental Plastic Flow Theory

Incremental theory, also known as flow theory, is a theory that describes the relationship between stress and strain increment or strain rate when the material is in a plastic state. It determines the instantaneous strain increment according to each instantaneous stress state in the loading process. In other words, the increment theory holds that in the plastic state, it is the relationship between plastic strain increment (or strain rate) and stress and stress increment (or stress rate). The main contents are as follows: 1) Lévy–Mises theory, which is proposed and perfected by M. Lévy and R. von Mises respectively, it is suitable for rigid-plastic materials [8, 9]; 2) Prandtl–Reuss theory, proposed by L. Prandtl and A. Ruess respectively, considers elastic deformation and is suitable for elastoplastic materials [10, 11].

In the process of plastic deformation, there is no one-to-one correspondence between stress and strain, but in a given state, there is a stress increment; accordingly, there must be a unique strain increment. Therefore, under the condition of general plastic deformation, the relationship between stress and strain increment can only be established. This kind of constitutive relation expressed in incremental form is called incremental theory or flow theory [3]. Its starting point is that only the theory established in the form of increment can track the whole loading path and solve the plastic problems.

Shield and Ziegler pointed out that the basic elements of the plastic constitutive relationship include the following three aspects:

1. Initial yield condition, which can determine when plastic deformation begins and divide the range of plastic zone and elastic zone, so that different constitutive relations can be used to analyze it.
2. Loading function, that is, the strengthening condition to describe the strengthening characteristics of materials.
3. A flow law associated with the initial yield and the subsequent yield surface, that is, the qualitative relationship between stress and strain (or their increment). This relationship includes the direction relationship (that is, the relationship between the two principal axes) and the distribution relationship (that is, the proportional relationship between the two). In fact, it is to study the relationship between their deviations.

Among the three elements discussed above, Elements (1) and (2) have been examined in detail in previous chapters. The plastic constitutive relation will now be developed based on the discussion of Element (3), namely, the flow law.

For ideal elastoplastic materials, the subsequent yield surface coincides with the initial yield surface. If Mises' yield condition is adopted:

$$\sigma_i = \sigma_s \tag{3.33}$$

So,

$$f\left(\sigma_{ij}\right) = \sigma_i^2 - \sigma_s^2 \text{ or } J_2 - \frac{1}{3}\sigma_s^2 = 0 \tag{3.34}$$

where

$$f\left(\sigma_{ij}\right) = J_2 - \frac{1}{3}\sigma_s^2 \tag{3.35}$$

Known by $J_2 = \frac{1}{2}S_{ij}S_{ij} = -(S_1S_2 + S_2S_3 + S_3S_1)$

$$\frac{\partial J_2}{\partial \sigma_{ij}} = \frac{\partial J_2}{\partial S_{ij}} = S_{ij} \tag{3.36}$$

Then

$$d\varepsilon_{ij}^{p} = d\lambda \cdot S_{ij} \tag{3.37}$$

Equation (3.37) is called the Mises flow rule.
Note $d\varepsilon_{ij}^{p} = d\varepsilon_{m}^{p}\delta_{ij} + de_{ij}^{p}$ and plastic incompressibility (i.e. $d\varepsilon_{ii}^{p} = 0$), defined as:

$$d\varepsilon_{ij}^{p} = de_{ij}^{p} \tag{3.38}$$

Then

$$de_{ij}^{p} = d\lambda \cdot S_{ij} \tag{3.39}$$

This means that the direction of de_{ij}^{p} is the same as that of S_{ij}, and the ratio between the corresponding components of the two tensors is the same, that is

$$\frac{de_x^p}{S_x} = \frac{de_y^p}{S_y} = \frac{de_z^p}{S_z} = \frac{de_{xy}^p}{\tau_{xy}} = \frac{de_{yz}^p}{\tau_{yz}} = \frac{de_{zx}^p}{\tau_{zx}} = d\lambda \tag{3.40}$$

Equation (3.40) shows that the increment of plastic strain deviation is proportional to the stress deviation.

1. The constitutive relation of ideal elastoplastic material, Prandtl–Reuss relation.
   For ideal elastoplastic materials, the total increment of strain bias is

$$de_{ij}^{e} = \begin{cases} de_{ij}^{e} + de_{ij}^{p} \\ \dfrac{1}{2\mu} dS_{ij} + d\lambda \cdot S_{ij} \end{cases} \tag{3.41}$$

   Because there is $de_{ii} = 0$, only five of the above six expressions are independent, so it is necessary to supplement the relation about $d\varepsilon_{ij}$. In this way, the incremental constitutive relation of ideal elastoplastic materials can be summarized as follows:

$$\begin{cases} de_{ij} = \dfrac{1}{2\mu} dS_{ij} + d\lambda \cdot S_{ij} \\ d\varepsilon_{ii} = \dfrac{1}{3K}\, d\sigma_{ii} \end{cases} \tag{3.42}$$

   Equation (3.42) is called Prandtl–Reuss relation. The proportional coefficient $d\lambda$ needs to be determined in combination with the yield condition. According to the increment of the strain-specific energy function

$$\begin{aligned} dW &= \sigma_{ij}\, d\varepsilon_{ij} = (\sigma_m\delta_{ij} + S_{ij})(d\varepsilon_m\delta_{ij} + de_{ij}) \\ &= \sigma_m d\varepsilon_m\delta_{ij}\delta_{ij} + d\varepsilon_m S_{ij}\delta_{ij} + \sigma_m de_{ij}\delta_{ij} + S_{ij}de_{ij} \end{aligned} \tag{3.43}$$

But

$$\delta_{ij}\delta_{ij} = 3 \tag{3.44}$$
$$S_{ij}\delta_{ij} = S_{ii} = 0$$
$$\delta_{ij}\, de_{ij} = de_{ii} = 0$$

So

$$dW = 3\sigma_m d\varepsilon_m + S_{ij} de_{ij} \tag{3.45}$$

The first term on the right of Eq. (3.45) is the increment of volume strain energy, and the second term is the increment of shape strain energy:

$$dW_d = S_{ij} \cdot de_{ij} \tag{3.46}$$

By substituting the Prandtl–Reuss relation, it can obtain:

$$dW_d = S_{ij}\left(\frac{1}{2\mu} dS_{ij} + d\lambda S_{ij}\right) = \frac{1}{2\mu} S_{ij} dS_{ij} + d\lambda S_{ij} S_{ij} \tag{3.47}$$

According to the yield condition $J_2 = \frac{1}{3}\sigma_s^2$ and using $\frac{\partial J_2}{\partial S_{ij}} = S_{ij}$ and $dJ_2 = S_{ij} dS_{ij}$, the differential of the yield condition equation is as follows:

$$dJ_2 = S_{ij} dS_{ij} = d\left(\frac{1}{3}\sigma_s^2\right) = 0 \tag{3.48}$$

Add $S_{ij}S_{ij} = \frac{2}{3}\sigma_i^2$, $\sigma_i = \sqrt{\frac{3}{2}}\sqrt{S_x^2 + S_y^2 + S_z^2 + 2\left(S_{xy}^2 + S_{yz}^2 + S_{zx}^2\right)} = \sqrt{\frac{3}{2}}\sqrt{S_{ij}S_{ij}}$
So

$$d\lambda = \frac{3dW_d}{2\sigma_i^2} = \frac{3dW_d}{2\sigma_s^2} \tag{3.49}$$

And then substituted into the constitutive equation, there is:

$$\begin{cases} de_{ij} = \dfrac{1}{2\mu} dS_{ij} + \dfrac{3dW_d}{2\sigma_s^2} S_{ij} \\ d\varepsilon_{ii} = \dfrac{1}{3K}\, d\sigma_{ii} \end{cases} \tag{3.50}$$

Since $d\varepsilon_{ij} = d\varepsilon_m \delta_{ij} + de_{ij}$, Eq. (3.50) can also be written as:

$$d\varepsilon_{ij} = \frac{1-2\nu}{E}\, d\sigma_m \delta_{ij} + \frac{1}{2\mu} dS_{ij} + \frac{3dW_d}{2\sigma_s^2} S_{ij} \tag{3.51}$$

There is another representation of $d\lambda$ which is derived below.

$$J_2 = \frac{1}{2}\left[S_x^2 + S_y^2 + S_z^2 + 2\left(\tau_{xy}^2 + \tau_{yz}^2 + \tau_{zx}^2\right)\right] = \frac{1}{2} S_{ij} S_{ij} \tag{3.52}$$

Substituting $d\varepsilon_{ij}^p = d\lambda S_{ij}$ into Eq. (3.52) yields:

$$\frac{1}{2}\frac{d\varepsilon_{ij}^p d\varepsilon_{ij}^p}{(d\lambda)^2} = J_2 = \frac{1}{3}\sigma_s^2 \tag{3.53}$$

In the plastic state:

$$d\lambda = \frac{\sqrt{\frac{3}{2}}\sqrt{d\varepsilon_{ij}^{p} d\varepsilon_{ij}^{p}}}{\sigma_s} = \frac{3d\varepsilon_i^p}{2\sigma_s} \tag{3.54}$$

Namely,

$$d\lambda = \frac{3}{2}\frac{d\varepsilon_i^p}{\sigma_s} \tag{3.55}$$

Equation (3.55) shows that in the process of plastic deformation, the proportional coefficient $d\lambda$ is not only related to the yield limit of the material, but also to the degree of deformation. However, at a certain moment of deformation, the ratio of each component of the strain deviation increment to the corresponding stress deviation component is the same $d\lambda$.

2. The constitutive relation of ideal rigid-plastic material Lévy–Mises constitutive relation.

For rigid-plastic materials, elastic deformation can be omitted

$$d\varepsilon_{ij} = d_{ij}^{p} = d\lambda \cdot S_{ij} \tag{3.56}$$

Equation (3.56) is the Lévy–Mises flow rule (also known as Lévy–Mises equation), written in the form of components

$$\frac{d\varepsilon_x}{S_x} = \frac{d\varepsilon_y}{S_y} = \frac{d\varepsilon_z}{S_z} = \frac{d\varepsilon_{xy}}{\tau_{xy}} = \frac{d\varepsilon_{yz}}{\tau_{yz}} = \frac{d\varepsilon_{zx}}{\tau_{zx}} = d\lambda \tag{3.57}$$

Lévy–Mises equation shows that for ideal rigid-plastic materials, the strain increment is proportional to the stress deviation. In history, the formula was obtained first in Eq. (3.57) and then established in Eq. (3.56). Multiply both sides of Eq. (3.57), there are

$$d\varepsilon_{ij}\, d\varepsilon_{ij} = (d\lambda)^2\, S_{ij}S_{ij} \tag{3.58}$$

According to Mises' yield condition

$$\sigma_i = \sqrt{\frac{3}{2}}\sqrt{S_{ij}S_{ij}} = \sigma_s \tag{3.59}$$

and strain increment strength (noting that this is distinct from the total differential of strain strength ε_i).

$$d\varepsilon_i = \sqrt{\frac{2}{3}}\sqrt{d\varepsilon_{ij}\, d\varepsilon_{ij}} \tag{3.60}$$

So,

$$d\lambda = \frac{3d\varepsilon_i}{2\sigma_s} \tag{3.61}$$

Compared with the formula of ideal elastoplastic material in Eq. (3.58). There is

$$d\varepsilon_{ij} = \frac{3d\varepsilon_i}{2\sigma_s} S_{ij} \tag{3.62}$$

The Lévy–Mises constitutive relation can also be briefly discussed here.

1. Can stress be determined by a known strain increment? For ideal rigid-plastic materials, due to the incompressibility condition, knowledge of the strain increment $d\varepsilon_{ij}$ allows determination of S_{ij} from the constitutive relation; however, σ_m remains indeterminate, and thus the stress state cannot be fully resolved.
2. Can the strain increment be determined at a given stress? If the stress σ_{ij} is given, that is, S_{ij} is known, the strain increment $d\lambda$ cannot be fully determined. Only the ratios among its components can be obtained, while the actual magnitudes remain undetermined.

To sum up, the assumptions in Lévy–Mises's plastic increment theory are as follows:

1. The material is ideal rigid and plastic, $\varepsilon_{ij} = \varepsilon_{ij}^{p}$;
2. The material is incompressible, $e_{ij} = \varepsilon_{ij}$;
3. The material satisfies Mises yield condition, $\sigma_i = \sigma_s$;
4. The increment of strain deviation is proportional to the stress deviation, $d\varepsilon_{ij} \propto S_{ij}$.

The flow rule associated with the Tresca yield condition, the Tresca yield surface has an irregular sharp corner, and the plastic strain at the smooth part can still be solved by the same method as above, but the outer normal direction of the yield surface at the sharp angle is not unique. How to determine the plastic strain increment at this time is a problem that must be solved when applying the Tresca condition.

In elasticity, there is the following relation between strain and elastic strain complementary energy, namely the Castigliano formula:

$$\varepsilon_{ij} = \frac{\partial W_c\left(\sigma_{ij}\right)}{\partial \sigma_{ij}} \tag{3.63}$$

where W_c is the complementary energy of elastic strain per unit volume. For an ideal elastic body, it is a positive definite potential function, which is called the elastic potential. If $W_c\left(\sigma_{ij}\right) = C$ (C is constant) is regarded as an equipotential surface in the stress space, then Eq. (3.63) indicates that the direction of the strain vector aligns with the gradient of the elastic potential, that is, the outward normal to the equipotential surface.

Similarly, von Mises proposed the plastic potential theory in 1928. The plastic potential function is not only related to the stress state, but also to the loading history. If the loading history is represented by the strengthening parameter K, the plastic potential function can be expressed as

$$g = g\left(\sigma_{ij}, K\right) \tag{3.64}$$

Then

$$\varepsilon_{ij}^{p} = d\lambda \frac{\partial g\left(\sigma_{ij}, K\right)}{\partial \sigma_{ij}} \tag{3.65}$$

If $g = C$, it represents the equipotential surface in the stress space, and Eq. (3.65) indicates that the normal direction of the plastic strain increment vector is consistent with the gradient direction of the plastic potential, that is, the outer normal direction of the equipotential surface.

It can be seen that the inference of the Drucker axiom is that the yield function f is regarded as the plastic potential function g. In this way, the yield condition, strengthening condition, and plastic strain increment are related. The flow rule obtained by combining the yield condition with the constitutive relation is called the joint flow rule or the flow rule associated with the yield condition (associated flow rule), which is suitable for stable materials in accordance with the Drucker axiom. The flow law of $g \neq f$ is called non-joint flow law or nonassociated flow law, which is often used in geotechnical materials and some composite materials.

When the yield surface consists of n regular functions $f_s = 0$. When constituting, the generalized plastic potential theory holds that

$$d\varepsilon_{ij}^{p} = \sum_{s=1}^{n} d\lambda_s \frac{\partial f_s}{\partial \sigma_{ij}}$$
$$d\lambda_s \begin{cases} = 0, & f_s < 0 \text{ or } f_s = 0, df_s < 0 \\ > 0, & f_s = 0, df_s = 0 \end{cases} \quad (s = 1, 2, \ldots, n) \tag{3.66}$$

Take the generalized plastic potential function as the Tresca yield function, as shown in Figure 3.6. Taking the AB and BC planes in the graph as an example, their equations are:

$$\begin{aligned} AB &: \sigma_1 - \sigma_2 = \sigma_s \\ BC &: \sigma_1 - \sigma_3 = \sigma_s \end{aligned} \tag{3.67}$$

The phase plasticity potential function is:

$$\begin{aligned} AB &: f_1 = \sigma_1 - \sigma_2 - \sigma_s = 0 \\ BC &: f_2 = \sigma_1 - \sigma_3 - \sigma_s = 0 \end{aligned} \tag{3.68}$$

For AB faces, there are:

$$\begin{cases} d\varepsilon_1^p = d\lambda_1 \dfrac{\partial f_1}{\partial \sigma_1} = d\lambda_1 \\ d\varepsilon_2^p = d\lambda_1 \dfrac{\partial f_1}{\partial \sigma_2} = -d\lambda_1 \\ d\varepsilon_3^p = d\lambda_1 \dfrac{\partial f_1}{\partial \sigma_3} = 0 \end{cases} \tag{3.69}$$

Therefore, the flow rule on the AB surface is:

$$d\varepsilon_1^p : d\varepsilon_2^p : d\varepsilon_3^p = 1 : (-1) : 0 \tag{3.70}$$

It is also obtained that the flow rule on the BC surface is:

$$d\varepsilon_1^p : d\varepsilon_2^p : d\varepsilon_3^p = 1 : 0 : (-1) \tag{3.71}$$

The proportional relationship between the above $d\varepsilon_{ij}^{p}$ can only be obtained here, but the stress deviation S_{ij} cannot be determined, because any point on the same yield surface has the same external normal.

At point B, because the direction of the outer normal is not unique, the direction of the plastic strain increment is uncertain. However, from the Drucker inequality, it can be proved that the plastic strain increment in any direction within the angle range of the normal directions on both sides (the shadow area in Figure 3.8) satisfies the Drucker postulate. The plastic strain increment at the corner can be obtained by a linear combination of plastic strain increments on the surface. Multiply Eq. (3.70) by any coefficient $0 \le \mu \le 1$, and Eq. (3.71) by $1 - \mu$, and then add the two to get the flow law of point B.

$$d\varepsilon_1^p : d\varepsilon_2^p : d\varepsilon_3^p = 1 : (-\mu) : -(1-\mu) \tag{3.72}$$

It can be proved that once the plastic strain increment $d\varepsilon_{ij}^{p}$ is given, the plastic work increment $dW_p = \sigma_{ij}\, d\varepsilon_{ij}^{p}$ is a single value.

For strengthening materials, if the isotropic strengthening model is adopted and the Mises yield condition is selected, the proportional coefficient $d\lambda$ in Eq. (3.37) can be determined by the strengthening condition

$$\sigma_i = H\left(\int d\varepsilon_i^p\right) \tag{3.73}$$

According to Prandtl–Reuss 's flow rule, that is, the previous relationship:

$$d\varepsilon_{ij}^{p} = d\lambda \cdot S_{ij} \tag{3.74}$$

Put the above into the definition of the incremental strength of plastic strain:

$$d\varepsilon_i^p = \sqrt{\frac{2}{3}\, d\varepsilon_{ij}^{p} d\varepsilon_{ij}^{p}} \tag{3.75}$$

$$d\varepsilon_i^p = \frac{2}{3}\, d\lambda \cdot \sqrt{\frac{3}{2} S_{ij} S_{ij}} = \frac{2}{3} d\lambda \cdot \sigma_i \tag{3.76}$$

$$d\lambda = \frac{3}{2}\frac{d\varepsilon_i^p}{\sigma_i} \tag{3.77}$$

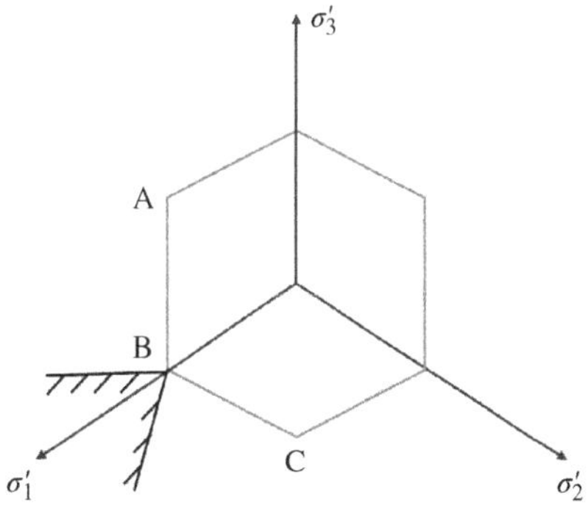

Figure 3.8 Flow laws associated with Tresca conditions.

Then defined by Eq. (3.73),

$$H' = \frac{d\sigma_i}{d\varepsilon_i^p} \tag{3.78}$$

The slope of the curve is the slope of $\sigma_i - \int d\varepsilon_i^p$ (Figure 3.9).

$$d\lambda = \frac{3}{2}\frac{d\varepsilon_i^p}{\sigma_i} = \frac{3d\sigma_i}{2H'\sigma_i} \tag{3.79}$$

Replace Eq. (3.79) back to the Prandtl–Reuss flow rule, that is:

$$de_{ij}^p = d\lambda \cdot S_{ij} \tag{3.80}$$

Then,

$$de_{ij}^p = \frac{3d\sigma_i}{2H'\sigma_i}S_{ij} \tag{3.81}$$

By adding the elastic strain deviator, the total strain deviator increment can be obtained.

$$de_{ij} = de_{ij}^e + de_{ij}^p = \frac{1}{2\mu}dS_{ij} + \frac{3}{2H'}\frac{d\sigma_i}{\sigma_i}S_{ij} \tag{3.82}$$

Further consideration of the volume change part of the elasticity can be obtained

$$d\varepsilon_{ij} = \frac{1-2\nu}{E}\, d\sigma_m\delta_{ij} + \frac{1}{2\mu}dS_{ij} + \frac{3}{2H'}\frac{d\sigma_i}{\sigma_i}S_{ij} \tag{3.83}$$

Equation (3.83) is equivalent to the following relation

$$\begin{cases} d\varepsilon_{ii} = \dfrac{1-2\nu}{E}d\sigma_{ii} \\ de_{ij} = \dfrac{1}{2\mu}dS_{ij} + \dfrac{3}{2H}\dfrac{d\sigma_i}{\sigma_i}S_{ij} \end{cases} \tag{3.84}$$

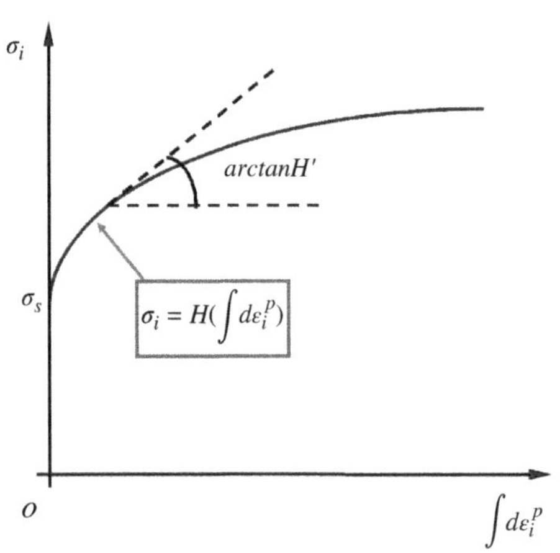

Figure 3.9 Enhancement curve.

This is the incremental constitutive equation of the reinforced material. If a certain instantaneous stress and stress increment are given, the strain increment can be uniquely determined, and the total strain can be determined by superimposing these strain increments sequentially along the strain path.

There is no strict theoretical basis for the specific relationship between the plastic potential function g and the yield function f. However, they are often assumed to be identical, which constitutes the associated flow rule, also known as the orthogonal flow rule. This rule is derived from the maximum plastic work principle, which applies to Drucker-stable materials.

The following theoretical results are also the basis for the correctness of the correlated flow rule.

1. The correlation flow rule can be used to prove the uniqueness of stress distribution in deformed solid under plastic deformation [5].
2. According to the Drucker postulate, the plastic flow can be guaranteed to be stable by adopting the correlated flow law.
3. Using the correlated flow rule, it can be proved that the strain state is unique when returning to the yield surface along any closed path under neutral variable load.

However, the above conditions are sufficient but not necessary to prove that the associated flow law is valid in the existing plastic deformation theory. In other words, whether the correlated flow rule is valid or not is not a necessary condition for the above results, because some models using non-correlated flow rules can also satisfy the results of the above three aspects.

In fact, an increasing number of experimental results show that the influence of hydrostatic pressure on the plastic deformation of many materials cannot be ignored, and the yield condition that relies solely on the second invariant of the stress deviator J_2 is not sufficient to describe its initial yield and strengthening behavior. On the contrary, if some nonassociated flow laws are adopted, they can more reasonably describe its plastic response.

In summary, the associated flow law is not entirely correct, as it assumes that the plastic potential function and yield function are equal, which is actually a restrictive condition or constraint. Nonassociated flow law models can increase the possibility of differentiating the plastic potential function and yield function in different ways, which is very beneficial for accurately describing the complex plastic response of materials. In particular, for materials exhibiting initial anisotropy, porosity, or other structural complexities, the yield conditions are often highly nonlinear. In such cases, the use of a nonassociated flow rule provides an effective modeling approach and helps to avoid unrealistic predictions that contradict experimental observations.

3.5 Total Plastic Deformation Theory

In plasticity theory, because stress is not only related to strain but also to its deformation history, especially because the loading and unloading processes in plasticity have different laws, the plastic constitutive relationship should be incremental in nature. However, due to the mathematical complexity of tracking the deformation history, the incremental theory is not convenient for practical applications. While developing incremental constitutive relationships, people have not abandoned the effort to establish total constitutive relationships under special loading histories. For example, in the case of a simply supported beam subjected to a monotonic tensile load without unloading, a one-to-one correspondence exists between stress and strain. This scenario is equivalent to a nonlinear elastic problem and does not require the iterative solution procedure associated with the incremental theory.

Now consider the case where the principal axis of stress remains unchanged and the stress components increase proportionally during deformation, known as proportional loading (also known as simple loading). It means that at each point within the material, the individual stress components increase proportionally during the loading process. Assuming that the reference stress state σ_{ij}^0 at a given moment is nonzero, the subsequent stress state can be expressed as

$$\sigma_{ij} = \alpha\,(t)\,\sigma_{ij}^0 \tag{3.85}$$

In the equation: $\alpha\,(t)$ is a monotonically increasing function of time.

1. Hencky's Total Theory [12]
In the case of simple loading, all stress components grow proportionally, i.e.

$$S_{ij} = \alpha(t)\,S_{ij}^0,\ \sigma_i = \alpha\,(t)\,\sigma_i^0,\ dS_{ij} = S_{ij}^0 d\alpha(t) \tag{3.86}$$

According to the Prandtl–Reuss constitutive relationship, there is a change in the strain increment.

$$de_{ij} = \frac{1}{2\mu} dS_{ij} + d\lambda \cdot S_{ij} = \frac{1}{2\mu} S_{ij}^0 d\alpha(t) + d\lambda \cdot S_{ij}^0 \alpha(t) \tag{3.87}$$

Integration of Eq. (3.87) yields the strain components.

$$e_{ij} = S_{ij}^0 \left[\frac{1}{2\mu}\alpha\,(t) + \int \alpha\,(t)\,d\lambda\right] \tag{3.88}$$

Because of $d\lambda = \frac{3d\varepsilon_i^p}{2\sigma_i}$. Therefore, Eq. (3.88) can be written as:

$$\begin{aligned} e_{ij} &= S_{ij}^0 \left[\frac{1}{2\mu}\alpha\,(t) + \int \alpha\,(t)\,\frac{3d\varepsilon_i^p}{2\sigma_i}\right] \\ &= S_{ij}^0 \alpha\,(t) \left[\frac{1}{2\mu} + \frac{1}{\alpha\,(t)} \int \alpha\,(t)\,\frac{3d\varepsilon_i^p}{2\alpha\,(t)\,\sigma_i^0}\right] \\ &= S_{ij}^0 \alpha\,(t) \left[\frac{1}{2\mu} + \frac{3}{2\alpha\,(t)\,\sigma_i^0} \int d\varepsilon_i^p\right] = \frac{S_{ij}}{2\mu} + \frac{3\varepsilon_i^p}{2\sigma_i} S_{ij} \end{aligned} \tag{3.89}$$

If $\lambda = \frac{3\varepsilon_i^p}{2\sigma_i}$, then Eq. (3.88) can also be written as:

$$e_{ij} = e_{ij}^e + e_{ij}^p = \frac{1}{2\mu} S_{ij} + \lambda S_{ij} \tag{3.90}$$

or

$$e_{ij} = e_{ij}^e + e_{ij}^p = \frac{1+\varphi_S}{2\mu} S_{ij} \tag{3.91}$$

Here, it is $\varphi = \frac{3\mu\varepsilon_i^p}{\sigma_i}$. Then the following strain expression is given:

$$\varepsilon_{ij} = \frac{1-2\nu}{E}\sigma_m\delta_{ij} + \frac{1}{2\mu}S_{ij} + \lambda S_{ij} \tag{3.92}$$

It is assumed that λ is positive when loaded and zero in all other cases. Then, the plastic part of the strain can be expressed as

$$\varepsilon_{ij}^p = e_{ij}^p = \lambda S_{ij} \tag{3.93}$$

2. Ilyushin's Total Elastic-plastic Theory
For the small elastoplastic deformation of the reinforced material, Ilyushin proposed the following assumptions about the basic elements according to generalized Hooke's law [13].
1. When the material is deformed, the volume change is elastic, that is

$$\varepsilon_{ii} = \frac{1-2\nu}{E}\sigma_{ii} \text{ or } \sigma_m = \frac{E}{1-2\nu}\cdot\varepsilon_m \tag{3.94}$$

2. The strain partial tensor is proportional to the stress partial tensor, i.e.

$$e_{ij} = \lambda \cdot S_{ij} \tag{3.95}$$

Equation (3.95) gives the qualitative relationship between strain and stress, that is, the direction relationship is that the principal axis of strain deflection and the principal axis of stress deflection coincide, that is, the principal axis of strain and the principal axis of stress are consistent. The distribution relation is that the strain deviant component is proportional to the stress deviant component. However, it should be noted that Eq. (3.95) is only similar to generalized Hooke's law in form; the difference is that the scale coefficient λ here is not a constant; it is related to the load level and coordinate position. Lambda is different at different points in the object; λ is also different at different load levels at the same point. Therefore, such a relationship is actually nonlinear, and correspondingly, the problem of plastic mechanics is much more complex than that of elastic mechanics.

Further by

$$\sigma_i = \sqrt{\frac{3}{2}}\sqrt{S_{ij}S_{ij}},\ \varepsilon_i = \sqrt{\frac{2}{3}}\sqrt{e_{ij}e_{ij}} \tag{3.96}$$

Double multiplication:

$$e_{ij}e_{ij} = \lambda^2 S_{ij}S_{ij} \tag{3.97}$$

Specifically,

$$\frac{3}{2}\varepsilon_i^2 = \lambda^2\cdot\frac{2}{3}\sigma_i^2 \tag{3.98}$$

It is available.

$$\lambda = \frac{3\varepsilon_i}{2\sigma_i} \tag{3.99}$$

Return, get:

$$e_{ij} = \frac{3\varepsilon_i}{2\sigma_i} S_{ij} \tag{3.100}$$

Or it can be written as:

$$S_{ij} = \frac{2\sigma_i}{3\varepsilon_i} e_{ij} \tag{3.101}$$

3. The stress strength is a definite function of the strain strength, i.e.

$$\sigma_i = \Phi\left(\varepsilon_i\right) \tag{3.102}$$

This is the strengthening condition established by the single curve hypothesis.
To sum up, the total plastic constitutive equation is

$$\begin{cases} \varepsilon_{ii} = \dfrac{1-2\nu}{E}\sigma_{ii} \\ e_{ij} = \dfrac{3\varepsilon_i}{2\sigma_i} S_{ij} \\ \sigma_i = \Phi\left(\varepsilon_i\right) \end{cases} \tag{3.103}$$

It is important to note that Eq. (3.103) only describes the elastoplastic deformation law during loading. The sign of loading is monotonically growing, i.e. $d\sigma_i > 0$, $d\sigma_i < 0$ when it belongs to unloading, it obeys the law of elasticity, i.e.

$$\begin{cases} d\varepsilon_{ii} = \dfrac{1-2\nu}{E} d\sigma_{ii} \\ de_{ij} = \dfrac{1}{2\mu} dS_{ij} \end{cases} \tag{3.104}$$

Note that they are in incremental form. As can be seen from Eq. (3.101), when strain ε_{ij} is given, ε_i, σ_i, and σ_m can be obtained in turn, and then stress σ_{ij} can be obtained. Conversely, if a stress σ_{ij} is given, ε_i, σ_i, and ε_m can be obtained in turn, and then the strain ε_{ij} can be obtained.

It is also important to note that Ilyushin assumes that the total strain is proportional to the stress bias.

3.6 Characteristics of Constitutive Relations in Plastic Deformation

As shown in Figure 3.10, for an ideal plastic material, the same yield stress σ_s may correspond to any strain. For a strain-hardening material, if the material is loaded to σ_a, the corresponding strain is ε_e, if it is then unloaded to σ_e, the corresponding strain is $\varepsilon_f^{'}$. Although $\varepsilon_f^{'} \neq \varepsilon_e$, this indicates that under the same stress state, different strain states may occur. This implies that the stress–strain relationship is no longer uniquely defined.

Figure 3.11 presents the stress–strain relationship of a strain-hardening plastic material under uniaxial tension and pure shear. Figure 3.12 illustrates the combined loading path

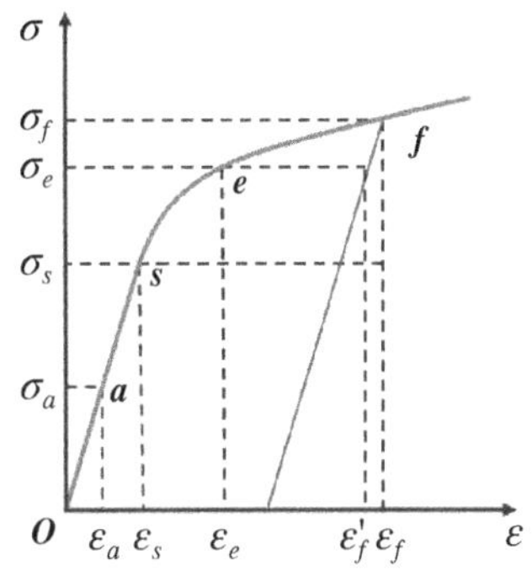

Figure 3.10 Schematic stress–strain curve for a material under uniaxial tension.

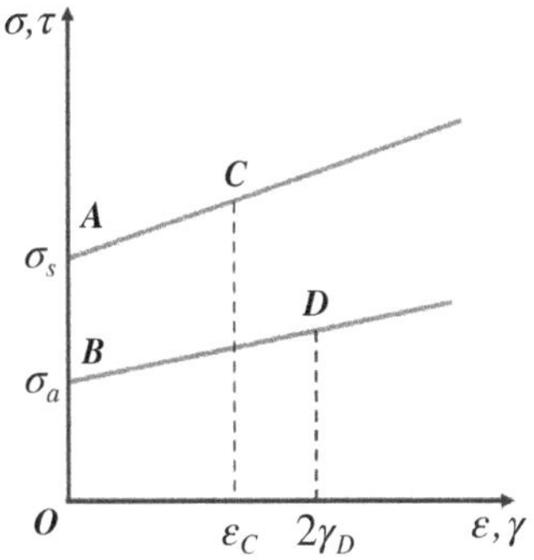

Figure 3.11 Stress–strain relationship of rigid-plastic strain-hardening material under tension and shear.

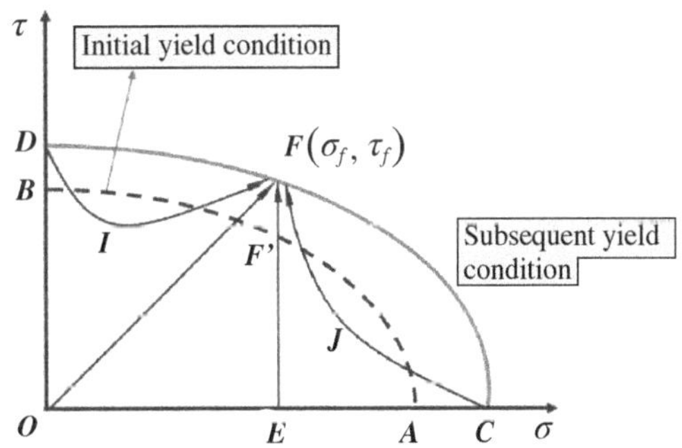

Figure 3.12 Schematic diagram of the yield path.

in the $\sigma - \tau$ plane. In Figure 3.12, segment AB represents the initial yield trajectory, and segment CD represents the subsequent yield trajectory. When the material is first loaded uniaxially to the initial yield point A in Figure 3.11 and then further loaded to point C, point C lies on the yield surface represented by segment CD in Figure 3.12. The stress in the material at point C is σ_c, and the corresponding strains are: $\varepsilon_1 = \varepsilon_c$, $\varepsilon_2 = \varepsilon_3 = -\frac{1}{2}\varepsilon_c$.

Due to the irreversibility of plastic deformation, the strains $\varepsilon_1, \varepsilon_2, \varepsilon_3$ are all non-recoverable and remain in the material. Therefore, if unloading occurs at point E in Figure 3.12, the strain remains unchanged. Then, upon applying a subsequent shear load, the state moves along the yield surface to point F. At this point, the stress state is (σ_f, τ_f). Since point F and point C lie on the same yield surface and correspond to the same equivalent stress without any increase, the material undergoes no further deformation, and the strain state remains the same as that at point C. This indicates that although the stress states at points F and C are different, they correspond to the same strain state. Similarly, in Figure 3.12, any reloading within segment CD of the yield surface will result in the same situation.

This clearly demonstrates that the relationship between plastic strain and stress is not one-to-one but is determined by the loading path. Therefore, it is impossible to establish a complete and unique constitutive relationship solely based on the current stress state.

The essential characteristics of plastic constitutive relations can be summarized as follows:

1. The relationship between plastic strain and stress is nonlinear, so the strain components are not linearly aligned with the stress axes.
2. Plastic strain cannot be considered a state function. Thus, $\varepsilon_v^p = 0, d\varepsilon_v^p = 0, \dot{\varepsilon}_v^p = 0$ are invalid.
3. For strain-hardening materials, the yield stress under reloading is the same as the previous initial yield stress before unloading.
4. Plastic deformation is irreversible and depends on the loading history, so the stress–strain relationship is no longer uniquely defined.

References

1 Duvaut, G. and Lions, J. L. (1976). *Inequalities in Mechanics and Physics*. Berlin, Heidelberg: Springer Berlin Heidelberg.

2 Shang, F. (2018). *Fundamentals of Plasticity Theory*, 3e. Xi 'an: Xi' an Jiaotong University Press.

3 Yu, T. and Xue, P. (2022). *Introduction to Engineering Plasticity*. Elsevier.

4 Hill, R. A. (1948). Theory of the yielding and plastic flow of anisotropic metals. *Proceedings of the Royal Society A* 193: 281–297.

5 Hill, R. (1949). The plastic yielding of notched bars under tension. *Quarterly Journal of Mechanics and Applied Mathematics* 2: 40–52.

6 Barlat, F., Brem J. C., Yoon, J. W. et al. (2003). Plane stress yield function for aluminum alloy sheets - part 1: theory. *International Journal of Plasticity* 19: 1297–1319.

7 Barlat, F. and Lian, J. (1989). Plastic behavior and stretchability of sheet metals .1. A yield function for orthotropic sheets under plane-stress conditions. *International Journal of Plasticity* 5: 51–66.

8 Lévy, M. (1871). Extrait du Mémoire sur les équations générales des mouvements intérieurs des corps solides ductiles au delà des limites où l'élasticité pourrait les ramener à leur premier état; présenté le 20 juin 1870. *Journal de Mathématiques pures et Appliquées* 16: 369–372.

9 Mises, R. V. (1986). Mechanics of solids in the plastically deformable state. *NASA Technical Memorandum* 1913: 88488.

10 Prandtl, L. (1921). Mutual influence of wings and propeller (No. NACA-TN-74).

11 Reuss, A. (1930). Berücksichtigung der elastischen Formänderung in der Plastizitätstheorie. *Journal of Applied Mathematics and Mechanics/Zeitschrift für Angewandte Mathematik und Mechanik* 10 (3): 266–274.

12 Hencky, H. (1924). Zur Theorie plastischer Deformationen und der hierdurch im Material hervorgerufenen Nachspannungen. *Journal of Applied Mathematics and Mechanics/Zeitschrift für Angewandte Mathematik und Mechanik* 4 (4): 323–334.

13 Starov, A. and Kalashnikov, S. (2020). A.A. Ilyushin s final relation, alternative equivalent relations and versions of its approximation in problems of plastic deformation of plates and shells part 1: A.A. Ilyushin's final relation. *International Journal for Computational Civil and Structural Engineering* 16: 106–129.

4

Damage Constitutive Principles and Methods

4.1 Overview of Material Damage

4.1.1 Continuum Damage Mechanics

Defects such as dislocations, voids, and microcracks are commonly found in practical engineering materials. Research on the mechanical effects of these internal defects is typically conducted at two scales: microscale and macroscale. At the microscale, micromechanics seeks to establish constitutive models that describe the macroscopic evolution of material behavior based on the interactions among microstructural constituents (e.g. matrix, particles, voids). This approach has led to the development of a dedicated branch of mechanics focusing on microscale material behavior. At the macroscale, two principal theoretical models are employed: fracture mechanics and damage mechanics.

Fracture mechanics deals with bodies containing one or more discrete cracks. It analyzes the stress, strain, displacement, and energy release near crack tips to determine criteria for crack initiation, stable propagation, and catastrophic failure. However, this framework does not address the formation and growth of microdefects prior to the appearance of visible cracks, nor their cumulative effect on a material's overall mechanical response. In many cases, especially under fatigue or creep loading, a large number of microdefects, such as microvoids, first develop at the weakest regions of a material. Macroscopic cracks eventually emerge as a result of the growth and coalescence of these microdefects—a process that often dominates the material's entire service life.

To study this process, a different approach is required. Instead of modeling discrete cracks, continuum damage mechanics (CDM) introduces the concept of a damage field—a continuous internal variable field that represents the degradation of material stiffness due to microstructural changes. Analogous to stress or strain fields, the damage field enables the description of the initiation, evolution, and interaction of microdefects and their impact on macroscopic mechanical behavior. CDM thus provides a powerful framework for predicting material degradation and failure due to distributed internal damage.

During manufacturing, transportation, assembly, and service, engineering components inevitably undergo microstructural changes due to external loading. Some of these changes—such as the initiation of microcracks or voids—are irreversible and can lead to the degradation of mechanical performance. Under continued loading, such microdamage evolves and merges, forming macroscopic defects that may severely reduce structural integrity and ultimately result in failure.

In the context of continuum mechanics, any irreversible microstructural change that impairs mechanical performance is referred to as damage. If the damage originates during manufacturing or processing, it is classified as initial damage. The life cycle of a material or structure can be divided into two stages: The initiation and growth of microdefects until they evolve into macrocracks or voids (the specific definition of "macro" depends on the application). The propagation and coalescence of macro-defects, leading to ultimate failure.

Typically, the first stage occupies the majority of the service life, making it essential to understand, characterize, and predict the evolution of damage during this phase. This was the original focus of damage mechanics when it emerged in the late 1970s, while the second stage was addressed using fracture mechanics methods.

In reality, once damage occurs, it continuously affects the material's mechanical behavior—altering stress and strain distributions near crack tips and even influencing other physical fields such as deformation and temperature. Therefore, the development of damage-coupled constitutive models enables more accurate and reliable predictions of deformation and failure in materials and structures.

4.1.2 Relationship Between Material Damage and Microstructure

During the service life of solid materials, various forms of microstructural evolution inevitably occur, such as the formation of voids, microcracks, slip bands, twins, and the degradation of phase boundaries [1–4]. The accumulation of these microscopic phenomena ultimately leads to the deterioration of macroscopic properties, manifested as reductions in strength, stiffness, and toughness, and may eventually result in fracture failure. Therefore, the essence of material damage can be regarded as the evolution of microdefects across multiple scales. One of the core goals in the study of damage mechanics is to effectively map these microscale mechanisms into macroscopic constitutive models.

In metallic materials, damage typically originates from the local accumulation of dislocations, the formation of vacancy clusters, or the decohesion at second-phase particle interfaces. Taking void coalescence as an example, plastic deformation under tensile loading often induces the nucleation of microcracks at grain boundaries or second-phase regions. These microcracks grow and gradually link up, eventually forming a macroscopic crack that leads to fracture failure [1, 2]. This evolution process can be modeled using micromechanical constitutive theories such as the Gurson—Tvergaard—Needleman (GTN) model, which explicitly introduces porosity as an internal damage variable to capture the influence of microstructural changes on macroscopic mechanical behavior.

In quasi-brittle materials such as concrete, damage primarily manifests as the nucleation, propagation, and coalescence of microcracks. Stress concentrations at the interface between aggregate and mortar are typical initiation sites for cracking [3]. Microscopic observations have shown that these cracks tend to nucleate at weak interfaces during loading and subsequently develop into a random yet predictable crack network governed by mesoscale heterogeneities. To simulate this process, researchers have developed micromechanical approaches—such as meso-damage models and stochastic interfacial network methods—that treat the material as a multiphase system with explicitly represented microstructures, enabling a more realistic reproduction of the damage initiation mechanism and its macroscopic consequences.

In recent years, increasing efforts have been devoted to employing experimental techniques such as digital image correlation (DIC), X-ray computed tomography (XCT), and electron backscatter diffraction (EBSD) to capture real-time microstructural evolution during loading. These experimental methods provide crucial data for defining damage variables and formulating evolution equations. They also reveal the multiscale coupling between damage and factors such as stress state, crystallographic texture, grain orientation, and interface quality, thereby promoting a shift from empirical damage models to physically based constitutive models rooted in microstructural evolution.

4.1.3 Types of Material Damage

The multi-angle microstructure imaging enables classification of material damage based on three characteristic dimensions: physical mechanism, morphological features, and scale-dependent behavior.

1. Microcrack Damage
   Microcrack damage is widely present in brittle materials (such as ceramics, concrete, and rocks). Its characteristics are: a large number of tiny cracks are formed inside the material in the stress concentration area, and these cracks continue to germinate, expand, and connect with each other during the loading process, eventually leading to macroscopic damage.
2. Void Growth and Coalescence Damage
   Void damage mainly occurs during the plastic deformation of metal materials, especially in the tensile state. Microscopic voids (originating from inclusions, second-phase interfaces, or atomic vacancies) gradually grow and aggregate during plastic flow, eventually forming cracks.
3. Crystal Slip and Dislocation Damage
   In crystalline materials, the proliferation, slip, interaction, and entanglement of dislocations lead to local strain inhomogeneity, which in turn triggers microscale damage accumulation. This type of damage usually manifests as hardening, brittleness, and local instability of the material. The microscopic manifestation is the increase in dislocation density and the evolution of dislocation structure.
4. Interfacial Damage
   In composite materials, multiphase materials, or coating systems, the degradation of the bond strength between different phase interfaces or materials leads to interfacial cracking, debonding, or slippage, which in turn causes the overall performance to deteriorate. From a mechanical point of view, stress concentration occurs at the interface, and cracks initiate at or near the interface.
5. Environment-assisted Damage
   The environment (such as corrosion, hydrogen embrittlement, high temperature, oxidation, irradiation, etc.) and mechanical loads work together to cause material performance degradation. This type of damage often has complex spatiotemporal evolution characteristics.
6. Creep Damage
   Under high temperature and long-term loading conditions, microscopic holes, grain boundary sliding, or local strain accumulation inside the material cause damage. The main microscopic characteristics are that holes grow preferentially at grain boundaries and grain boundaries fracture.

4.2 Research Methods of Damage Mechanics

Current research methodologies in damage mechanics can be broadly categorized into two main approaches. The first approach is microstructure-based, wherein micro-voids or other defects within a representative volume element (RVE) are treated as inclusions or inhomogeneities, and their influence on macroscopic mechanical properties is studied using micromechanical methods such as self-consistent homogenization. The second approach is phenomenological, where the discontinuities induced by damage are disregarded, and damage is treated as a continuous field variable, analogous to deformation or temperature. Based on experimental observations and mechanistic understanding, relationships are then established between the damage field and field quantities such as stress and strain to analyze the mechanisms and evolution of damage.

The elastoplastic constitutive theories discussed in Chapters 2 and 3 do not account for damage; they assume that materials are ideal and that mechanical properties remain unchanged during deformation. Consequently, these models cannot capture or predict the degradation in mechanical performance or the failure behavior of materials or structures due to damage evolution. To realistically and accurately describe the influence of damage and its progression during deformation, it is necessary to develop damage-coupled constitutive equations that integrate the theories and methods of damage mechanics, allowing the mutual interaction between deformation and damage to be characterized.

Damage is not an independent physical property but rather a macroscopic manifestation of various internal physical degradation processes in engineering materials. It is highly dependent on the material type and service environment. For instance:

1. In terms of materials, damage can occur in metals, polymers, rocks, concrete, and composites.
2. In terms of deformation mechanisms, it may involve elastic damage, plastic damage, fatigue damage, or creep damage.
3. In terms of loading conditions, damage can occur under static, dynamic, ambient, or high-temperature environments.
4. In terms of spatial distribution, damage may be isotropic or anisotropic.

The following sections establish the fundamental thermodynamic framework of CDM based on deformation mechanisms of various engineering materials [5, 6]:

1. Elastic Damage
   Within the elastic analysis framework, damage variables are introduced to represent material degradation. This category applies to materials such as high-strength low-toughness metals and alloys, high-strength concrete, rocks, and ceramics, where no significant irreversible deformation occurs. Therefore, this type of damage is often referred to as brittle damage.
2. Elastoplastic Damage
   Using an elastoplastic framework, damage variables are introduced to analyze materials such as low-strength high-toughness metals and alloys, medium-strength concrete, fiber-reinforced concrete, composites, and polymers, where damage is accompanied by observable permanent deformation. At room or elevated temperatures, damage in

metals undergoing large deformations is typically of this type, also known as ductile or plastic damage.

3. Creep Damage
   For metallic materials operating at temperatures above approximately one-third of their melting temperature (in Kelvin) or under time-dependent loading due to viscoplasticity, creep deformation occurs. A creep analysis framework incorporating damage variables enables the characterization of degradation behavior at moderate to high temperatures.
4. Fatigue Damage
   Within a fatigue analysis framework, damage variables are introduced to assess cyclic degradation in metallic materials under various stress amplitudes.

In actual engineering practice, numerous other degradation modes exist. These include dynamic damage (caused by impact or high-rate loading, such as spalling) and corrosion-induced damage. Regardless of the specific form, CDM models all such degradation mechanisms as a continuously distributed field, the damage field, and considers the material as a damaged continuum [7]. By introducing appropriate damage variables, the macroscopic properties of the damaged material can be characterized, enabling the formulation of damage evolution laws which describe the initiation, growth, and coalescence of defects.

While fracture mechanics, micromechanics, and damage mechanics all address irreversible failure processes, they do so at different scales:

Fracture mechanics focuses on macroscopic cracks or voids (typically ≥1 mm) and analyzes their propagation until failure, without considering the mechanisms by which they form. Micromechanics investigates the behavior of microstructural constituents (e.g. features visible under optical or electron microscopy) using multiscale continuum theories that incorporate statistical averaging of microstructural features.

Damage mechanics, particularly CDM, does not resolve individual microdefects such as dislocations, microvoids, or microcracks. Instead, it introduces damage variables to represent the collective effect of distributed microdefects (typically ~0.01 mm) on macroscopic mechanical degradation [8]. Macroscopic (or phenomenological) methods establish analytical models based on observable behavior. As a phenomenological framework grounded in continuum mechanics, damage mechanics offers a systematic means to account for material degradation and its impact on mechanical properties.

4.3 Definition and Basic Assumptions of Damage

4.3.1 Variable Selection

As mentioned in Section 4.2, the damage variable characterizes the deterioration of the internal mechanical properties of the engineering material. In an intuitive physical sense, it can be understood as the percentage of defects such as microcracks and voids in the entire engineering material. At the micro and mesoscopic scales, the number, length, area, and volume of microdefects can be selected as the damage variables to measure the deterioration of the mechanical properties of materials. On the macroscopic phenomenological scale, the elastic coefficient, yield stress, and density can be selected as the damage variables to measure the deterioration of the mechanical properties of materials. In continuum mechanics, the damage variable is defined as an internal variable (from the thermodynamic point of view, the internal variable reflects the irreversible process of the material structure), and other parameters (physical and mechanical quantities) will be affected by the damage variable.

In a continuum, the field variable that describes the damage state in a material is called the damage variable. For continuous damage mechanics, the damage variables can be selected as scalars, vectors, and tensors as required. For example, for small irregular void distributions or spherical voids with the same anisotropic distribution, the damage variable can be selected as a scalar ω; For the crack in the small distribution plane, the vector $\boldsymbol{\omega}$ perpendicular to the crack plane can be used as the damage variable. However, in general, the damage variable is taken as a tensor $\boldsymbol{\omega}$, which is generally of the second or fourth order.

As shown in Figure 4.1, according to the Kachanov—Rabotnov classical damage theory, three states are constructed for unidirectional tensile specimens (unidirectional uniform stress states): initial damage-free states, damage states, and virtual damage-free states [5, 6]. In the initial damage-free state (current configuration) subjected to tensile force **P**, the internal stress of the rod is σ; in the damaged state (in the current configuration), subjected to the same tensile force **P**, the bearing capacity of the rod decreases due to damage

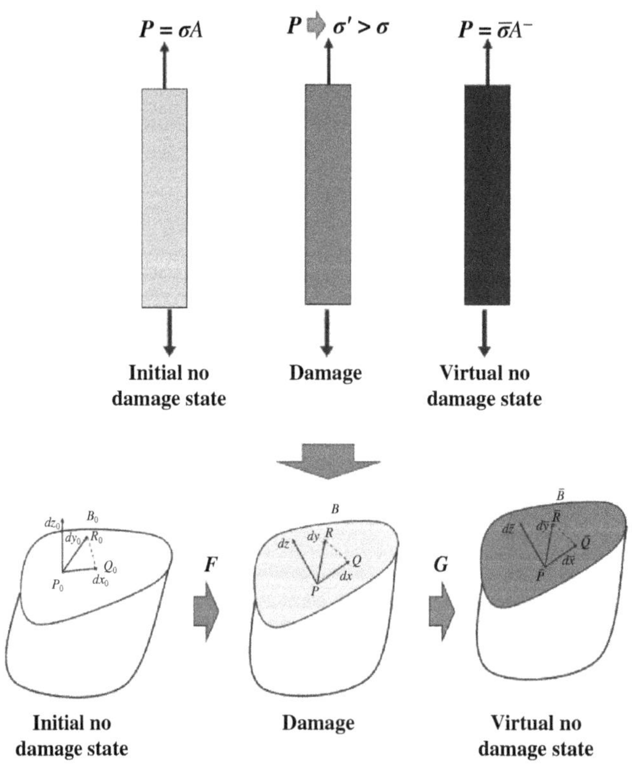

Figure 4.1 The relationship between the three states of one-way tensile specimens.

(if the damage is measured by the cross-sectional area, it is equivalent to a decrease in the effective bearing area, i.e. $A < A^{-}$), and the stress is $\sigma' > \sigma$.

4.3.2 Definition of Damage

In the process of damage initiation and propagation, it is accompanied by irreversible dissipation of energy. From a thermodynamic point of view, the evolution of damage is an irreversible process of change in the internal structure of matter, so damage is an internal variable. The definition of damage variables varies depending on the object or problem under study, and, in general, as a quantitative representation of irreversible changes in the structure of matter, it cannot be directly measured like the elastic constant, and is generally determined by indirect means. For example, there is only one damage of micropores with volume V in volume, and the damage variable D can be defined as the ratio of the pores to the total volume, i.e. $D = V_d/V$. Obviously, when there are no holes, i.e. $V_d = 0$, the damage variable $D = 0$, means that there is no damage. When the hole begins to germinate and expands with external loading, the ultimate limit case is that the hole occupies all the volume, i.e. $D = 1$, indicating that the material has completely failed. Generally, when the damage variable reaches 1, the component has lost the ability to bear or continue to serve, so the critical damage variable D_{cr} is often used to represent the value of the damage variable when the component actually fails. The above definition assumes that the damage is uniformly distributed and spread uniformly within volume V. If the anisotropy of damage development is to be considered, the damage variable needs to be defined as tensors.

Another common definition of impairment is described below. According to the classical damage theory proposed by Kachanov and Rabotnov, three states are constructed for an object in a state of unidirectional uniform stress [9, 10]. Assuming that there is no initial damage to the component, there is no damage before the external action is loaded. The other state is that after deformation, damage occurs, so it is called the damage state. At the same time, a virtual damage-free state was constructed. In the non-damage state, the stress of the specimen is σ under the action of tensile force P. If the cross-sectional area of the specimen is A_0, there is

$$P = \sigma A_0 \tag{4.1}$$

In the case of bearing the same load P, the bearing capacity of the material decreases due to the existence of damage in the damaged state. Assuming that the damage is present in the cross-section as a microcrack, the effective area of the cross-section under the load is less than the initial area A_0 due to the presence of the crack. Stress in the damaged state is denoted by σ'. Comparative Eq. (4.1) shows that the stress increases due to the decrease in the effective cross-sectional area that can bear the load in the damage state. In order to characterize this change in stress due to damage, it is assumed that the specimen is also in a non-damaged state under the action of uniaxial uniform stress, and its load is the same as the damage state, and the uniform stress in the cross-section is equal to the stress in the damaged state, i.e. $\tilde{\sigma} = \sigma'$. The cross-sectional area that can bear the load in the virtual damage-free state is A_{eff}, and since the load in this state is the same as the load in the initial damage-free state, therefore,

$$P = \tilde{\sigma} A_{eff} \tag{4.2}$$

It is assumed that there is a relationship between the cross-sectional area of the damage-free state and the cross-section of the virtual damage-free state as follows:

$$A_{eff} = A_0 - DA_0 \tag{4.3}$$

where D is the damage factor,

$$D = \frac{(A_0 - A_{eff})}{A_0} \tag{4.4}$$

With the initiation and development of the lesion, A_{eff} gradually decreases from A_0 to 0, and the damage factor D also changes from 0 to 1.

It can be obtained from Eq. (4.1) to Eq. (4.4).

$$\tilde{\sigma} = \frac{P}{A_{eff}} = \frac{P}{(1 - D) A_0} = \frac{\sigma}{1 - D} \tag{4.5}$$

where $\tilde{\sigma}$ is the equivalent stress.

It should be pointed out that the above derivation is obtained in a uniform unidirectional stress state, and it is assumed that the cracks are uniformly distributed and developed uniformly in the cross-section. If Eq. (4.5) is to be extended to complex three-dimensional stress states, and there is directionality in both damage distribution and development, the damage variables of the scalar will no longer be able to meet the requirements. In this case, the damage variable needs to be defined as a tensor.

Assuming that the strain in the damaged state is only related to the effective stress, the following assumptions are made:

Under the action of external forces, the constitutive relationship of the material containing the damage can be used in the form of no damage, and only the stress in it needs to be replaced by an equivalent stress. This is the strain equivalence assumption. That is, the strain calculated by the Cauchy stress in the damaged state is considered to be equal to the strain calculated by the equivalent stress in the virtual non-damaged state. This assumption is the basis for coupling damage into the constitutive model. According to this assumption, in the elastic deformation of a uniaxial line, according to Hooke's law:

$$\varepsilon = \frac{\tilde{\sigma}}{E} = \frac{\sigma}{(1 - D) E} \tag{4.6}$$

where E is the modulus of elasticity. If the effective modulus of elasticity is defined as E_{eff}:

$$E_{eff} = (1 - D) E \tag{4.7}$$

It is possible to obtain the formula for calculating the damage variables that are commonly used in experiments

$$D = 1 - \frac{E_{eff}}{E} \tag{4.8}$$

If the effective modulus of elasticity is regarded as the slope of the linear segment of the stress—strain curve in the uniaxial tensile test, its evolution with deformation can be obtained through experiments, and then the evolution of damage with deformation can be obtained.

The damage variable can also be expressed in other forms, for example, for fatigue problems, N represents the actual cycle, and N_f represents the ultimate fatigue life, and the damage factor can be defined according to the evolution law of the specific injury:

$$D = f\left(\frac{N}{N_f}\right) \tag{4.9}$$

The bearing capacity of the material decreases linearly with the fatigue cycle; that is, the damage increases linearly with the fatigue cycle, so the damage factor can be defined as:

$$D = \frac{N}{N_f} \tag{4.10}$$

Equation (4.10) is the simplest fatigue damage evolutionary relationship.

If the definition has an effect, it becomes

$$\tilde{\varepsilon} = \varepsilon\,(1 - D) \tag{4.11}$$

Contrasting Eq. (4.6), it can be obtained

$$\sigma = E\tilde{\varepsilon} \tag{4.12}$$

Equation (4.12) can be expressed using the stress equivalence assumption:

Under the action of external forces, the constitutive relation of the material containing the damage can be formed in the absence of damage, and only the strain in it needs to be transformed into an equivalent effect.

The strain equivalence assumption is often used in constitutive damage coupling models with strain as the fundamental variable, while the stress equivalence assumption can be used for constitutive models with stress as the fundamental variable.

4.4 Damage Evolution Equation

4.4.1 State Variables

A group of particles that are surrounded by boundaries in space is called a system, and the area occupied by the particles and the boundaries of that region are represented by B and ∂B, respectively. A system consisting of a set of fixed particles is called a closed system. Let's assume that a closed system, a collection of particles, or a fixed mass occupying region B would be called System B.

The thermodynamic state of a system can be described by a set of macroscopic variables that characterize the state of the system in a targeted manner. Of these variables, the one that is specified only by the current state of the system is called a state variable. In addition, state variables are divided into external variables (or observable variables) and internal variables (or hidden variables) according to whether they can be observed externally.

A thermodynamic system is in a state of equilibrium when its macroscopic state variables (e.g. pressure, temperature, and composition) remain constant over time in the absence of external perturbations. The temporal evolution of these variables due to external interactions defines a thermodynamic process. A process is considered reversible if the system can be restored to its initial equilibrium state by precisely reversing all energy and mass transfers involved, leaving no net change in either the system or its surroundings. Processes that do not satisfy this criterion are inherently irreversible, typically due to dissipative effects such as friction, heat conduction, or chemical reactions proceeding at finite rates [4].

The state variables that serve as independent control parameters for thermodynamic state changes are termed independent variables. Variables that are uniquely determined as mathematical functions of these independent variables are classified as dependent variables. Notably, the selection of independent variables is nonunique and depends on the specific thermodynamic representation being employed [4].

4.4.2 Damage Local State Principle

During deformation of a continuum, field quantities such as strain and temperature exhibit spatial and temporal variations. Consequently, the thermodynamic state of the continuum becomes nonuniform and departs from equilibrium, rendering the process thermodynamically irreversible [4]. Classical thermodynamics, developed under the assumptions of uniformity and equilibrium (e.g. treatment of thermostats), cannot adequately describe such nonequilibrium processes. To address this fundamental limitation, several theoretical frameworks have been developed. Among these, the most widely adopted approach in continuum thermodynamics is the principle of local states [7, 8], which is also known as the local equilibrium hypothesis [9, 10]:

> *Take a small element of material at a point in an object and assume that the thermodynamic state of that element at any given time is represented entirely by a set of state variables.*
>
> *Then, even if the element is in nonequilibrium, the state variables of the element at any moment are determined by the same thermodynamic relationship as the equilibrium case.*

This hypothesis assumes that the material elements in the nonequilibrium state in the continuum exhibit the same thermodynamic response as the material elements in the corresponding equilibrium state. Therefore, this assumption always holds true if the response time of the material element reaching its equilibrium state is short compared to the characteristic time of the kinematic and thermodynamic evolution of the continuum. That is, this assumption implies that the thermodynamic process of the material elements in the moving continuum is a continuous process of its equilibrium state.

4.4.3 Basic Principles of Irreversible Thermodynamics

Unlike traditional equilibrium thermodynamics, which mainly deals with static or quasistatic processes, irreversible thermodynamics focuses on actual processes with driving forces and dissipation mechanisms, such as heat conduction, viscous flow, diffusion, plasticity, damage, chemical reactions, etc. In continuum mechanics and material constitutive theory, the introduction of irreversible thermodynamic principles is the basis for constructing an evolutionary model with consistent dissipation and controllable stability. Irreversible thermodynamics is generally centered on the following basic principles:

1. Local Equilibrium Hypothesis
   It assumes that in the nonequilibrium process, each tiny area in the system can still be described by local thermodynamic state quantities (such as temperature, specific entropy, chemical potential, etc.), that is, the "local state function" still holds.
2. Clausius—Duhem Inequality
   In the nonequilibrium process, the total entropy $\bar{s}_{tot}$ cannot be reduced, and the entropy generation rate $\bar{s}_p$ is nonnegative, that is:

$$\bar{s}_{tot} = \bar{s}_r + \bar{s}_p, \qquad \bar{s}_p \geq 0 \tag{4.13}$$

   In continuous media, the corresponding form is the Clausius—Duhem inequality, which is the key constraint condition for constructing dissipative constitutive conditions.
3. Thermodynamic Forces and Fluxes
   Each dissipation process can be represented as a set of "flow-force" pairs: heat flow corresponds to temperature gradient, diffusion flow corresponds to chemical potential gradient, and stress and strain rate correspond to mechanical dissipation. These "flow-force" pairs follow linear relationships (such as Fourier's law, d'Alembert—Owen relationship, etc.) when approaching equilibrium.

4.4.4 Basis of Thermodynamic Damage

As noted earlier, the initiation and development of damage are an irreversible process. Therefore, the description of damage evolution should conform to the laws of irreversible thermodynamics. The relevant thermodynamic basis is provided in Chapter 2, and here it is discussed in elasticity. Forms in inelastic deformations.

When an object is plastically or viscoplastically deformed, it needs to dissipate energy and generate heat. The energy dissipated per unit time during deformation is:

$$\sigma_V : \varepsilon + \sigma_p : \varepsilon_p \tag{4.14}$$

where σ_V denotes the inelastic stress tensor, ε_p denotes the viscoplastic strain tensor, and σ_p denotes the stress tensor that causes the viscoplastic deformation.

All of the energy dissipated during deformation is converted into heat, and all of this heat is absorbed by the object except some of it is lost, resulting in entropy. For a microelement of volume V:

$$\int_V \rho \dot{s} dV = \int_V \frac{1}{T}\left[\sigma_V : \dot{\varepsilon} + \sigma_p : \dot{\varepsilon}_p + A_i \dot{\alpha}_i - \nabla \boldsymbol{q}\right] dV \tag{4.15}$$

where T is the temperature of the microelement, α_i denotes the internal state variable that causes various physical or chemical changes, and A_i denotes the energy dissipated by the internal variable α_i to produce a unit change. $\nabla \boldsymbol{q}$ is the heat lost.

From Eq. (4.15), it follows that:

$$\nabla \boldsymbol{q} = \boldsymbol{\sigma}_V : \dot{\varepsilon} + \boldsymbol{\sigma}_p : \dot{\varepsilon}_p + \dot{A_i \alpha_i} - \rho T \dot{s} \tag{4.16}$$

And according to the conservation of energy, combined with Eq. (4.16), there is:

$$\rho \dot{e} - (\sigma - \boldsymbol{\sigma}_V) : \dot{\varepsilon} + \boldsymbol{\sigma}_p : \dot{\varepsilon}_p + \dot{A_i \alpha_i} - \rho T \dot{s} = 0 \tag{4.17}$$

where e denotes the density of internal energy.

For processes with comparable changes of state, the energy state of matter can be expressed as a free energy density function. The free energy density function can be expressed as:

$$\Phi = e - Ts \tag{4.18}$$

Deriving both sides of Eq. (4.18) with respect to time yields:

$$\dot{\Phi} = \dot{e} - \dot{T}s - T\dot{s} \tag{4.19}$$

From Eq. (4.16) and (4.18) it follows that:

$$\rho \dot{\Phi} = (\sigma - \boldsymbol{\sigma}_V) : \dot{\varepsilon} - \boldsymbol{\sigma}_p : \dot{\varepsilon}_p - \dot{A_i \alpha_i} - \rho T \dot{s} \tag{4.20}$$

The free energy is a function of the elastic strain tensor, the temperature T, and the internal variables. From the relationship between the total strain and the elastic and viscoplastic strains, the free energy density function can be expressed as a function of the total and viscoplastic strains, as well as the temperature and other internal variables:

$$\Phi = \Phi\left(\varepsilon, \boldsymbol{\sigma}_p, T, \alpha_i\right) \tag{4.21}$$

$$\dot{\Phi} = \frac{\partial \Phi}{\partial \varepsilon} : \dot{\varepsilon} + \frac{\partial \Phi}{\partial \varepsilon_p} : \dot{\varepsilon}_p + \frac{\partial \Phi}{\partial \alpha_i} \dot{\alpha}_i + \frac{\partial \Phi}{\partial T} \dot{T} \tag{4.22}$$

Equations (4.20) and (4.21) allow derivation of:

$$(\sigma - \boldsymbol{\sigma}_V) = \rho \frac{\partial \Phi}{\partial \varepsilon} \tag{4.23}$$

$$\varepsilon_p = -\rho \frac{\partial \Phi}{\partial \varepsilon_p} \tag{4.24}$$

$$s = -\frac{\partial \Phi}{\partial T} \tag{4.25}$$

$$A_i = -\rho \frac{\partial \Phi}{\partial \alpha_i} \tag{4.26}$$

The above is the equation between each state variable and its generalized force expressed as a function of the free energy density.

4.5 Helmholtz Free Energy (HFE)

4.5.1 Physical Significance of HFE

HFE is a state function in thermodynamics, and its general expression is usually:

$$\psi = U - TS \tag{4.27}$$

Among them, U is the total energy of the system, including molecular kinetic energy, potential energy, etc.; T is the absolute temperature; S is entropy, which measures the degree of disorder of the system. The physical meaning of HFE is that under constant temperature conditions, the reduction of free energy is equal to the maximum useful work that the system can do, such as mechanical deformation, internal phase change of materials, etc. At the same time, in continuum mechanics, it can be used as a potential function to describe the constitutive behavior of materials.

In damage mechanics, the free energy contained in a unit volume is called the HFE density ψ. It can be expressed as a function of strain ε and damage d:

$$\psi = \psi(\varepsilon, d) \tag{4.28}$$

4.5.2 Tensor Decomposition of HFE

According to the basic principle of irreversible thermodynamics, in terms of expression, the form of the stress—strain constitutive relation is only related to the elastic HFE potential of the material. However, since the damage constitutive relation contains two types of internal variables—damage variables and plastic variables, even if the stress—strain constitutive relation expression is the same, the different evolution laws of the internal variables will lead to completely different stress—strain constitutive relations.

Different damage models often use different damage variables to describe the effect of damage on the degradation of the macroscopic mechanical properties of the material, resulting in completely different specific expressions of the elastic HFE potential defined by each model. In fact, even if the same damage variable is used, the elastic HFE potential defined by each model is different. In order to describe the unilateral effect of certain materials, the definition of the elastic HFE potential requires the decomposition of a certain physical quantity, such as concrete damage. Different decomposition methods not only lead to different constitutive models, but also greatly affect the computational efficiency of the model numerical algorithm. Regarding the evolution of the free energy potential function, Wu listed them in his study. This study employs the modified material elastic Helmholtz free energy (HFE) potential function proposed by Faria et al. in 1998 [11]:

$$\psi = \psi^e = \frac{1}{2}(1 - d^+)\overline{\sigma}^+ : \boldsymbol{C}_0^{-1} : \overline{\sigma}^+ + \frac{1}{2}(1 - d^-)\overline{\sigma}^- : \boldsymbol{C}_0^{-1} : \overline{\sigma}^- \tag{4.29}$$

Here, the effective stress tensor $\overline{\sigma}$ is decomposed into two parts, positive and negative, $\overline{\sigma}^+$ and $\overline{\sigma}^-$, and it is only related to the elastic strain tensor, so the free energy potential function can be shown as an explicit expression of strain. $\boldsymbol{C}_0$ is the elastic stiffness tensor. This decomposition method is more obvious and direct than the previous one, which greatly improves the calculation efficiency. However, the influence of plasticity is still not considered, and the total free energy potential is only considered to be equivalent to the elastic free energy.

4.5.3 Damage Criteria Based on Energy Dissipation

The core of the damage constitutive model is to give an accurate and reasonable criterion for the evolution of damage variables. Since Sidoroff first proposed the concept of damage surface and used the orthogonal law to describe the evolution of damage variables, the key to the problem has begun to transform into how to determine the damage criterion (i.e. damage surface) of solid materials [12]. Compared with other methods, the damage criterion based on the damage energy release rate has a clearer physical meaning and a more solid irreversible thermodynamic foundation because it is conjugate with the energy of damage change in the thermodynamic sense.

At the same time, plastic properties have a greater impact on the initiation, development, and evolution of cracks [13]. Therefore, it is incorrect to ignore the impact of damage on plastic properties in the physical sense. From the previous research results of Mazars, it can be seen that the prediction results and experimental results obtained by establishing the damage criterion based solely on the elastic damage energy release rate are often quite different. Therefore, Wu et al. [14] incorporated plastic-free energy into the construction of the damage model to more accurately express the stress—strain relationship of concrete materials.

Like the decomposition of the strain tensor, the total elastic-plastic HFE potential can be defined as the sum of the elastic component ψ^e and the plastic component ψ^p given in Eq. (4.29), which can be decomposed as follows:

$$\psi\left(\varepsilon^{e}, d^{\pm}, \kappa\right)=\psi^{e}\left(\varepsilon^{e}, d^{\pm}\right)+\psi^{p}\left(d^{\pm}, \kappa\right) \tag{4.30}$$

Among them, κ represents the plastic hardening function, which should be determined based on the plastic yield function and hardening criterion. Different materials have different plastic behaviors, so it is necessary to select appropriate yield functions and hardening criteria.

In the process of force in which the material does not exchange energy with the outside world, the second law of thermodynamics [15] states that any irreversible process should satisfy the Clausius—Duhem inequality, which is simplified as follows:

$$\dot{\gamma}=-\dot{\psi}+\sigma : \dot{\varepsilon} \geq 0 \tag{4.31}$$

Differentiating Eq. (4.30) with respect to time yields:

$$\dot{\psi}=\frac{\partial \psi^{e}}{\partial \varepsilon^{e}} : \dot{\varepsilon}^{e}+\frac{\partial \psi}{\partial d^{+}} \dot{d}^{+}+\frac{\partial \psi}{\partial d^{-}} \dot{d}^{-}+\frac{\partial \psi^{p}}{\partial \kappa} : \dot{\kappa} \tag{4.32}$$

Based on the standard thermodynamic arguments, Wu et al. [14] assumed that damage and plastic unloading are elastic processes and that the following conditions must be satisfied:

$$\sigma=\frac{\partial \psi^{e}}{\partial \varepsilon^{e}} \tag{4.33}$$

$$\dot{\gamma}^d = -\left(\frac{\partial \psi}{\partial d^+}\dot{d}^+ + \frac{\partial \psi}{\partial d^-}\dot{d}^-\right) \geq 0 \tag{4.34}$$

$$\dot{\gamma}^p = \boldsymbol{\sigma} : \dot{\boldsymbol{\varepsilon}}^p - \frac{\partial \psi^p}{\partial \boldsymbol{\kappa}} : \dot{\boldsymbol{\kappa}} \geq 0 \tag{4.35}$$

From Eq. (4.33), it can be seen that the Cauchy stress tensor depends only on the elastic HFE potential, which is an improvement on the model of Faria et al. [11], in which the total HFE potential is considered. Combined with Eqs. (4.29) and (4.33) can be rewritten as:

$$\boldsymbol{\sigma} = (1 - d^+)\boldsymbol{P}^+ : \overline{\boldsymbol{\sigma}} + (1 - d^-)\boldsymbol{P}^- : \overline{\boldsymbol{\sigma}} = \overline{\boldsymbol{\sigma}} - (d^+\boldsymbol{P}^+ + d^-\boldsymbol{P}^-) : \overline{\boldsymbol{\sigma}} \tag{4.36}$$

This leads to the following expression for the Cauchy stress tensor:

$$\boldsymbol{\sigma} = (\boldsymbol{I} - \boldsymbol{D}) : \overline{\boldsymbol{\sigma}} = (\boldsymbol{I} - \boldsymbol{D}) : \boldsymbol{C}_0 : (\varepsilon - \varepsilon^p) \tag{4.37}$$

Among them, $\boldsymbol{I}$ is the fourth-order unit tensor; the fourth-order damage tensor $\boldsymbol{D}$ is expressed as:

$$\boldsymbol{D} = d^+\boldsymbol{P}^+ + d^-\boldsymbol{P}^- \tag{4.38}$$

From Eqs. (4.34) and (4.36), the damage energy release rate (DERR) Y can be defined as the damage variable as follows:

$$Y = -\frac{\partial \psi}{\partial d} = -\left(\frac{\partial \psi^e}{\partial d} + \frac{\partial \psi^p}{\partial d}\right) \tag{4.39}$$

These results demonstrate that the material damage evolution rate is governed by the total free energy function. To develop an accurate damage model, however, the selection of an appropriate plastic HFE expression must be guided by the specific yield characteristics or hardening behavior of the material.

According to the research of Wu et al. [14], the damage variable and the energy release rate are a pair of conjugate variables. It is proposed that the evolution of material damage can be controlled by the damage criterion function G:

$$G(Y, r) = Y - r \leq 0 \tag{4.40}$$

where r is the threshold of the damage energy release rate at any time. Only when the damage energy release rate Y is greater than r, the material will be damaged and begin to enter the inelastic stage from the linear stage.

References

1 Lavenstein, S., Gu, Y., Madisetti, D., and El-Awady J. A. (2020). The heterogeneity of persistent slip band nucleation and evolution in metals at the micrometer scale. *Science* 370: eabb2690.

2 Jamali, A., Ma, A., and Llorca, J. (2022). Influence of grain size and grain boundary misorientation on the fatigue crack initiation mechanisms of textured AZ31 Mg alloy. *Scripta Materialia* 207: 114304.

3 Li, Y., Yang, E-H., and Tan, K. H. (2019). Effects of heating followed by water quenching on strength and microstructure of ultra-high performance concrete. *Construction and Building Materials* 207: 403–411.
4 Murakami, S. (1987). Anisotropic aspects of material damage and application of continuum damage mechanics. In: *Continuum Damage Mechanics Theory and Application* (ed. D. Krajcinovic and J. Lemaitre), 91–133. Vienna: Springer.
5 Liu, X. and Hao, J. (2011). *Damage Mechanics of Continuum Media*. Beijing: National Defense Industry Press.
6 Lou, Z. (1991). *Fundamentals of Damage Mechanics*. Xi'an: Xi'an Jiaotong University Press.
7 Stuart, E. B., Gal-Or, B., Brainard, A. J., et al. (1970). *A Critical Review of Thermodynamics*, 275–298. Baltimore, MD: Mono Book Corp.
8 Germain, P. (1973). *Cours de Mécanique des Milieux Continus*. Paris: Masson et Cie.
9 de Groot, S. R. and Mazur, P. (1962). *Non-equilibrium Thermodynamics*. Amsterdam: North-Holland.
10 Glansdorff, P. and Prigogine, I. (1971). *Thermodynamic Theory of Structure, Stability and Fluctuations*. London: Wiley.
11 Faria, R., Oliver, J., and Cervera, M. (1998). A strain-based plastic viscous-damage model for massive concrete structures. *International Journal of Solids and Structures* 35: 1533–1558.
12 Sidoroff, F. (1981). Description of anisotropic damage application to elasticity. In: *Physical Non-Linearities in Structural Analysis* (ed. J. Hult and J. Lemaitre), 237–244. Berlin, Heidelberg: Springer.
13 Chow, C. L. and Wang, J. (1987). An anisotropic theory of continuum damage mechanics for ductile fracture. *Engineering Fracture Mechanics* 27: 547–558.
14 Wu, J. Y., Li, J., and Faria, R. (2006). An energy release rate-based plastic-damage model for concrete. *International Journal of Solids and Structures* 43: 583–612.
15 Lubliner, J., Oliver, J., Oller, S., and Oñate, E. (1989). A plastic-damage model for concrete. *International Journal of Solids and Structures* 25: 299–326.

5

Basics of Micromechanics

5.1 Basic Concepts of Micromechanics

Micromechanics is one of the important scientific research achievements in the field of mechanics in the 20th century, and it is a discipline that combines continuum mechanics and materials science. Continuum mechanics assumes that matter is continuously distributed, and that the constitutive relations of mechanics apply to any volume of the object. However, the mesostructure of both natural and artificial materials is heterogeneous. Therefore, the description of continuum mechanics is only an approximation and cannot establish a connection between the mesostructure and the macroscopic properties of the material.

Micromechanics studies media that are macroscopically homogeneous but exhibit mesoscopic inhomogeneity. By utilizing information about the mesostructure of materials, it enables the determination of effective properties at the macroscopic scale. The fundamental concept underlying micromechanics is homogenization [1–3]. Building upon the framework of continuum mechanics, micromechanics introduces internal variables to characterize mesoscopic structures and damage. It establishes homogenization approaches to bridge the mesoscopic and macroscopic scales, thereby enabling the quantitative correlation between microstructure and macroscopic mechanical behavior. This forms a new theoretical framework for multiscale modeling. Applications of micromechanics span a wide range of thermomechanical phenomena, including plasticity, fracture, fatigue, composite materials, and constitutive modeling of polycrystalline solids. Specifically, by integrating elasticity and plasticity theories, this approach examines how defects, inclusions, and material inhomogeneities influence the macroscopic response of crystals, alloys, and composites. Consequently, it facilitates the predictive modeling of novel materials, reducing the need for extensive physical experimentation.

Micromechanics provides a powerful tool for the design, fabrication, and analysis of materials that can be used in polycrystalline materials, composites, geotechnical materials, biomaterials, and electronic materials. The transition from macroscopic to microscopic reflects the modern development of materials science. Through micro, mesoscopic, and macroscopic multiscale methods, the phenomenon and properties of materials of different sizes are increasingly studied with precision and effectiveness.

Constitutive Models of Solid Materials: From Mechanical Principles to Engineering Applications, First Edition.

5.1.1 Introduction to Representative Volume Elements

In general, engineering materials are nonuniform in their mesostructure. However, whether or not it is called a heterogeneous material depends on the scale used in the observation. Certain components or phases within a material can only be identified at or below a specific length scale. While these components may appear homogeneous at that scale, they can exhibit inhomogeneities when examined at a finer resolution. In the case of polycrystalline copper, a common metallic material, the material is homogeneous under visual observation (resolution of about 10 μm). However, observations using the electron backscatter diffraction technique (with a resolution of about 1 μm) show that the polycrystalline copper is composed of many copper grains with different grain orientations. At this scale, although each grain can be regarded as a homogeneous material, polycrystalline copper composed of grains can no longer be regarded as a homogeneous material, and its material properties are determined by the mesostructural characteristics such as grain size, grain orientation, and grain boundaries. When observed on a smaller scale, it is found that each copper grain also has its own mesostructure, such as voids, dislocations, etc., and can no longer be considered a homogeneous material.

Therefore, when studying the properties of real materials, it is necessary to define scales that are directly related to the properties of the research. The microstructural features that cannot be observed at this scale are negligible. In practice, only the average properties of the microstructure are of concern. Therefore, when studying inhomogeneous materials, the overall properties can be used to represent the average properties of the material over a certain volume.

When studying materials using mesoscopic forces, two scales are generally considered: macroscopic and mesoscopic. At the macroscopic scale, a continuum consists of numerous material points, each associated with a mesoscopic region known as the representative volume element (RVE). The RVE is a fundamental micromechanical concept that must balance scale duality: it should be sufficiently small to approximate uniform macroscopic stress and strain fields within it, yet large enough to encompass numerous mesoscopic constituents and capture statistically representative microstructural features of the local continuum. The mesoscopic stress–strain fields, on the other hand, only have an effect on the macroscopic properties of the material through their volume average. There is no fixed length size for RVE, and how to choose RVE is related to the material being studied and the characteristics of the material being studied. In general, the selection of RVE must satisfy the duality of scale. The characteristic dimensions of the heterogeneous material components (e.g. the average size of defects, reinforcing phases, or grains) are denoted by d. The dimensions of RVE are denoted by l. The characteristic dimensions of the macrostructure are indicated by L. And these three parameters need to meet the following relationships:

1. $l \cdot L$: The RVE is small enough relative to the macroscopic structure to be regarded as a point of matter, so it satisfies the conditions for the use of continuum mechanics.
2. $l \cdot d$: The relative feature size of RVE is large enough to contain enough mesostructure information to represent the statistically averaged properties of the local volume.

5.1.2 Localization

If body forces are neglected, it is assumed that either uniform stress or uniform strain boundary conditions are applied on the boundary S of the RVE. The average stress within the Ω is defined as:

$$\dot{\sigma}_{ij} = \frac{1}{|\Omega|}\int_{\Omega} \sigma_{ij} dV \; or \; \bar{\boldsymbol{\sigma}} = \frac{1}{|\Omega|}\int_{\Omega} \boldsymbol{\sigma} dV \tag{5.1}$$

Similarly, the average strain within Ω is defined as:

$$\dot{\varepsilon}_{ij} = \frac{1}{|\Omega|}\int_{\Omega} \varepsilon_{ij} dV \; or \; \bar{\varepsilon} = \frac{1}{|\Omega|}\int_{\Omega} \varepsilon dV \tag{5.2}$$

This definition applies to either mean or non-mean materials.

1. **Uniform Stress Boundary Conditions**
   The Average Stress Theorem: Let $\sum_{ij}$ be a given constant stress tensor, the outer normal of the RVE boundary S is n_j (Figure 5.1), if a uniform stress acts on the boundary S:

$$\sigma_{ij} n_j\big|_S = P_j^0 = \sum_{ij} n_j \; or \; \boldsymbol{\sigma}\cdot\boldsymbol{n}|_S = \boldsymbol{P}^0 = \boldsymbol{\Sigma}\cdot\boldsymbol{n} \tag{5.3}$$

   Then, the average stress within Ω is:

$$\dot{\sigma}_{ij} = \Sigma_{ij} \; or \; \bar{\boldsymbol{\sigma}} = \boldsymbol{\Sigma} \tag{5.4}$$

   To prove the theorem, the average stress is considered as:

$$\dot{\sigma}_{ij} = \frac{1}{|\Omega|}\int_{\Omega} \sigma_{ij} dV = \frac{1}{|\Omega|}\int_{\Omega} \sigma_{ik}\delta_{jk} dV \tag{5.5}$$

   Since $x_{j,k} = \delta_{jk}$,

$$\dot{\sigma}_{ij} = \frac{1}{|\Omega|}\int_{\Omega} \sigma_{ik} x_{j,k} dV = \frac{1}{|\Omega|}\int_{\Omega} \left[\left(\sigma_{ik} x_j\right)_{,k} - \sigma_{ik,k}\right] dV \tag{5.6}$$

   When body forces are neglected, then $\sigma_{ik,k} = 0$. Combined with the sum of the divergence theorem and Eq. (5.3), it can be derived:

$$\dot{\sigma}_{ij} = \frac{1}{|\Omega|}\int_{S} \sigma_{ik} x_j n_k dS = \sum_{ik} \frac{1}{|\Omega|}\int_{S} x_j n_k dS \tag{5.7}$$

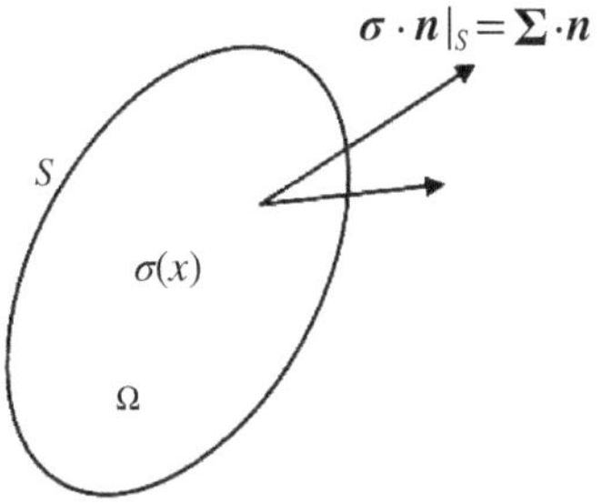

Figure 5.1 Uniform stress boundary condition.

It can be obtained by applying the divergence theorem again,

$$\dot{\sigma}_{ij} = \sum_{ik} \frac{1}{|\Omega|} \int_{\Omega} x_{j,k} dV = \sum_{ik} \frac{1}{|\Omega|} \int_{\Omega} \delta_{jk} dV = \sum_{ij} \tag{5.8}$$

The above is the derivation process of the mean stress theorem.

The mean stress theorem shows that when an object is subjected to the stress boundary condition of Eq. (5.6) and $\sum_{ij}$ is a constant stress tensor; no matter how complex the local stress field is, the mean stress inside it is equal to the macroscopic stress $\sum_{ij}$. Accordingly, the uniform stress boundary condition Ω may be represented as:

$$\sigma_{ij} n_j\big|_S = \dot{\sigma}_{ij} n_j \; or \; \boldsymbol{\sigma} \cdot \boldsymbol{n}|_S = \bar{\boldsymbol{\sigma}} \cdot \boldsymbol{n} \tag{5.9}$$

where $\bar{\boldsymbol{\sigma}}$ is the average stress tensor within Ω.

2. **Uniform Strain Boundary Conditions**

Average Strain Theorem: Let E_{ij} be the given constant strain tensor, the outer normal of their boundary S is n_{ij} (Figure 5.2), and there is the uniform strain on the boundary S, so:

$$u_i|_s = E_{ij} x_j \quad or \quad \boldsymbol{u}|_s = \boldsymbol{E} \cdot \boldsymbol{x} \tag{5.10}$$

Then, the average strain within Ω is:

$$\dot{\varepsilon}_{ij} = E_{ij} \quad or \quad \bar{\boldsymbol{\varepsilon}} = \boldsymbol{E} \tag{5.11}$$

To prove this theorem, the average strain is considered as:

$$\begin{aligned} \dot{\varepsilon}_{ij} &= \frac{1}{|\Omega|} \int_{\Omega} \varepsilon_{ij} dV = \frac{1}{2|\Omega|} \int_{\Omega} \left(u_{i,j} + u_{j,i}\right) dV \\ &= \frac{1}{2|\Omega|} \int_{\Omega} u_{i,j} dV + \frac{1}{2|\Omega|} \int_{\Omega} u_{j,i} dV \end{aligned} \tag{5.12}$$

Applying the divergence theorem, Eq. (5.12) becomes

$$\dot{\varepsilon}_{ij} = \frac{1}{2|\Omega|} \int_S u_i n_j \, dS + \frac{1}{2|\Omega|} \int_S u_j n_i dS \tag{5.13}$$

Substituting Eq. (5.10) into Eq. (5.13):

$$\dot{\varepsilon}_{ij} = \frac{1}{2|\Omega|} \int_S E_{ik} x_k n_j \, dS + \frac{1}{2|\Omega|} \int_S E_{ik} x_k n_i dS \tag{5.14}$$

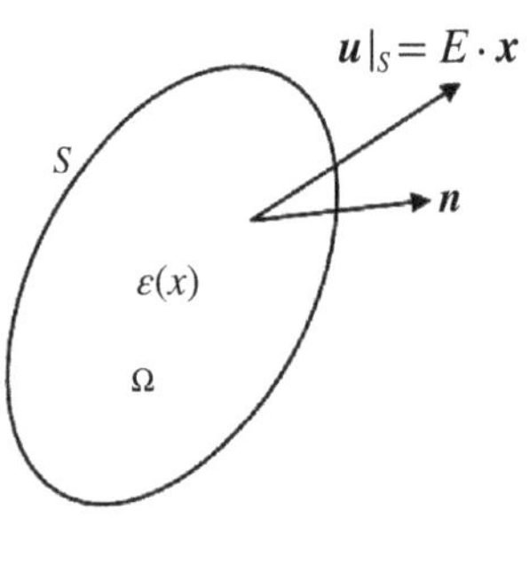

Figure 5.2 Uniform strain boundary condition.

Then apply the divergence theorem, and get:

$$\dot{\varepsilon}_{ij} = \frac{1}{2|\Omega|}\int_V E_{ik}x_{k,j}\,dV + \frac{1}{2|\Omega|}\int_V E_{jk}x_{k,i}dV = \frac{1}{2}\left(E_{ij} + E_{ji}\right) = E_{ij} \tag{5.15}$$

The above is the proof process of the mean strain theorem.

According to the average strain theorem, when an object is subjected to the uniform strain boundary condition of Eq. (5.10) and E_{ij} is the constant strain tensor, no matter how complex the local strain field is, the average strain in Ω is equal to the macroscopic strain E_{ij}. Therefore, the uniform strain boundary condition can be written as

$$u_i|_S = \dot{\varepsilon}_{ij}x_j \quad or \quad \boldsymbol{u}|_S = \bar{\varepsilon}\cdot\boldsymbol{x} \tag{5.16}$$

where $\dot{\varepsilon}_{ij}$ is the average strain tensor within Ω.

5.2 Eshelby Eigenstrain Theory and Equivalent Inclusion Theory

5.2.1 Eigenstrain Theory

An infinitely large homogeneous elastomer medium, shown in Figure 5.3a, c with a region represented as Ω. When a local area I shown in Figure 5.3b produces strain ε^*_{ij} in the absence of constraints due to some physical or chemical reasons (such as phase change, temperature change, or plastic deformation), this strain is called eigenstrain. However, the material in the region I in Figure 5.3a shows an infinite elastomer in reality, constrained by the surrounding medium, and cannot be freely deformed. Its deformation is limited due to the forces of the surrounding medium. For small strain deformations, when there are both elastic and eigenstrain in the continuum, the total strain is the sum of the two

$$\varepsilon_{ij} = e_{ij} + \varepsilon^*_{ij} \tag{5.17}$$

where ε_{ij} is the total strain, e_{ij} is elastic strain, and ε^*_{ij} is an eigenstrain.

Consider the Eshelby problem of uniform eigenstrain ε^*_{ij} in an ellipsoidal subdomain Ω in an infinite domain with a material stiffness tensor of L_{ijkl}. Assume that the ellipsoidal subdomain (Figure 5.4) is:

$$\Omega = \left\{x_1, x_2, x_3 \left| \left(\frac{x_1}{a_1}\right)^2 + \left(\frac{x_2}{a_2}\right)^2 + \left(\frac{x_3}{a_3}\right)^2 \right| 1\right\} \tag{5.18}$$

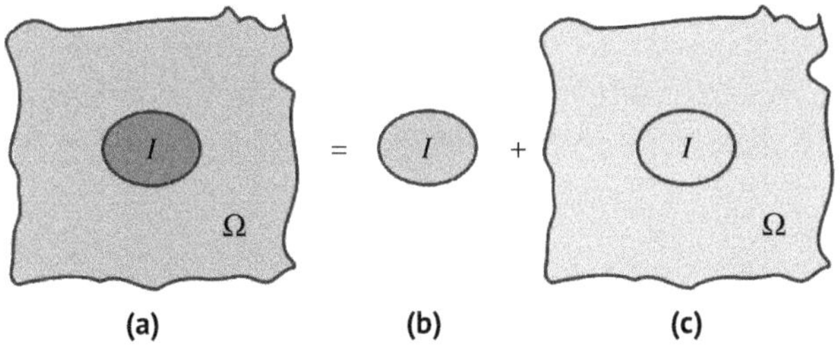

Figure 5.3 Eigenstrain in infinite medium: (a) infinite medium; (b) free body I; and (c) infinite medium with holes Ω.

where a_1, a_2, and a_3 are the semi-axes of the ellipsoid.

According to the Green function expression solved by the eigenstrain problem, there is:

$$\varepsilon_{ij}(\boldsymbol{x}) = S_{ijkl}(\boldsymbol{x})\,\varepsilon_{kl}^{*} \tag{5.19}$$

where the fourth-order tensor S_{ijkl} is the Eshelby tensor. Eq. (5.18) is the strain solution for the Eshelby ellipsoid. And

$$S_{ijkl} = S_{jikl} = S_{ijlk} \tag{5.20}$$

However, Eshelby tensors do not have diagonal symmetry, so $S_{ijkl} \neq S_{kiij}$.

Eshelby tensor S_{ijkl} is only related to the properties of the matrix material and the shape of the ellipsoid. For anisotropic materials, the integral can be found numerically S_{ijkl}. For isotropic materials, this integral can be expressed as an elliptic integral. In some specific cases, you can also get an explicit expression. The Eshelby tensor S_{ijkl} of the spherical region in an isotropic material can be expressed as

$$\begin{gathered} S_{ijkl} = \gamma\delta_{ij}\delta_{kl} + \delta\left(\delta_{ik}\delta_{jl} + \sigma_{ii}\delta_{jk} - \frac{2}{3}\delta_{ij}\delta_{kl}\right) \\ \gamma = \frac{k}{3k+4\mu} = \frac{1+\upsilon}{9(1-\upsilon)},\ \delta = \frac{3(k+2\mu)}{5(3k+4\mu)} = \frac{4-5\upsilon}{15(1-\upsilon)} \end{gathered} \tag{5.21}$$

For special shapes, the expression of the Eshelby tensor S_{ijkl} for an isotropic material with a Poisson's ratio υ is given below.

1. Ellipsoids $(a_1 > a_2 > a_3)$

$$\begin{cases} S_{ijkl} = S_{jkl} = S_{ijk} \\ S_{1111} = \dfrac{3}{8\pi(1-\upsilon)}a_1^2 I_{11} + \dfrac{1-2\upsilon}{8\pi(1-\upsilon)}I_1 \\ S_{1122} = \dfrac{1}{8\pi(1-\upsilon)}a_2^2 I_{12} - \dfrac{1-2\upsilon}{8\pi(1-\upsilon)}I_1 \\ S_{1133} = \dfrac{1}{8\pi(1-\upsilon)}a_3^2 I_{13} - \dfrac{1-2\upsilon}{8\pi(1-\upsilon)}I_1 \\ S_{1212} = \dfrac{a_1^2+a_2^2}{16\pi(1-\upsilon)}I_{12} + \dfrac{1-2\upsilon}{16\pi(1-\upsilon)}(I_1+I_2) \end{cases} \tag{5.22}$$

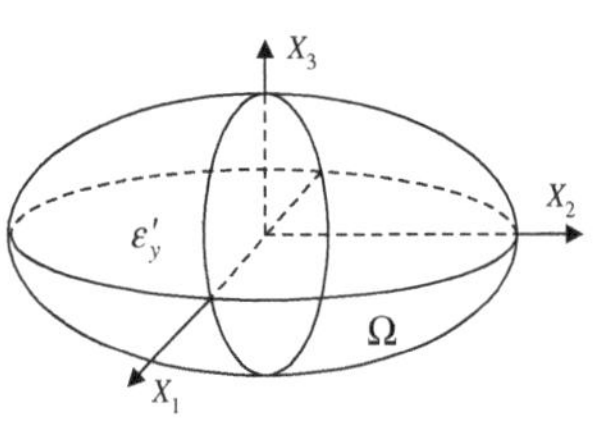

Figure 5.4 Ellipsoidal subdomain **Ω**.

All nonzero components can be obtained by cyclic permutation of the indices (1, 2, 3). Other components are zero, such as: $S_{1112} = S_{1223} = S_{1232} = 0$. In Eq. (5.22), constant I_1, I_2, I_{11}, I_{12}, and I_{13} can be expressed as:

$$\begin{cases} I_1 = \dfrac{4\pi a_1 a_2 a_3}{(a_1^2 - a_2^2)(a_1^2 - a_3^2)} \{F(\theta, \kappa) - E(\theta, \kappa)\} \\ I_3 = \dfrac{4\pi_{a1} a_2 a_3}{(a_2^2 - a_3^2)(a_1^2 - a_3^2)^{\frac{1}{2}}} \left\{ \dfrac{a_2 (a_1^2 - a_3^2)^{\frac{1}{2}}}{a_1 a_3} - E(\theta, \kappa) \right\} \end{cases} \tag{5.23}$$

where elliptic integral is defined as:

$$\begin{cases} F(\theta, \kappa) = \displaystyle\int_0^\theta \frac{\mathrm{dw}}{\left(1 - \kappa^2 \sin^2 w\right)^{\frac{1}{2}}} \\ E(\theta, \kappa) = \displaystyle\int_0^\theta \left(1 - \kappa^2 \sin^2 w\right)^{\frac{1}{2}} \mathrm{dw} \\ \theta = \arcsin\left(1 - \dfrac{a_3^2}{a_1^2}\right)^{\frac{1}{2}}, \kappa = \left\{\dfrac{(a_1^2 - a_2^2)}{(a_1^2 - a_3^2)}\right\}^{\frac{1}{2}} \end{cases} \tag{5.24}$$

and

$$\begin{cases} I_1 + I_2 + I_3 = 4\pi \\ 3I_{11} + I_{12} + I_{13} - \dfrac{4\pi}{a_1^2} \\ 3a_1^2 I_{11} + a_2^2 I_{12} + a_3^2 I_{13} = 3I_1 \\ I_{12} = \dfrac{I_2 - I_1}{a_1^2 - a_2^2} \end{cases} \tag{5.25}$$

2. Sphere ($a_1 = a_2 = a_3 = a$)

$$\begin{cases} I_1 = I_2 = I_3 = \dfrac{4\pi}{3} \\ I_{11} = I_{22} = I_{33} = I_{12} = I_{23} = I_{31} = \dfrac{4\pi}{5a^2} \end{cases} \tag{5.26}$$

$$\begin{cases} S_{1111} = S_{2222} = S_{3333} = \dfrac{7 - 5\upsilon}{15(1 - \upsilon)} \\ S_{1122} = S_{2233} = S_{3311} = S_{1133} = S_{2211} = S_{3322} = \dfrac{5\upsilon - 1}{15(1 - \upsilon)} \\ S_{1212} = S_{2323} = S_{3131} = \dfrac{4 - 5\upsilon}{15(1 - \upsilon)} \end{cases} \tag{5.27}$$

3. Ellipsoidal Cylinder ($a_3 \to \infty$)

$$\begin{cases} I_1 = \dfrac{4\pi a_2}{a_1 + a_2}, I_2 = \dfrac{4\pi a_1}{a_1 + a_2}, I_3 = 0 \\ I_{12} = \dfrac{4\pi}{(a_1 + a_2)^2}, 3I_{11} = \dfrac{4\pi}{a_1^2} - I_{12} \\ 3I_{22} = \dfrac{4\pi}{a_2^2} - I_{12}, I_{13} = I_{23} = I_{33} = 0 \\ a_3^2 I_{13} = I_1, a_3^2 I_{23} = I_2, a_3^2 I_{33} = 0 \\ S_{1111} = \dfrac{1}{2(1-v)} \left\{ \dfrac{a_2^2 + 2a_1 a_2}{(a_1 + a_2)^2} + (1 - 2v) \dfrac{a_2}{a_1 + a_2} \right\} \\ S_{2222} = \dfrac{1}{2(1-v)} \left\{ \dfrac{a_1^2 + 2a_1 a_2}{(a_1 + a_2)^2} + (1 - 2v) \dfrac{a_1}{a_1 + a_2} \right\}, \\ S_{3333} = 0 \\ S_{1122} = \dfrac{1}{2(1-v)} \left\{ \dfrac{a_2^2}{(a_1 + a_2)^2} - (1 - 2v) \dfrac{a_2}{a_1 + a_2} \right\} \\ S_{2323} = \dfrac{1}{2(a_1 + a_2)}, S_{3131} = \dfrac{2v a_2}{a_1 + a_2}, S_{1133} = \dfrac{1}{2(1-v)} \\ S_{2211} = \dfrac{1}{2(1-v)} \left\{ \dfrac{a_1^2}{(a_1 + a_2)^2} - (1 - 2v) \dfrac{a_1)}{a_1 + a_2} \right\} \\ S_{3322} = 0, S_{1212} = \dfrac{1}{2(1-v)} \left\{ \dfrac{a_1^2 + a_2^2}{2(a_1 + a_2)^2} + \dfrac{1 - 2v}{2} \right\} \end{cases} \tag{5.28}$$

4. Disc (Coin-shaped) ($a_1 = a_2 >> a_3$)

$$\begin{cases} I_1 = I_2 = \dfrac{\pi^2 a_3}{a_1}, I_3 = 4\pi - \dfrac{2\pi^2 a_3}{a_1} \\ I_{12} = I_{21} = \dfrac{3\pi^2 a_3}{4a_1^3} \\ I_{13} = I_{23} = I_{31} = I_{32} - 3\left(\dfrac{4}{3}\pi - \dfrac{\pi^2 a_3}{a_1}\right) / a_1^2 \\ I_{11} = I_{22} = \dfrac{3\pi^2 a_3}{4a_1^3}, I_{33} = \dfrac{\frac{4}{3}\pi}{a_3^2} \end{cases} \tag{5.29}$$

$$
\begin{cases}
S_{1111} = S_{2222} = \dfrac{13-8v}{32(1-v)}\pi\dfrac{a_3}{a_1}, S_{3333} = 1 - \dfrac{1-2v}{1-v}\dfrac{\pi}{4}\dfrac{a_3}{a_1} \\
S_{1122} = S_{2211} = \dfrac{8v-1}{32(1-v)}\pi\dfrac{a_3}{a_1}, S_{2233} = \dfrac{2v-1}{8(1-v)}\pi\dfrac{a_3}{a_1} \\
S_{3311} = S_{3322} = \dfrac{v}{1-v}\left(1 - \dfrac{4v+1}{8v}\pi\dfrac{a_3}{a_1}\right) \\
S_{1212} = \dfrac{7-8v}{32(1-v)}\pi\dfrac{a_3}{a_1}, S_{1313} = S_{2323} = \dfrac{1}{2}\left(1 + \dfrac{v-2}{1-v}\dfrac{\pi}{4}\dfrac{a_3}{a_1}\right) \\
S_{kk11} = S_{kk22} = \dfrac{1-2v}{1-v}\dfrac{\pi}{4}\dfrac{a_3}{a_1} + \dfrac{v}{1-v}, S_{kk33} = 1 - \dfrac{1-2v}{1-v}\dfrac{\pi}{2}\dfrac{a_3}{a_1}
\end{cases}
\tag{5.30}
$$

when $a_3 = 0$,

$$
\begin{cases}
I_1 = I_2 = 0, I_3 = 4\pi \\
I_{12} = 0, I_{23} = \dfrac{4\pi}{a_2^2}, I_{31} = \dfrac{4\pi}{a_1^2} \\
I_{11} = I_{22} = 0, a_3^2 I_{33} = \dfrac{4\pi}{3} \\
S_{2323} = S_{3131} = \dfrac{1}{2} \\
S_{3311} = S_{3322} = \dfrac{v}{1-v} \\
S_{3333} = 1, \text{其余} S_{ijkl} = 0
\end{cases}
\tag{5.31}
$$

5. Elliptical Discs ($a_1 > a_2 >> a_3$)

$$
\begin{cases}
I_1 = \dfrac{4\pi a_2 a_3 \{F(\kappa) - E(\kappa)\}}{a_1^2 - a_2^2} \\
I_2 = \dfrac{4\pi a_3 E(\kappa)}{a_2} - \dfrac{4\pi a_2 a_3 \{F(\kappa) - E(\kappa)\}}{a_1^2 - a_2^2} \\
I_3 = 4\pi - \dfrac{4\pi a_3 E(\kappa)}{a_2} \\
I_{12} = \dfrac{\left[\frac{4\pi a_3 E(\kappa)}{a_2} - \frac{8\pi a_2 a_3 \{F(\kappa)-E(\kappa)\}}{a_1^2-a_2^2}\right]}{a_1^2 - a_2^2} \\
I_{23} = \dfrac{\left[4\pi - \frac{8\pi a_3 E(\kappa)}{a_2} + \frac{4\pi a_2 a_3 \{F(\kappa)-E(\kappa)\}}{a_1^2-a_2^2}\right]}{a_2^2} \\
I_{31} = \dfrac{\left[4\pi - \frac{4\pi a_2 a_3 \{F(\kappa)-E(\kappa)\}}{a_1^2-a_2^2} - \frac{4\pi a_3 E(\kappa)}{a_2}\right]}{a_1^2} \\
I_{33} = \dfrac{4\pi}{3a_3^2}
\end{cases}
\tag{5.32}
$$

where $E(\varkappa)$ and $F(\varkappa)$ are perfect elliptic integrals of the first and second kinds, respectively:

$$\begin{cases} E(\varkappa) = \int_0^{\frac{\pi}{2}} \left(1 - \varkappa^2 \sin^2 \phi\right)^{\frac{1}{2}} d\phi \\ F(\varkappa) = \int_0^{\frac{\pi}{2}} \left(1 - \varkappa^2 \sin^2 \phi\right)^{-\frac{1}{2}} d\phi \\ \varkappa^2 = \dfrac{a_1^2 - a_2^2}{a_1^2} \end{cases} \tag{5.33}$$

6. Rotate the Flat Sphere ($a_1 = a_2 > a_3$)

$$\begin{cases} I_1 = I_2 = \dfrac{2\pi a_1^2 a_3}{\left(a_1^2 - a_3^2\right)^{\frac{3}{2}}} \left\{ \arccos \dfrac{a_3}{a_1} - \dfrac{a_3}{a_1}\left(1 - \dfrac{a_3^2}{a_1^2}\right)^{\frac{1}{2}} \right\} \\ I_3 = 4\pi - 2I_1, I_{11} = I_{22} = I_{12} \\ I_{12} = \dfrac{\pi}{a_1^2} - \dfrac{1}{4} I_{13} = \dfrac{\pi}{a_1^2} - \dfrac{I_1 - I_3}{4\left(a_3^2 - a_1^2\right)} \\ I_{13} = I_{23} = \dfrac{I_1 - I_3}{a_3^2 - a_1^2}, 3I_{33} = \dfrac{4\pi}{a_3^2} - 2I_{13} \end{cases} \tag{5.34}$$

7. Spheroid ($a_1 > a_2 = a_3$)

$$\begin{cases} I_2 = I_3 = \dfrac{2\pi a_1 a_3^2}{\left(a_1^2 - a_3^2\right)^{3/2}} \left\{ \dfrac{a_1}{a_3}\left(\dfrac{a_1^2}{a_3^2} - 1\right)^{1/2} - \cosh^{-1} \dfrac{a_1}{a_3} \right\} \\ I_1 = 4\pi - 2I_2, I_{12} = \dfrac{I_2 - I_1}{a_1^2 - a_2^2} \\ 3I_{11} = \dfrac{4\pi}{a_1^2} - 2I_{12}, I_{22} = I_{33} = I_{23} \\ 3I_{22} = \dfrac{4\pi}{a_2^2} - I_{23} - \dfrac{I_2 - I_1}{a_1^2 - a_2^2} \\ I_{23} = \dfrac{\pi}{a_2^2} - \dfrac{I_2 - I_1}{4\left(a_1^2 - a_2^2\right)} \end{cases} \tag{5.35}$$

5.2.2 Equivalent Inclusion Theory

Consider the following two scenarios:

1. As shown in Figure 5.5a: Assume that the stiffness tensor of an infinitely large uniform elastic medium is $\boldsymbol{L}_0$ and the space region is Ω. A uniform eigenstrain $\boldsymbol{\varepsilon}^*$ occurs in the ellipsoidal region I, and the uniform strain boundary condition is at infinity $\boldsymbol{u} = \boldsymbol{E} \cdot \boldsymbol{x}$.

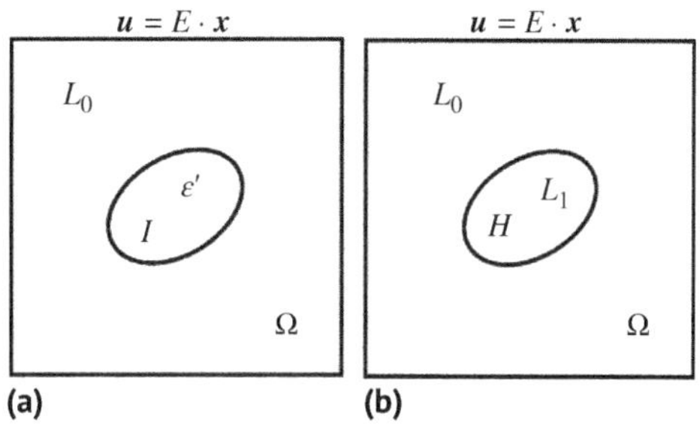

Figure 5.5 Eigenstrain problems and inclusion problems in homogeneous media: (a) eigenstrain problems and (b) inclusion problems.

2. As shown in Figure 5.5(b): Assume an infinite uniform elastic medium with stiffness tensor $\boldsymbol{L}_0$ and space region Ω. An ellipsoidal inclusion $\boldsymbol{H}$ with stiffness tensor $\boldsymbol{L}_1$ is embedded in it, uniform strain boundary condition $\boldsymbol{u} = \boldsymbol{E} \cdot \boldsymbol{x}$ is at infinity, and displacement is continuous on the interface between region $\boldsymbol{x} \in (\Omega - H)$ and inclusion $\boldsymbol{x} \in H$.

The above two cases are the eigenstrain problem and the inclusion problem in a homogeneous medium, respectively. The governing equation for the eigenstrain problem of a homogeneous medium is:

$$\begin{cases} \nabla \cdot (\boldsymbol{L}_0 : \boldsymbol{\varepsilon}) + \nabla \cdot (-\boldsymbol{L}_0 : \boldsymbol{\varepsilon} *) = 0 & \boldsymbol{x} \in I \\ \nabla \cdot (\boldsymbol{L}_0 : \boldsymbol{\varepsilon}) = 0 & \boldsymbol{x} \in (\Omega - I) \\ \boldsymbol{u} = \boldsymbol{E} \cdot \boldsymbol{x} & |\boldsymbol{x}| \to \infty \end{cases} \tag{5.36}$$

And the governing equation for the inclusion problem is:

$$\begin{cases} \nabla \cdot (\boldsymbol{L}_0 : \boldsymbol{\varepsilon}) + \nabla \cdot [(\boldsymbol{L}_1 - \boldsymbol{L}_0) : \boldsymbol{\varepsilon}] = 0 & \boldsymbol{x} \in H \\ \nabla\, (\boldsymbol{L}_0 : \boldsymbol{\varepsilon}) = 0 & \boldsymbol{x} \in (\Omega - H) \\ \boldsymbol{u} = \boldsymbol{E} \cdot \boldsymbol{x} & |\boldsymbol{x}| \to \infty \end{cases} \tag{5.37}$$

It can be seen that the only difference between the two problems is the second term of the first equation when the geometries of regions I and H are the same. However, the eigenstrain $\boldsymbol{\varepsilon}^*$ is known in region I, but $(\boldsymbol{L}_1 - \boldsymbol{L}_0) : \boldsymbol{\varepsilon}$ is unknown in region H.

In order to make an analogy between the above two types of problems, it is assumed that regions I and H are ellipsoids of the same shape. The solution of Eq. (5.36) can be decomposed into the Eshelby solution when eigenstrain occurs in region I, and then the uniform strain $\boldsymbol{E}$ is superimposed, that is,

$$\boldsymbol{\varepsilon}^I = \boldsymbol{E} + \boldsymbol{S} : \boldsymbol{\varepsilon}^* \tag{5.38}$$

If $\boldsymbol{\varepsilon}^*$ meet the following conditions

$$-\boldsymbol{L}_0 : \boldsymbol{\varepsilon}^* = (\boldsymbol{L}_1 - \boldsymbol{L}_0) : \boldsymbol{\varepsilon}^I \tag{5.39}$$

The solution to the governing equation of the homogeneous medium problem also satisfies the governing equation of the inclusion problem.

Similarly, if the governing equations for the inclusion problem with an ellipsoidal region H are solved, the solution also satisfies the governing equations for the homogeneous medium problem, provided that the chosen $\boldsymbol{\varepsilon}^*$ satisfies Eq. (5.29).

The solutions to Eqs. (5.36) and (5.37) are both unique. To further illustrate this problem, substitute Eq. (5.39) into Eq. (5.38), eliminate $\boldsymbol{\varepsilon}^*$, and obtain:

$$\boldsymbol{\varepsilon}^I = \boldsymbol{E} - \boldsymbol{S} : \boldsymbol{L}_0^{-1} : (\boldsymbol{L}_1 - \boldsymbol{L}_0) : \boldsymbol{\varepsilon}^I \tag{5.40}$$

By substituting $\boldsymbol{\varepsilon}^I$ and $\boldsymbol{\varepsilon}^H$, Eq. (5.40) can be rewritten as

$$\boldsymbol{\varepsilon}^H = \left[\boldsymbol{I} + \boldsymbol{S} : \boldsymbol{L}_0^{-1} : (\boldsymbol{L}_1 - \boldsymbol{L}_0)\right]^{-1} : \boldsymbol{E} \tag{5.41}$$

Therefore, the case where an infinite elastic medium (stiffness tensor is $\boldsymbol{L}_0$) contains an ellipsoidal inclusion (stiffness tensor is $\boldsymbol{L}_1$) and is subjected to a uniform strain (E) at infinity is equivalent to the problem where an ellipsoidal region of the same geometric shape in an infinite uniform medium undergoes a eigenstrain $\boldsymbol{\varepsilon}^*$ and is subjected to a uniform strain at infinity. At this time, substituting Eq. (5.39) into Eq. (5.38), eliminating $\boldsymbol{\varepsilon}^I$, the equivalent eigenstrain $\boldsymbol{\varepsilon}^*$ is

$$\boldsymbol{\varepsilon}^* = -\left[\boldsymbol{S} + (\boldsymbol{L}_1 - \boldsymbol{L}_0)^{-1} : \boldsymbol{L}_0\right]^{-1} : \boldsymbol{E}.$$

5.3 Hill Theorem

Consider a RVE with volume V and boundary S. Hill lemma states that under applied force boundary conditions or displacement boundary conditions, any stress field and strain field at a given point of the RVE can be obtained by the following relationship:

$$\overline{\sigma_{ij}\varepsilon_{ij}} - \bar{\sigma}_{ij}\bar{\varepsilon}_{ij} = \frac{1}{D}\int_S (u_i - x_j\bar{\varepsilon}_{ij})(\sigma_{ik}n_k - \bar{\sigma}_{ik}n_k)\, dS \tag{5.42}$$

The upper horizontal line corresponds to the volume-averaged value, exemplified by

$$\overline{\sigma_{ij}\varepsilon_{ij}} = \frac{1}{V}\int_V \sigma_{ij}\varepsilon_{ij} dv. \tag{5.43}$$

The proof of the Hill lemma requires expansion of the surface integral integrand in Eq. (5.42). This preliminary step involves:

$$\begin{aligned}&\int_S (u_i - x_j\bar{\varepsilon}_{ij})(\sigma_{ik}n_k - \bar{\sigma}_{ik}n_k)\, dS \\ &= \int_S \left(u_i\sigma_{ik}n_k - u_i n_k\bar{\sigma}_{ik} - \sigma_{ik}n_k x_j\bar{\varepsilon}_{ij} + x_j n_k\bar{\varepsilon}_{ij}\bar{\sigma}_{ik}\right) dS.\end{aligned} \tag{5.44}$$

The surface integral on the right side of Eq. (5.44) can be easily found

$$\int_S u_i\sigma_{ik}n_k dS = \int_V u_{i,j}\sigma_{ij}dV = \int_V \varepsilon_{ij}\sigma_{ij}dV = D\overline{\sigma_{ij}\varepsilon_{ij}},$$
$$\int_S u_i n_k \bar{\sigma}_{ik} dS = \bar{\sigma}_{ik}\int_V u_{i,k}dV = D\bar{\sigma}_{ik}\bar{\varepsilon}_{ik},$$
$$\int_S \sigma_{ik}n_k x_j dS = \int_V \sigma_{ik}\delta_{ik}dV = D\bar{\sigma}_{ij},$$
$$\int_S x_j n_k dS = \int_V \delta_{jk}dV = D\delta_{jk}.$$

When the solution is substituted into Eq. (5.44), the result is:

$$\frac{1}{D}\int_S (u_i - x_j\bar{\varepsilon}_{ij})(\sigma_{ik}n_k - \bar{\sigma}_{ik}n_k)\,dS = \overline{\sigma_{ij}\varepsilon_{ij}} - \bar{\varepsilon}_{ik}\bar{\sigma}_{ik} - \bar{\sigma}_{ij}\bar{\varepsilon}_{ij} + \bar{\varepsilon}_{ik}\bar{\sigma}_{ik}$$
$$= \overline{\sigma_{ij}\varepsilon_{ij}} - \bar{\varepsilon}_{ij}\bar{\sigma}_{ij} \tag{5.45}$$

Thus, the Hill lemma is proved.

A corollary of Hill's lemma is that if a stress boundary condition $\sigma_{ij}n_j\big|_S = \bar{\sigma}_{ij}n_j$ or a displacement boundary condition $u_i|_S = \bar{\varepsilon}_{ij}x_j$ is applied in the RVE, then any stress field and strain field at a given point have the following relationship:

$$\overline{\sigma_{ij}\varepsilon_{ij}} = \bar{\varepsilon}_{ij}\bar{\sigma}_{ij}, \tag{5.46}$$

In other words, for a statically allowable stress field $\sigma_{ij}n_j\big|_S = \bar{\sigma}_{ij}n_j$, or a kinematically allowable displacement field $u_i|_S = \bar{\varepsilon}_{ij}x_j$, the volume average of the product $\overline{\sigma_{ij}\varepsilon_{ij}}$ is equal to the product of the volume averages of $\bar{\varepsilon}_{ij}$ and $\bar{\sigma}_{ij}$. Eq. (5.46) is also called the Hill macroscopic homogeneity condition or the Mandel–Hill condition. Note that $\sigma_{ij}\varepsilon_{ij}$ is twice the strain energy density. Equation (5.45) indicates that the volume-averaged strain energy density of a heterogeneous material can be expressed through the volume averages of stress and strain. Accordingly, the process of homogenization can be interpreted as identifying a homogeneous medium that is energetically equivalent to the original heterogeneous microstructure. This principle forms the basis for defining the equivalent macroscopic properties of heterogeneous materials [4, 5].

5.4 Mean Field Method Based on Eshelby Equivalent Inclusion Theory

5.4.1 Basic Process of Homogenization Method

As a branch of micromechanics, mesomechanics systematically derives effective continuum properties for heterogeneous materials through rigorous averaging of microstructural features and their mechanical interactions [1–3]. The homogenization procedure ensures thermodynamic consistency while accounting for dominant deformation mechanisms at relevant length scales.

As shown in Figure 5.6, a composite material region Ω is introduced, which consists of the matrix phase region Ω_0 and the ellipsoidal inhomogeneous region $\Omega_1, \Omega_2, \dots, \Omega_N$.

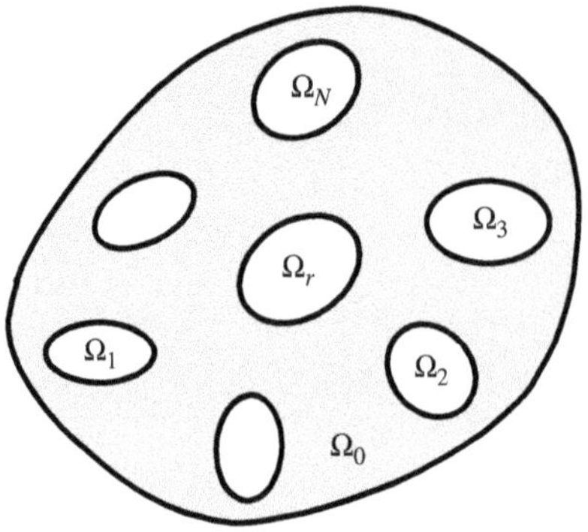

Figure 5.6 Multiphase composite material.

The volume fraction of the inhomogeneous region Ω_r is $c_r = |\Omega_r| \,/\, |\Omega|$, and $\sum_{r=0}^{N} c_r = 1$. Let the material stiffness of the region be $\boldsymbol{L}_r$ $(r = 0, 1, \dots, N)$.

Based on the homogenization concept, the equivalent stiffness of the composite material in Figure 5.6 is defined as

$$\bar{\sigma} = \bar{\boldsymbol{L}}\bar{\varepsilon} \tag{5.47}$$

$\bar{\sigma}$ and $\bar{\varepsilon}$ are the average stress and strain tensors of the composite material, respectively. The fourth-order tensor $\bar{\boldsymbol{L}}$ is the equivalent modulus or equivalent stiffness tensor of the composite material. The average strain of RVE is defined as

$$\bar{\varepsilon} = \sum_{r=0}^{N} c_r \bar{\varepsilon}_r \tag{5.48}$$

$\bar{\varepsilon}$ is the average strain of the composite material, and $\bar{\varepsilon}_r$ is the average strain of the rth phase inhomogeneity, so the strain concentration tensor $\boldsymbol{A}_r$ is introduced to link the two variables

$$\bar{\varepsilon}_r = \boldsymbol{A}_r \bar{\varepsilon} \tag{5.49}$$

The average strain of the composite material is defined as

$$\bar{\sigma} = \sum_{r=0}^{N} c_r \bar{\sigma}_r = c_0 \boldsymbol{L}_0 \bar{\varepsilon}_0 + \sum_{r=1}^{N} c_r \boldsymbol{L}_r \bar{\varepsilon}_r, \tag{5.50}$$

where $\bar{\varepsilon}_r$ is the average strain of the rth phase inhomogeneity, and Eq. (5.50) contains Hooke's law for each phase. From the above definitions, we can deduce

$$\bar{\sigma} = \left[\boldsymbol{L}_0 + \sum_{r=1}^{N} c_r \left(\boldsymbol{L}_r - \boldsymbol{L}_0 \right) \boldsymbol{A}_r \right] \bar{\varepsilon} \tag{5.51}$$

Then

$$\bar{\boldsymbol{L}} = \boldsymbol{L}_0 + \sum_{r=1}^{N} c_r \left(\boldsymbol{L}_r - \boldsymbol{L}_0 \right) \boldsymbol{A}_r \tag{5.52}$$

Because each phase heterogeneity of the composite is embedded in the matrix, a local concentration tensor is introduced to relate the average strain of the rth phase heterogeneity to the average strain of the matrix.

$$\bar{\varepsilon}_r = \boldsymbol{G}_r \bar{\varepsilon}_0 \tag{5.53}$$

The relationship between the global strain concentration tensor and the local strain concentration tensor is obtained by derivation:

$$\boldsymbol{A}_r = \boldsymbol{G}_r \left[c_0 \boldsymbol{I} + \sum_{r=1}^{N} c_r \boldsymbol{G}_r \right]^{-1} \tag{5.54}$$

According to the above relationship, as long as one of the strain concentration tensors is determined, it is sufficient to determine the equivalent stiffness tensor of the composite material. Therefore, the problem of determining the equivalent stiffness tensor of the composite material is transformed into solving the strain concentration tensor.

For the multiphase composite material in Figure 5.6, given the displacement boundary condition

$$\boldsymbol{u}|_S = \boldsymbol{\varepsilon}^0 \boldsymbol{x} \tag{5.55}$$

or stress boundary condition

$$\boldsymbol{\sigma} \cdot \mathbf{n}|_S = \boldsymbol{\sigma}^0 \mathbf{n}, \tag{5.56}$$

These two boundary conditions need to be met

$$\boldsymbol{\sigma}^0 = \boldsymbol{L}_0 \boldsymbol{\varepsilon}^0 \tag{5.57}$$

If region Ω is uniform and filled with matrix material with a stiffness tensor of $\boldsymbol{L}_0$, the strain and stress fields in region Ω are $\boldsymbol{\varepsilon}^0$ and $\boldsymbol{\sigma}^0$, respectively. In composite materials, there are multiphase heterogeneities, and the strains and stresses in the matrix and heterogeneity will no longer be $\boldsymbol{\varepsilon}^0$ and $\boldsymbol{\sigma}^0$. Mori-Tanaka et al. proposed a method for calculating the average strain in heterogeneity based on physical significance. For the rth phase heterogeneity, the influence of other heterogeneities on the rth phase can be regarded as the rth phase being embedded in another equivalent matrix different from the matrix, that is, as an isolated heterogeneity embedded in a fictitious matrix with a prestrain $\widehat{\boldsymbol{\varepsilon}}^0$, and the stiffness tensor of the fictitious matrix is $\widehat{\boldsymbol{L}}_0$.

Therefore, the above problem becomes an ellipsoidal inhomogeneity Ω_r with a stiffness tensor of $\boldsymbol{L}_r$ embedded in a homogeneous matrix with a stiffness tensor of $\widehat{\boldsymbol{L}}_0$ and a prestrain $\widehat{\boldsymbol{\varepsilon}}^0$. Therefore, only by solving the values of $\widehat{\boldsymbol{\varepsilon}}^0$ and $\widehat{\boldsymbol{L}}_0$ can the two problems be equivalent. When $\widehat{\boldsymbol{\varepsilon}}^0$ and $\widehat{\boldsymbol{L}}_0$ are known, the Eshelby method and equivalent inclusion theory can be used to determine the stress and strain of a single-phase ellipsoidal inhomogeneity in an infinite elastic body. The equivalent inclusion equation is

$$\boldsymbol{L}_r \left(\widehat{\boldsymbol{\varepsilon}}^0 + \boldsymbol{\varepsilon}_r^{pt} \right) = \widehat{\boldsymbol{L}}_0 \left(\widehat{\boldsymbol{\varepsilon}}^0 + \boldsymbol{\varepsilon}_r^{pt} - \boldsymbol{\varepsilon}_r^* \right) \tag{5.58}$$

According to Eshelby's method, there is

$$\boldsymbol{\varepsilon}_r^{pt} = \widehat{\boldsymbol{S}}_r \boldsymbol{\varepsilon}_r^* \tag{5.59}$$

where $\widehat{S}_r$ is the Eshelby tensor, calculated from the stiffness tensor $\widehat{L}_0$ and the geometry of the rth phase ellipsoidal inhomogeneity Ω_0, and the eigenstrain is

$$\varepsilon_r^* = \left[(L_r - \widehat{L}_0)\, \widehat{S}_r + \widehat{L}_0 \right]^{-1} (L_r - L_0)\, \widehat{\varepsilon}^0 \tag{5.60}$$

Therefore, the strain of the rth phase ellipsoidal inhomogeneity can be obtained as

$$\varepsilon_r = \widehat{\varepsilon}^0 + \varepsilon_r^{pt} = \widehat{\varepsilon}^0 + \widehat{S}_r \varepsilon^* = \widehat{T}_r \widehat{\varepsilon}^0 \tag{5.61}$$

where

$$\widehat{T}_r = \left[I + \widehat{S}_r \widehat{L}_0^{-1} (L_r - \widehat{L}_0) \right]^{-1} \tag{5.62}$$

Therefore, the stress of the rth phase ellipsoidal anisotropy is

$$\sigma_r = L_r \varepsilon_r = L_r \widehat{T}_r \widehat{\varepsilon}^0 \tag{5.63}$$

Now, the strain and stress fields of the rth phase inhomogeneity have been related to the fictitious matrix material of unknown strain field $\widehat{\varepsilon}^0$ and stiffness $\widehat{L}_0$. Micromechanics uses various approximate methods to choose $\widehat{\varepsilon}^0$ and $\widehat{L}_0$. Once $\widehat{\varepsilon}^0$ and $\widehat{L}_0$ are given, the concentrated tensor can be solved to determine the equivalent properties of the composite material.

5.4.2 Sparse Methods

The sparse method is also called the Eshelby method. It is applicable to the case where the distance between heterogeneities is far and the interaction can be ignored. The sparse method assumes that the average strain of the heterogeneity is approximately equal to the strain when the isolated heterogeneity is embedded in an infinite matrix, without considering the influence of other heterogeneities contained in the matrix. Therefore, the rth phase heterogeneity can be regarded as an ellipsoidal heterogeneity in a uniform matrix with a stiffness of L_0. The matrix has a uniform strain ε^0 before the heterogeneity is embedded, that is:

$$\widehat{L}_0 = L_0 \ \& \ \widehat{\varepsilon}^0 = \varepsilon^0 \tag{5.64}$$

Since the solution of the Eshelby tensor is only related to the matrix stiffness L_0 and the inhomogeneous geometry,

$$\widehat{S}_r = S_r \tag{5.65}$$

According to Eq. (5.61), the strain in the rth phase inhomogeneity can be obtained:

$$\varepsilon_r = \varepsilon^0 + \varepsilon_r^{pt} = \varepsilon^0 + S_r \varepsilon_r^* = T_r \varepsilon^0 \tag{5.66}$$

where

$$T_r = \left[I + \widehat{S}_r \widehat{L}_0^{-1} (L_r - \widehat{L}_0) \right]^{-1} \tag{5.67}$$

It is worth noting that although the domain of T_r is $r > 0$, it can be seen from Eq. (5.67) that $T_0 = I$.

Now assume that the composite material is subjected to the following displacement boundary conditions:

$$\boldsymbol{u}|_S = \boldsymbol{\varepsilon}^0 \boldsymbol{x} \tag{5.68}$$

Under uniform strain boundary conditions, according to the average strain theorem, the average strain of the composite material is equal to $\boldsymbol{\varepsilon}^0$, that is:

$$\bar{\boldsymbol{\varepsilon}} = \frac{1}{V}\int_V \boldsymbol{\varepsilon} dV = \boldsymbol{\varepsilon}^0 \tag{5.69}$$

Therefore, Eq. (5.66) can be written as:

$$\boldsymbol{\varepsilon}_r = \boldsymbol{T}_r \bar{\boldsymbol{\varepsilon}} \tag{5.70}$$

By comparing Eqs. (5.49) and (5.70), the overall strain concentration tensor of the rth phase inhomogeneity is:

$$\boldsymbol{A}_r = \boldsymbol{T}_r \tag{5.71}$$

Finally, the equivalent stiffness tensor of the composite material is obtained:

$$\begin{aligned}\bar{\boldsymbol{L}} &= \boldsymbol{L}_0 + \sum_{r=1}^{N} c_r (\boldsymbol{L}_r - \boldsymbol{L}_0)\, \boldsymbol{T}_r \\ &= \boldsymbol{L}_0 + \sum_{r=1}^{N} c_r (\boldsymbol{L}_r - \boldsymbol{L}_0)\, [\boldsymbol{I} + \boldsymbol{S}_r(\boldsymbol{M}_0\boldsymbol{L}_r - \boldsymbol{I})]^{-1}\end{aligned} \tag{5.72}$$

where $\bar{\boldsymbol{L}}$ is the sparse estimate of the composite material equivalent stiffness tensor. This equivalent stiffness tensor is based on the assumption that the heterogeneities in the composite material are far apart and cannot affect each other. Therefore, the sparse estimate is only applicable when the heterogeneity volume fraction is very low.

5.4.3 Mori–Tanaka Method

Since the sparse method does not consider the interaction between heterogeneities, some scholars believe that the average strain of the heterogeneity is equal to the strain of a single heterogeneity in an infinite matrix [4–7]. Therefore, Mori and Tanaka et al. proposed the Mori–Tanaka method by changing the distant strain or stress to consider the interaction between heterogeneities [8].

The mean strain and mean stress in a composite material are defined as:

$$\bar{\varepsilon} = \frac{1}{D}\int_D \varepsilon dV = \sum_{r=0}^{N} \bar{\varepsilon}_r, \quad \bar{\sigma} = \frac{1}{D}\int_D \sigma dV = \sum_{r=0}^{N} \bar{\sigma}_r, \tag{5.73}$$

where $\bar{\boldsymbol{\varepsilon}}_r$ and $\bar{\boldsymbol{\sigma}}_r$ are the average values of the heterogeneous strain tensor and stress tensor of the rth phase, respectively.

$$\bar{\varepsilon}_r = \frac{1}{\Omega_r}\int_{\Omega_r} \varepsilon dV, \quad \bar{\sigma}_r = \frac{1}{\Omega_r}\int_{\Omega_r} \sigma dV_r, \tag{5.74}$$

And

$$\bar{\sigma}_r = L_r : \bar{\varepsilon}_r \tag{5.75}$$

Consider the composite material shown in Figure 5.6. For a typical inhomogeneity of stiffness L_r (r > 0), the effects of other inhomogeneities are mutually transmitted through the strain and stress fields of the surrounding matrix. When a large number of inhomogeneities exist and are randomly distributed in the matrix, although the strain and stress fields in the matrix have different distributions at different locations, the average values of the matrix strain $\bar{\varepsilon}_0$ and stress $\bar{\sigma}_0$ and can be well approximated to the true field in the matrix surrounding each inhomogeneity. Therefore, it can be assumed that removing an inhomogeneity from the matrix will not affect the overall elastic behavior of the composite. In other words, when the *r*th inhomogeneity is removed and replaced by the matrix material, the average strain and average stress of the matrix will remain unchanged. For the *r*th inhomogeneity, it can be considered as an ellipsoidal inhomogeneity with a stiffness tensor L_r embedded in a uniform matrix with a stiffness tensor of L_0, which has a uniform strain $\bar{\varepsilon}_0$ before the inhomogeneity is embedded. Therefore, in Eq. (5.72),

$$\hat{L}_0 = L_0 \ \& \ \hat{\varepsilon}^0 = \bar{\varepsilon}_0 \tag{5.76}$$

It can be seen that the Eshelby tensor:

$$\hat{S}_r = S_r \tag{5.77}$$

By substituting Eq. (5.77) into Eq. (5.51), the strain field of the *r*th phase inhomogeneity can be determined.

$$\varepsilon_r = \bar{\varepsilon}_0 + \varepsilon_r^{pt} = \bar{\varepsilon}_0 + S_r \varepsilon_r^* = T_r \bar{\varepsilon}_0 \tag{5.78}$$

T_r can be given by Eq. (5.67). Although Eqs. (5.62) and (5.78) are similar in form; Eq. (5.84) represents the relationship between the inhomogeneous strain and the average strain of the composite material, while Eq. (5.78) represents the relationship between the inhomogeneous strain and the average strain of the surrounding matrix. Therefore, the tensor T_r in Eq. (5.71) is the global strain concentration tensor, while the tensor T_r in Eq. (5.78) is the local strain concentration tensor,

$$G_r = T_r. \tag{5.79}$$

The overall strain concentration tensor is obtained:

$$A_r = G_r \left[c_0 I + \sum_{n=1}^{N} c_n G_n \right]^{-1} = T_r \left[\sum_{n=0}^{N} c_n T_n \right]^{-1} \tag{5.80}$$

where $T_0 = I$. Therefore, the equivalent stiffness tensor of the composite material is:

$$\bar{L} = L_0 + \sum_{r=1}^{N} c_r (L_r - L_0) A_r = \sum_{r=0}^{N} c_r L_r T_r \left[\sum_{n=0}^{N} c_n T_n \right]^{-1} \tag{5.81}$$

where $\bar{L}$ is the calculation result of the equivalent stiffness tensor of the composite material using the Mori–Tanaka method.

5.4.4 Self-consistent Method for Polycrystalline Materials

In conventional self-consistent schemes for evaluating the effective properties of composites, a clear matrix phase is typically assumed, with other constituents treated as inclusions. However, this assumption is not applicable to certain heterogeneous materials like polycrystals, where no distinct matrix exists. Instead, each grain is considered as an inclusion embedded within the surrounding aggregate, and all grains are treated with equal significance.

Consider a polycrystalline material composed of N randomly oriented grains. In the case of a single-phase polycrystal, each grain represents a single crystal with a distinct crystallographic orientation. Assuming that the volume fraction of the rth grain is c_r, under uniform strain and stress boundary conditions, the average strain and average stress in the polycrystalline are:

$$\bar{\varepsilon} = \sum_{r=0}^{N} c_r \bar{\varepsilon}_r, \quad \bar{\sigma} = \sum_{r=0}^{N} c_r \bar{\sigma}_r \tag{5.82}$$

where $\bar{\varepsilon}_r$ and $\bar{\sigma}_r$ are the average values of the strain tensor and stress tensor of the rth grain, respectively, and the overall strain concentration tensor of the rth grain is

$$\bar{\varepsilon}_r = \boldsymbol{A}_r \bar{\varepsilon} \tag{5.83}$$

According to Hooke's law,

$$\bar{\sigma} = \sum_{r=1}^{N} c_r \bar{\sigma}_r = \sum_{r=1}^{N} c_r \boldsymbol{L}_r \bar{\varepsilon}_r = \sum_{r=1}^{N} c_r \boldsymbol{L}_r \boldsymbol{A}_r \bar{\varepsilon} \tag{5.84}$$

So,

$$\boldsymbol{L} = \sum_{r=1}^{N} c_r \boldsymbol{L}_r \boldsymbol{A} \tag{5.85}$$

Obviously, for polycrystalline materials, if the strain concentration tensor $\boldsymbol{A}_r$ is known, the equivalent stiffness tensor $\bar{\boldsymbol{L}}$ can be calculated according to Eq. (5.62). Generally speaking, unless the exact expression of $\boldsymbol{A}_r$ is given, it is difficult to directly calculate the equivalent stiffness tensor from these two equations.

References

1 Zhang, Y. and Han, L. (2014). *Fundamentals of Micromechanics*. Beijing: Science Press.
2 Zhang, Y. and Zhang, Z. (2008). *Micromechanics of Materials*. Beijing: Science Press.
3 Qu, J. and Cherkaoui, M. (2006). *Fundamentals of Micromechanics of Solids*. Hoboken: Wiley.
4 He, X., Liu, L., Zeng, T. et al. (2020). Micromechanical modeling of work hardening for coupling microstructure evolution, dynamic recovery and recrystallization: application to high entropy alloys. *International Journal of Mechanical Sciences* 177: 105567.
5 Liu, L., Yao, Y., and Zeng, T. (2019). A micromechanical analysis to the elasto-viscoplastic behavior of solder alloys. *International Journal of Solids and Structures* 159: 211–220.

6 Zecevic, M. and Knezevic, M. (2015). A dislocation density based elasto-plastic self-consistent model for the prediction of cyclic deformation: application to AA6022-T4. *International Journal of Plasticity* 72: 200–217.

7 Zeng, T., Shao, J. F., and Xu, W. (2015). A self-consistent approach for micro-macro modeling of elastic-plastic deformation in polycrystalline geomaterials: Polycrystalline model. *International Journal for Numerical and Analytical Methods in Geomechanics* 39: 1735–1752.

8 Mori, T. and Tanaka, K. (1973). Average stress in matrix and average elastic energy of materials with misfitting inclusions. *Acta Metallurgica* 21: 571–574.

6

Numerical Implementation of Constitutive Relations

The previous five chapters established a relatively complete constitutive theoretical framework. Through reasonable assumptions and deformation mechanism analysis, the complex behavior of materials under load can be described. Constitutive relations are generally expressed using partial differential equations (PDEs). In most cases, obtaining a theoretical solution to these PDEs (or systems of PDEs) is challenging, which necessitates the introduction of appropriate numerical algorithms [1–3].

6.1 Basic Concepts

To facilitate subsequent discussions, this section will focus on one-dimensional elastoplastic constitutive relations as an example, emphasizing key concepts and the numerical implementation process in numerical algorithms.

6.1.1 Numerical Solution of Differential Equations

Consider the following initial value problem:

$$\begin{cases} \dot{x}(t) = \dfrac{\mathrm{d}x}{\mathrm{d}t} = f(x(t)) \\ x(t=0) = x_0 \end{cases} \quad t \in [0, T] \tag{6.1}$$

In Eq. (6.1), $f(x)$ is known. According to numerical analysis knowledge [1], when the function $x(t)$ is sufficiently smooth, for example, if $x(t)$ satisfies the Lipschitz condition, it can theoretically be guaranteed that the solution to the initial value problem exists and is unique.

The so-called numerical solution refers to determining the function $x(t)$ at a series of discrete nodes

$$0 = t_0 < t_1 < \cdots < t_{n-1} < t_n = T \tag{6.2}$$

Thus, the approximate value of Eq. (6.1) is $x_1, x_2, \cdots, x_{n-1}, x_n$. Assume $x_n \cong x(t_n)$, where x_n represents the numerical solution approximating the exact solution $x(t_n)$. The interval between two adjacent time steps, $\Delta t = t_n - t_{n-1}$, is called the step size. Unless otherwise specified, the step size is assumed to be constant, $\Delta t = T/k$, with nodes

Constitutive Models of Solid Materials: From Mechanical Principles to Engineering Applications, First Edition.

$t_n = t_0 + n\Delta t, n = 0, 1, 2, \cdots, k$. Considering that nonlinear problems generally adopt incremental loading, the current stress state of the material depends on the material's loading history. This corresponds to the "step-by-step" solving strategy used in numerical methods, where the solution process advances sequentially, following the arrangement of the nodes. As long as the known information (historical information) $x_{n-1}, x_{n-2}, \cdots, x_0$ is provided, the value of x_n can be computed.

For differential equations similar to Eq. (6.1), the following single-parameter integration algorithm, known as the generalized midpoint method, is often used:

$$\begin{cases} x_{n+1} = x_n + \Delta t \cdot f(x_{n+\theta}) \\ x_{n+\theta} = (1-\theta)\, x_n + \theta x_{n+1} \end{cases} \quad \theta \in [0, 1] \tag{6.3}$$

Different values of θ determine the stability, accuracy, and convergence speed of the numerical algorithm. In practice, common choices for θ include the following:

$$\begin{cases} \theta = 0 & \text{Euler forward interpolation method (Explicit algorithm)} \\ \theta = \dfrac{1}{2} & \text{Midpoint Euler method (Implicit algorithm)} \\ \theta = 1 & \text{Euler post-interpolation method (Implicit algorithm)} \end{cases} \tag{6.4}$$

When $\theta = 0$, Eq. (6.3) simplifies to: $x_{n+1} = x_n + \Delta t \cdot f(x_n)$. This shows that if x_0 is known, the numerical solution at all future time steps can be obtained by traversing the entire time interval $[0, T]$, as shown in Figure 6.1. It can be seen that the forward Euler method is an explicit method, where the incremental direction is determined by the tangent at the previous time step. This method is computationally efficient and can be directly solved. However, its drawbacks are quite apparent. As illustrated in Figure 6.1, if the numerical solution is not corrected, it is prone to drift, making it an unstable solution method. To ensure accuracy, the step size often needs to be very small, which increases the computational cost.

When $\theta \neq 0$, Eq. (6.3) becomes:

$$x_{n+1} = x_n + \Delta t \cdot f(x_{n+\theta}) = x_n + \Delta t \cdot f((1-\theta)\, x_n + \theta x_{n+1}) \tag{6.5}$$

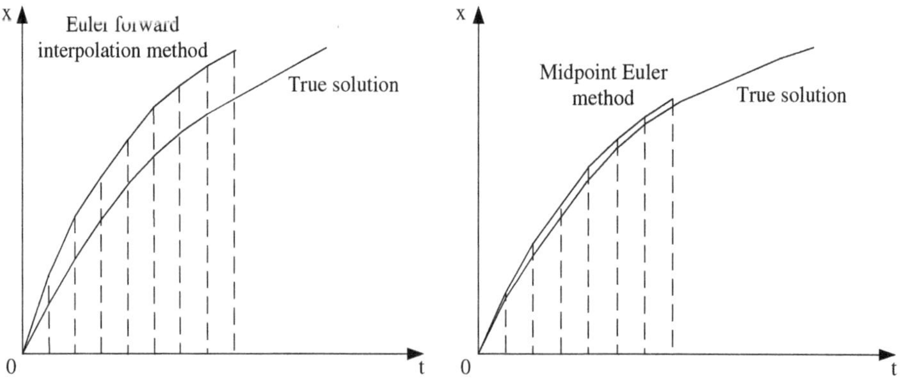

Figure 6.1 Numerical solution methods for differential equations.

Clearly, both sides of Eq. (6.5) contain x_{n+1}, making it a nonlinear equation that must be solved using an iterative method. Research shows that to ensure the stability of the numerical algorithm, θ should be no less than $1/2(\theta \geq 1/2)$. In practice, θ is often chosen as 1/2 or 1. When $\theta = 1/2$, the method corresponds to the midpoint Euler method, where the increment direction is taken along the tangent at the midpoint $[x_n, x_{n+1}]$. This method provides second-order accuracy. However, the trade-off of using this method is that it increases the storage of state variables and makes the numerical implementation more complex.

Another simpler and unconditionally stable method is to choose $\theta = 1$, which takes the tangent direction at the unknown time x_{n+1} as the increment direction. This method corresponds to the backward Euler method in numerical algorithms. Although it has the same first-order accuracy as the forward Euler method, the backward Euler method has the advantage that errors do not accumulate, and it is unconditionally stable. Moreover, the numerical implementation is simpler than that of the midpoint Euler method, making it more widely used in practice.

6.1.2 Newton's Iterative Method

The numerical integration of constitutive equations involves solving many nonlinear equations (or systems of equations). For computational efficiency, Newton's Iteration Method (Newton–Raphson Method, or NR Method) is commonly used. This method, proposed by Newton in the 17th century, is an approach to approximate the roots of equations in the real and complex domains. Its primary advantage is that it exhibits quadratic convergence near the root of the equation.

For a single equation $g(x) = 0$, assume the equation has several roots, with x^r being one of them. The basic steps of the NR method for solving x^r are as follows:

1. Choose an initial guess $x^{(0)}$ near the root x^r;
2. If $g\left(x^{(k)}\right) < \text{Tol}$ (where Tol is the permissible error tolerance, typically, choosing $10^{-3} \sim 10^{-5}$ provides good accuracy depending on the complexity of the problem), then $x^{(k)}$ can be considered as the root of the equation. The number in parentheses indicates the current iteration step k (starting with 0 as the initial value). If $g\left(x^{(k)}\right) \geq \text{Tol}$, further correction of $x^{(k)}$ is needed. Assume the corrected value is closer to the root. The corrected value can be obtained by expanding $g(x)$ at $x^{(k)}$ using a first-order Taylor expansion and setting it to zero:

$$g(x) \approx g\left(x^{(k)}\right) + \left.\frac{\mathrm{d}g}{\mathrm{d}x}\right|_{x=x^{(k)}} \left(x - x^{(k)}\right) = g\left(x^{(k)}\right) + g'\left(x^{(k)}\right)\left(x - x^{(k)}\right) \tag{6.6}$$

3. Let $g(x) = 0$, then the updated value of $x^{(k)}$ becomes

$$g(x) \approx g\left(x^{(k)}\right) + g'\left(x^{(k)}\right)\left(x - x^{(k)}\right) = 0 \Rightarrow x = x^{(k)} - \frac{g\left(x^{(k)}\right)}{g'\left(x^{(k)}\right)} \tag{6.7}$$

   Correction of iterative variables: $k + 1 \to k$, $x \to x^{(k)}$.
   Continue with step 2 until the convergence criterion is satisfied.

For a system of nonlinear equations $\boldsymbol{g}(\boldsymbol{x}) = [g_1(\boldsymbol{x}), g_2(\boldsymbol{x}), g_3(\boldsymbol{x}), \ldots, g_n(\boldsymbol{x})]^T = \boldsymbol{0}$, that is:

$$\boldsymbol{g}(\boldsymbol{x}) = \begin{bmatrix} g_1(x_1, x_2, x_3, \ldots, x_n) \\ g_2(x_1, x_2, x_3, \ldots, x_n) \\ g_3(x_1, x_2, x_3, \ldots, x_n) \\ \vdots \\ g_n(x_1, x_2, x_3, \ldots, x_n) \end{bmatrix} = \begin{bmatrix} 0 \\ 0 \\ 0 \\ \vdots \\ 0 \end{bmatrix} \tag{6.8}$$

where the unknowns form a vector $\boldsymbol{x} = [x_1, x_2, x_3, \ldots, x_n]^T$. The approach to solving a system of nonlinear equations is similar to that for a single equation. The steps are as follows:

1. Choose an initial guess $\boldsymbol{x}^{(0)}$;
2. If $\left\|\boldsymbol{g}(\boldsymbol{x}^{(k)})\right\| < \text{Tol}$, then $\boldsymbol{x}^r = \boldsymbol{x}^{(k)}$ is considered the solution. $\|\cdot\|$ represents the 2-norm of the vector. Otherwise, perform a first-order Taylor expansion of $\boldsymbol{g}(\boldsymbol{x})$ at $\boldsymbol{x}^{(k)}$ to obtain

$$\boldsymbol{g}(\boldsymbol{x}) \approx \boldsymbol{g}(\boldsymbol{x}^{(k)}) + \left.\frac{\partial(g_1, g_2, g_3, \ldots, g_n)}{\partial(x_1, x_2, x_3, \ldots, x_n)}\right|_{x=x^{(k)}} (\boldsymbol{x} - \boldsymbol{x}^{(k)}) = g(\boldsymbol{x}^{(k)}) + \left[\mathbf{J}^{(k)}\right](\boldsymbol{x} - \boldsymbol{x}^{(k)}) \tag{6.9}$$

$$\left[\mathbf{J}^{(k)}\right] = \left.\frac{\partial(g_1, g_2, g_3, \ldots, g_n)}{\partial(x_1, x_2, x_3, \ldots, x_n)}\right|_{x=x^{(k)}} = \begin{bmatrix} \frac{\partial g_1}{\partial x_1} & \frac{\partial g_1}{\partial x_2} & \frac{\partial g_1}{\partial x_3} & \cdots & \frac{\partial g_1}{\partial x_n} \\ \frac{\partial g_2}{\partial x_1} & \frac{\partial g_2}{\partial x_2} & \frac{\partial g_2}{\partial x_3} & \cdots & \frac{\partial g_2}{\partial x_n} \\ \frac{\partial g_3}{\partial x_1} & \frac{\partial g_3}{\partial x_2} & \frac{\partial g_3}{\partial x_3} & \cdots & \frac{\partial g_3}{\partial x_n} \\ \vdots & \vdots & \vdots & \ddots & \vdots \\ \frac{\partial g_n}{\partial x_1} & \frac{\partial g_n}{\partial x_2} & \frac{\partial g_n}{\partial x_3} & \cdots & \frac{\partial g_n}{\partial x_n} \end{bmatrix}$$

$\left[\mathbf{J}^{(k)}\right]$ is the Jacobian matrix at the kth iteration. Let $\boldsymbol{g}(\boldsymbol{x}) = 0$, then the correction value $\boldsymbol{x}^{(k)}$ can be obtained by solving:

$$\boldsymbol{g}(\boldsymbol{x}) \approx \boldsymbol{g}(\boldsymbol{x}^{(k)}) + \left[\mathbf{J}^{(k)}\right](\boldsymbol{x} - \boldsymbol{x}^{(k)}) = \boldsymbol{0} \Rightarrow \boldsymbol{x} = \boldsymbol{x}^{(k)} - \left[\mathbf{J}^{(k)}\right]^{-1} \boldsymbol{g}(\boldsymbol{x}^{(k)}) \tag{6.10}$$

Correct the iterative variables: $k + 1 \rightarrow k$, $x \rightarrow x^{(k)}$.

Continue with step 2 until the corresponding convergence criteria are satisfied.

The Newton–Raphson method has the advantage of fast convergence, but it also has drawbacks, such as its convergence depending on the choice of the initial guess. If the initial guess is not chosen properly, the numerical solution may diverge. Additionally, solving the Jacobian matrix can be complex, and its positive definiteness is not always guaranteed. In practical applications, improved versions of the Newton–Raphson method can be used. For example, Newton's downhill method can help prevent divergence, and methods like the secant method or parabolic method can utilize historical information from the iterative process to approximate the Jacobian matrix [1].

6.1.3 One-dimensional Elastoplastic Constitutive Model

In this chapter, we only consider small-strain cases. According to the principle of superposition, the total strain ε is the sum of the elastic strain ε^e and the plastic strain ε^p, which can be expressed as:

$$\varepsilon = \varepsilon^e + \varepsilon^p \tag{6.11}$$

The relationship between the Cauchy stress σ and the elastic strain ε^e is given by Hooke's law:

$$\sigma = E \cdot \varepsilon^e = E \cdot (\varepsilon - \varepsilon^p) \tag{6.12}$$

For the elastoplastic model, the absolute value of the Cauchy stress σ cannot exceed the material's yield stress σ_y, where the subscript y indicates yield. Assuming the material has the same tensile and compressive strength, the allowable range of Cauchy stress σ is $\left[-\sigma_y, \sigma_y\right]$. The stress that does not lead to material failure is called the admissible stress, and its set can be defined as

$$E_\sigma = \left\{\sigma \in \mathbb{R} | f(\sigma, \alpha) = |\sigma| - \sigma_y(\alpha) \le 0\right\} \tag{6.13}$$

f is the yield function, and α is an internal variable that reflects the hardening of the material, typically associated with plastic deformation.

When the absolute value of the stress σ is less than the yield stress, the material has not yet entered the plastic flow phase, meaning $\varepsilon^p = 0$. Once the stress reaches the yield stress ($f = 0$), the material undergoes plastic flow. According to the flow rule, plastic deformation can be expressed as

$$\dot{\varepsilon}^p = \dot{\gamma}\frac{\partial F}{\partial \sigma} \tag{6.14}$$

where $\dot{\gamma}$ is a nonnegative scalar known as the plastic multiplier, and F is the plastic potential function. When using the associated flow rule ($F = f$):

$$\dot{\varepsilon}^p = \dot{\gamma}\frac{\partial F}{\partial \sigma} = \dot{\gamma}\frac{\partial f}{\partial \sigma} = \begin{cases} +\dot{\gamma} & \text{if} \quad \sigma = \sigma_y > 0 \\ -\dot{\gamma} & \text{if} \quad \sigma = -\sigma_y < 0 \end{cases} \tag{6.15}$$

Introducing the sign function sign (x):

$$\operatorname{sign}(x) = \begin{cases} +1 & \text{if} \quad x > 0 \\ -1 & \text{if} \quad x < 0 \end{cases} \tag{6.16}$$

Equation (6.15) can then be rewritten as:

$$\dot{\varepsilon}^p = \dot{\gamma}\text{sign}(\sigma) \quad \text{if} \quad \sigma = \sigma_y \tag{6.17}$$

$\dot{\gamma}$ and σ must satisfy the following unilateral constraints:$f(\sigma) \leq 0$ and $\dot{\gamma} \geq 0$; $\dot{\gamma} \cdot f = 0$. This condition implies two things: 1) If $f < 0$, then $\dot{\gamma} = 0 \Rightarrow \dot{\varepsilon}^p = 0$; 2) If $f = 0$, then $\dot{\gamma} > 0$. These two conditions are known as the Kuhn–Tucker complementary conditions [2]. To determine $\dot{\gamma}$, we also need to introduce the consistency condition. This condition implies: 1) $\dot{f} = 0 \Rightarrow \dot{\gamma} > 0$; 2) $\dot{f} < 0 \Rightarrow \dot{\gamma} = 0$. It is important to note that $\dot{f}$ cannot be positive. The reasoning is as follows: since the current stress state lies on the yield surface i.e. $f = 0$. If $\dot{f} > 0$, it would result in $f > 0$ in the next loading step, which contradicts condition 1. This can be demonstrated using a Taylor expansion.

6.1.4 Rate Form of the Stress–Strain Relationship

The stress–strain rate form of a one-dimensional elastoplastic model can be determined using the Kuhn–Tucker complementary conditions and the consistency condition. There are two cases:

1. $\dot{\gamma} = 0$, the material remains in the elastic phase, and $\dot{\varepsilon}^p = 0$. Differentiating both sides of Eq. (6.12) with respect to time gives:

$$\dot{\sigma} = E \cdot (\dot{\varepsilon} - \dot{\varepsilon}^p) \tag{6.18}$$

Thus, the stress–strain relationship in the elastic state is:

$$\dot{\sigma} = E \cdot \dot{\varepsilon}^e \tag{6.19}$$

Assume that the material parameter E is constant and independent of the material state.

2. $f = 0, \dot{\gamma} > 0$. From the consistency condition, $\dot{f} = 0$, we obtain:

$$\dot{f} = \frac{\partial f}{\partial \sigma}\dot{\sigma} + \frac{\partial f}{\partial \sigma_y}\frac{\partial \sigma_y}{\partial \gamma_p}\dot{\gamma}_p = \frac{\partial f}{\partial \sigma} \cdot E \cdot (\dot{\varepsilon} - \dot{\varepsilon}^p) - \frac{\partial \sigma_y}{\partial \alpha}\dot{\alpha} = 0 \tag{6.20}$$

In general, material hardening is related to its plastic deformation. In the simplest case, assume $\dot{\alpha} = |\dot{\varepsilon}^p| = \dot{\gamma}$. Of course, more accurate functional forms are determined by the material's deformation mechanism. Considering Eqs. (6.17) and (6.20) simplifies to

$$\begin{aligned} \dot{f} &= \frac{\partial f}{\partial \sigma} \cdot E \cdot (\dot{\varepsilon} - \dot{\varepsilon}^p) - \frac{\partial \sigma_y}{\partial \alpha}\dot{\alpha} \\ &= \text{sign}(\sigma) \cdot E \cdot \dot{\varepsilon} - E \cdot \dot{\gamma} - \frac{\partial \sigma_y}{\partial \alpha}\dot{\gamma} = 0 \Rightarrow \dot{\gamma} = \frac{\text{sign}(\sigma) \cdot E \cdot \dot{\varepsilon}}{E + \frac{\partial \sigma_y}{\partial \alpha}} \end{aligned} \tag{6.21}$$

Substituting Eqs. (6.21) and (6.17) into Eq. (6.18) and simplifying yields

$$\dot{\sigma} = \frac{E \cdot \frac{\partial \sigma_y}{\partial \alpha}}{E + \frac{\partial \sigma_y}{\partial \alpha}} \cdot \dot{\varepsilon} = E^{ep} \cdot \dot{\varepsilon} \quad \text{where} \quad E^{ep} = \frac{E \cdot \frac{\partial \sigma_y}{\partial \alpha}}{E + \frac{\partial \sigma_y}{\partial \alpha}} \tag{6.22}$$

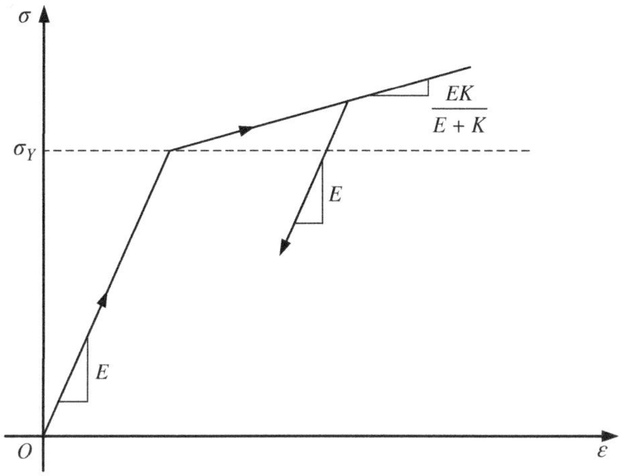

Figure 6.2 The stress–strain relationship of one-dimensional elastoplastic models.

Equation (6.22) represents the rate form of the stress–strain relationship after the material has yielded. The term E^{ep} is known as the continuum elastoplastic modulus. Consider two special cases:

1. Perfect plasticity (no hardening), i.e. $\partial\sigma_y/\partial\alpha = 0$, in this case $\dot{\sigma} = 0$.
2. Linear hardening, i.e. $\partial\sigma_y/\partial\alpha = K$, K is the isotropic hardening modulus, in this case E^{ep} becomes:

$$E^{ep} = \frac{E \cdot K}{E + K} \tag{6.23}$$

The stress–strain responses for these two cases are shown in Figure 6.2.

6.1.5 Stress–Strain Relationship with Kinematic Hardening in Rate Form

When a material is subjected to cyclic loading, the center of the yield surface is no longer fixed and moves in the direction of plastic flow, as illustrated in Figure 6.3. This movement of the yield surface center causes the material to exhibit different behavior under tension and compression. This phenomenon is known as the Bauschinger Effect (BE). To capture this effect, an additional variable q is introduced into the yield function:

$$f(\sigma, q, \alpha) = |\sigma - q| - \sigma_y(\alpha) \leq 0 \tag{6.24}$$

In this Eq. (6.24), q is called the back stress, which determines the position of the yield surface center. The magnitude of the back stress can be determined according to Ziegler's rule:

$$\dot{q} = H\dot{\varepsilon}^p = H \cdot \gamma \cdot \frac{\partial f}{\partial \sigma} = H \cdot \gamma \cdot \text{sign}(\sigma - q) \tag{6.25}$$

where H is the kinematic hardening modulus.

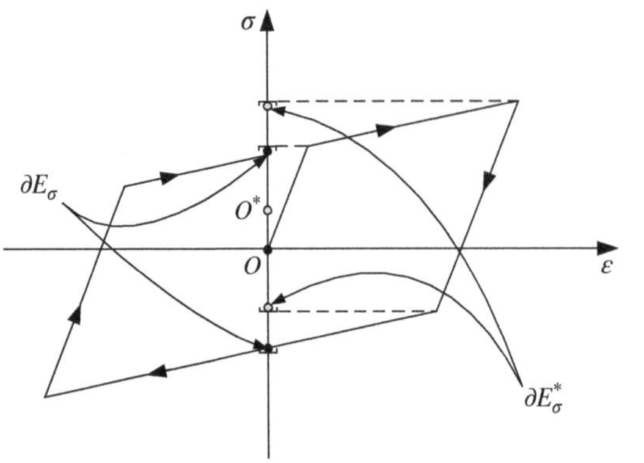

Figure 6.3 One-dimensional yield surface with motion hardening.

Referring to section 6.1.4, the derivation of the continuum elastoplastic tangent modulus with kinematic hardening is as follows:

$$\begin{aligned}\dot{f} &= \frac{\partial f}{\partial \sigma}\dot{\sigma} + \frac{\partial f}{\partial q}\dot{q} + \frac{\partial f}{\partial \sigma_y}\frac{\partial \sigma_y}{\partial \gamma_p}\dot{\gamma}_p \\ &= \text{sign}(\sigma - q)\cdot E\cdot(\dot{\varepsilon} - \dot{\varepsilon}^p) - \text{sign}(\sigma - q)\cdot H\cdot\dot{\gamma}\cdot\text{sign}(\sigma - q) - \frac{\partial \sigma_y}{\partial \alpha}\dot{\gamma} \\ &= \text{sign}(\sigma - q)\cdot E\cdot\dot{\varepsilon} - E\cdot\dot{\gamma} - H\cdot\dot{\gamma} - \frac{\partial \sigma_y}{\partial \alpha}\dot{\gamma} = 0 \Rightarrow \dot{\gamma} = \frac{\text{sign}(\sigma - q)\cdot E\cdot\dot{\varepsilon}}{E + H + \frac{\partial \sigma_y}{\partial \alpha}}\end{aligned} \tag{6.26}$$

Substituting Eqs. (6.30) and (6.17) into Eq. (6.18) and simplifying yields:

$$\dot{\sigma} = \frac{E\cdot\left(H + \frac{\partial \sigma_y}{\partial \alpha}\right)}{E + H + \frac{\partial \sigma_y}{\partial \alpha}}\cdot\dot{\varepsilon} = E^{\text{ep}}\cdot\dot{\varepsilon} \quad \text{where} \quad E^{\text{ep}} = \frac{E\cdot\left(H + \frac{\partial \sigma_y}{\partial \alpha}\right)}{E + H + \frac{\partial \sigma_y}{\partial \alpha}} \tag{6.27}$$

For materials with both linear hardening and kinematic hardening, the stress–strain response is shown in Figure 6.3:

$$E^{\text{ep}} = \frac{E\cdot(H + K)}{E + H + K} \tag{6.28}$$

6.1.6 Incremental Form of the Governing Equations

The governing equation for the one-dimensional elastoplastic model is as follows:

$$\begin{cases} \dot{\sigma} = E\cdot\dot{\varepsilon}^e = E\cdot(\dot{\varepsilon} - \dot{\varepsilon}^p) \\ \dot{\varepsilon}^p = \dot{\gamma}\,\text{sign}(\sigma - q) \\ \dot{\alpha} = \dot{\gamma} \\ \dot{q} = H\cdot\dot{\gamma}\,\text{sign}(\sigma - q) \\ f(\sigma, q, \alpha) = |\sigma - q| - \sigma_y(\alpha) \le 0 \\ \dot{\gamma} > 0 \quad f = 0 \quad \dot{\gamma} f = 0 \end{cases} \tag{6.29}$$

In finite element analysis (FEA), nonlinear problems generally require an incremental solution method. Suppose the total loading time is T, and we consider an incremental step within the time interval $t \in [0, T]$, $\Delta t = t_{n+1} - t_n$. Assume that at time t_n, all state variables involved in Eq. (6.29) are known, and the displacement increment $\Delta U = U_{n+1} - U_n$ due to external loading is applied, resulting in a corresponding strain increment $\Delta\varepsilon_{n+1} = \varepsilon_{n+1} - \varepsilon_n$. The objective of the numerical solution to the elastoplastic constitutive equation is to obtain all state variables at time t_{n+1}.

Using the backward Euler method (as mentioned in Section 6.1.2), Eq. (6.29) can be discretized as

$$\begin{cases} \dfrac{\sigma_{n+1} - \sigma_n}{\Delta t} = E \cdot \left(\dfrac{\varepsilon_{n+1} - \varepsilon_n}{\Delta t} - \dfrac{\varepsilon^p_{n+1} - \varepsilon^p_n}{\Delta t} \right) = E \cdot \left(\dfrac{\varepsilon_{n+1} - \varepsilon^p_{n+1} - \varepsilon^e_n}{\Delta t} \right) \\ \dfrac{\varepsilon^p_{n+1} - \varepsilon^p_n}{\Delta t} = \dfrac{\Delta\gamma}{\Delta t} \operatorname{sign}(\sigma_{n+1} - q_{n+1}) \\ \dfrac{\alpha_{n+1} - \alpha_n}{\Delta t} = \dfrac{\Delta\gamma}{\Delta t} \\ \dfrac{q_{n+1} - q_n}{\Delta t} = H \cdot \dfrac{\Delta\gamma}{\Delta t} \operatorname{sign}(\sigma_{n+1} - q_{n+1}) \\ f(\sigma_{n+1}, q_{n+1}, \alpha_{n+1}) = |\sigma_{n+1} - q_{n+1}| - \sigma_y(\alpha_{n+1}) \leq 0 \\ \Delta\gamma \geq 0 \quad f_{n+1} = 0 \quad \Delta\gamma f_{n+1} = 0 \end{cases} \tag{6.30}$$

In Eq. (6.30), $(\cdot)_{n+1} = (\cdot)_n + \Delta(\cdot)_{n+1}$ (The quantities in parentheses represent arbitrary state variables). The subscripts denote the values at time t_n and t_{n+1}, respectively. $\Delta\gamma = \dot{\gamma} \cdot \Delta t$. The first four equations in Eq. (6.30) can be further simplified as:

$$\begin{cases} \sigma_{n+1} = \sigma_n + E \cdot \left(\Delta\varepsilon_{n+1} - \Delta\varepsilon^p_{n+1}\right) \\ \varepsilon^p_{n+1} = \varepsilon^p_n + \Delta\gamma \operatorname{sign}(\sigma_{n+1} - q_{n+1}) \\ \alpha_{n+1} = \alpha_n + \Delta\gamma \\ q_{n+1} = q_n + H \cdot \Delta\gamma \operatorname{sign}(\sigma_{n+1} - q_{n+1}) \end{cases} \tag{6.31}$$

Correspondingly, the state variables σ_{n+1}, ε^p_{n+1}, α_{n+1}, and q_{n+1} must satisfy the yield criterion and the Kuhn–Tucker complementary conditions, as represented in the last two equations of Eq. (6.31).

6.1.7 Elastic Predictor/Plastic Corrector Solution Algorithm

Due to the presence of the yield criterion and the Kuhn–Tucker complementary conditions, the solution to the above nonlinear system of equations cannot follow the typical initial value problem strategy used for partial differential equations. Currently, the solution to Eq. (6.30) can be approached using the classical elastic predictor/plastic corrector algorithm. The main idea is as follows: 1) elastic predictor. When a strain increment $\Delta\varepsilon_{n+1}$ is applied, the material is assumed to remain in the elastic state, i.e. $\Delta\gamma = 0\ (\Delta\varepsilon_{n+1}^{p} = 0)$. The state variables at this stage are computed as follows: $\sigma_{n+1}^{\text{trial}} = \sigma_n + E \cdot \Delta\varepsilon_{n+1}$, $\varepsilon_{n+1}^{p,\text{trial}}$, $\alpha_{n+1}^{\text{trial}}$, q_{n+1}^{trial}. The superscript "trial" indicates a predicted state. If $f_{n+1}^{\text{trial}} = f\left(\sigma_{n+1}^{\text{trial}}, q_{n+1}^{\text{trial}}, \alpha_{n+1}^{\text{trial}}\right) < 0$, then the elastic predictor state is the true state of the material, i.e. $\sigma_{n+1} = \sigma_{n+1}^{\text{trial}}$, $\varepsilon_{n+1}^{p} = \varepsilon_{n+1}^{p,\text{trial}}$, $\alpha_{n+1} = \alpha_{n+1}^{\text{trial}}$, $q_{n+1} = q_{n+1}^{\text{trial}}$. 2) If $f_{n+1}^{\text{trial}} = \left(\sigma_{n+1}^{\text{trial}}, q_{n+1}^{\text{trial}}, \alpha_{n+1}^{\text{trial}}\right) > 0$, this means that the current stress state is outside the yield surface, and a plastic correction is required to bring the stress state back onto the yield surface, as illustrated in Figure 6.4.

The purpose of the plastic correction is to find state variables that satisfy both the yield criterion and the Kuhn–Tucker complementary conditions. To achieve this, the following nonlinear system of equations needs to be solved:

$$\begin{cases} \sigma_{n+1} = E \cdot \left(\varepsilon_{n+1} - \varepsilon_{n+1}^{p}\right) \\ \varepsilon_{n+1}^{p} = \varepsilon_n^p + \Delta\gamma \operatorname{sign}\left(\sigma_{n+1} - q_{n+1}\right) \\ \alpha_{n+1} = \alpha_n + \Delta\gamma \\ q_{n+1} = q_n + H \cdot \Delta\gamma \operatorname{sign}\left(\sigma_{n+1} - q_{n+1}\right) \\ f_{n+1} = \left|\sigma_{n+1} - q_{n+1}\right| - \sigma_y\left(\alpha_{n+1}\right) \end{cases} \Rightarrow \begin{cases} \sigma_{n+1} = E \cdot \left(\varepsilon_{n+1} - \varepsilon_{n+1}^{p}\right) \\ \varepsilon_{n+1}^{p} = \varepsilon_n^p + \Delta\gamma \operatorname{sign}\left(\xi_{n+1}\right) \\ \alpha_{n+1} = \alpha_n + \Delta\gamma \\ q_{n+1} = q_n + H \cdot \Delta\gamma \operatorname{sign}\left(\xi_{n+1}\right) \\ f_{n+1} = \left|\xi_{n+1}\right| - \sigma_y\left(\alpha_{n+1}\right) \end{cases} \tag{6.32}$$

In Eq. (6.32), $\varepsilon_{n+1} = \varepsilon_n + \Delta\varepsilon_{n+1}$, $\xi_{n+1} = \sigma_{n+1} - q_{n+1}$, σ_{n+1}, ε_{n+1}^{p}, α_{n+1}, q_{n+1}, and $\Delta\gamma$ are the five variables to be solved. Since the number of equations matches the number of unknowns, the system is closed and can be solved. The Newton–Raphson (NR) method is used to solve this nonlinear system of equations.

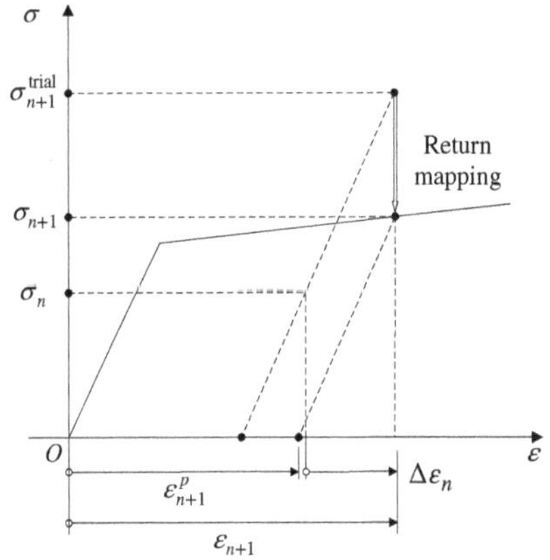

Figure 6.4 Classic Elastic predictor/plastic corrector algorithm.

The system of Eq. (6.32) can be rewritten as:

$$\begin{cases} g_1\left(\sigma_{n+1}, \varepsilon^p_{n+1}, \alpha_{n+1}, q_{n+1}, \Delta\gamma\right) = \dfrac{1}{E}\sigma_{n+1} - \varepsilon_{n+1} - \varepsilon^p_{n+1} = 0 \\ g_2\left(\sigma_{n+1}, \varepsilon^p_{n+1}, \alpha_{n+1}, q_{n+1}, \Delta\gamma\right) = \varepsilon^p_{n+1} - \varepsilon^p_n - \Delta\gamma \text{sign}\left(\xi_{n+1}\right) = 0 \\ g_3\left(\sigma_{n+1}, \varepsilon^p_{n+1}, \alpha_{n+1}, q_{n+1}, \Delta\gamma\right) = \alpha_{n+1} - \alpha_n - \Delta\gamma = 0 \\ g_4\left(\sigma_{n+1}, \varepsilon^p_{n+1}, \alpha_{n+1}, q_{n+1}, \Delta\gamma\right) = \dfrac{1}{H} q_{n+1} - \dfrac{1}{H} q_n - \Delta\gamma \text{sign}\left(\xi_{n+1}\right) = 0 \\ g_2\left(\sigma_{n+1}, \varepsilon^p_{n+1}, \alpha_{n+1}, q_{n+1}, \Delta\gamma\right) = \left|\xi_{n+1}\right| - \sigma_y\left(\alpha_{n+1}\right) = 0 \end{cases} \tag{6.33}$$

In Eq. (6.33), $g_i\left(\sigma_{n+1}, \varepsilon^p_{n+1}, \alpha_{n+1}, q_{n+1}, \Delta\gamma\right), i = 1, 2, 3, 4, 5$ are independent nonlinear equations. In a more compact form, it can be expressed as:

$$\boldsymbol{g}(\boldsymbol{x}) = 0 \tag{6.34}$$

and

$$\boldsymbol{g}(\boldsymbol{x}) = \left[g_1(\boldsymbol{x}), g_2(\boldsymbol{x}), g_3(\boldsymbol{x}), g_4(\boldsymbol{x}), g_5(\boldsymbol{x})\right]^T$$

$$\boldsymbol{x} = \left[x_1, x_2, x_3, x_4, x_5\right]^T = \left[\sigma_{n+1}, \varepsilon^p_{n+1}, \alpha_{n+1}, q_{n+1}, \Delta\gamma\right]^T$$

The superscript T denotes the transpose. To apply the Newton–Raphson method, the corresponding Jacobian matrix needs to be computed:

$$[\mathbf{J}(\boldsymbol{x})] = \begin{bmatrix} \dfrac{\partial g_1(\boldsymbol{x})}{\partial x_1} & \dfrac{\partial g_1(\boldsymbol{x})}{\partial x_2} & \dfrac{\partial g_1(\boldsymbol{x})}{\partial x_3} & \dfrac{\partial g_1(\boldsymbol{x})}{\partial x_4} & \dfrac{\partial g_1(\boldsymbol{x})}{\partial x_5} \\ \dfrac{\partial g_2(\boldsymbol{x})}{\partial x_1} & \dfrac{\partial g_2(\boldsymbol{x})}{\partial x_2} & \dfrac{\partial g_2(\boldsymbol{x})}{\partial x_3} & \dfrac{\partial g_2(\boldsymbol{x})}{\partial x_4} & \dfrac{\partial g_2(\boldsymbol{x})}{\partial x_5} \\ \dfrac{\partial g_3(\boldsymbol{x})}{\partial x_1} & \dfrac{\partial g_3(\boldsymbol{x})}{\partial x_2} & \dfrac{\partial g_3(\boldsymbol{x})}{\partial x_3} & \dfrac{\partial g_3(\boldsymbol{x})}{\partial x_4} & \dfrac{\partial g_3(\boldsymbol{x})}{\partial x_5} \\ \dfrac{\partial g_4(\boldsymbol{x})}{\partial x_1} & \dfrac{\partial g_4(\boldsymbol{x})}{\partial x_2} & \dfrac{\partial g_4(\boldsymbol{x})}{\partial x_3} & \dfrac{\partial g_4(\boldsymbol{x})}{\partial x_4} & \dfrac{\partial g_4(\boldsymbol{x})}{\partial x_5} \\ \dfrac{\partial g_5(\boldsymbol{x})}{\partial x_1} & \dfrac{\partial g_5(\boldsymbol{x})}{\partial x_2} & \dfrac{\partial g_5(\boldsymbol{x})}{\partial x_3} & \dfrac{\partial g_5(\boldsymbol{x})}{\partial x_4} & \dfrac{\partial g_5(\boldsymbol{x})}{\partial x_5} \end{bmatrix} \tag{6.35}$$

$$= \begin{bmatrix} 1/E & -1 & 0 & 0 & 0 \\ 0 & 1 & 0 & 0 & -\text{sign}\left(\xi_{n+1}\right) \\ 0 & 0 & 1 & 0 & -1 \\ 0 & 0 & 0 & 1/H & -\text{sign}\left(\xi_{n+1}\right) \\ \text{sign}\left(\xi_{n+1}\right) & 0 & -\dfrac{\partial\sigma_y}{\partial\alpha} & -\text{sign}\left(\xi_{n+1}\right) & 0 \end{bmatrix}$$

The initial value for NR method is $\boldsymbol{x}^{(0)} = \left[\sigma_{n+1}^{(0)}, \varepsilon_{n+1}^{p,(0)}, \alpha_{n+1}^{(0)}, q_{n+1}^{(0)}, \Delta\gamma^0\right] = \left[\sigma_{n+1}^{\text{trial}}, \varepsilon_n^p, \alpha_n, q_n, 0\right]$. For the $k+1$-th iteration, the corresponding solution is

$$\boldsymbol{x}_{n+1}^{(k+1)} = \boldsymbol{x}_{n+1}^{(k)} - \left[\mathbf{J}\left(\boldsymbol{x}_{n+1}^{(k)}\right)\right]^{-1} \boldsymbol{g}\left(\boldsymbol{x}_{n+1}^{(k)}\right) \tag{6.36}$$

If $\left\|\boldsymbol{g}\left(\boldsymbol{x}_{n+1}^{(k+1)}\right)\right\| \geq \text{Tol}$, where Tol is the permissible error tolerance, the iteration continues until the error satisfies $\left\|\boldsymbol{g}\left(\boldsymbol{x}_{n+1}^{(k+1)}\right)\right\| < \text{Tol}$. The iterative solution process follows these steps:

I. Elastic prediction

$$\varepsilon_{n+1}^{\text{trial}} = \varepsilon_n + \Delta\varepsilon_{n+1} \quad \varepsilon_{n+1}^{p,\text{trial}} = \varepsilon_n^p$$

$$\sigma_{n+1}^{\text{trial}} = E \cdot \left(\varepsilon_{n+1}^{\text{trial}} - \varepsilon_{n+1}^{p,\text{trial}}\right) \quad \alpha_{n+1}^{\text{trial}} = \alpha_n$$

$$q_{n+1}^{\text{trial}} = q_n \quad \varsigma_{n+1}^{\text{trial}} = \alpha_{n+1}^{\text{trial}} - q_{n+1}^{\text{trial}}$$

II. Plastic prediction

If $f_{n+1}^{\text{trial}} = f\left(\sigma_{n+1}^{\text{trial}}, q_{n+1}^{\text{trial}}, \alpha_{n+1}^{\text{trial}}\right) < 0$ then

$$(\cdot)_{n+1} = (\cdot)_{n+1}^{\text{trial}}$$

Go to step IV)
Else
Go to step III)
End

III. Plastic correction (stress iteration)

Do while $k < k^{\max}$ (where $k^{\max}$ is the maximum number of iterations)
Calculate the residual on the left side of Eq. (6.34)
If $\|g(x)\| < \text{Tol}$ then

$$(\cdot)_{n+1} = (\cdot)_{n+1}^{(k)}$$

Go to step IV)
End
Solve for the correction based on the Jacobian matrix and the residual
Update the state variables according to Eq. (6.36).
End

IV. Solve the consistent elastoplastic model and return to the main program.

6.1.8 Consistent Elastoplastic Modulus

In Section 6.1.7, we derived the continuum elastoplastic modulus. In this section, we will derive the consistent elastoplastic modulus. The term "consistent" refers to the fact that the modulus is derived in a manner consistent with the stress iteration algorithm of the constitutive model. Moreover, the consistent elastoplastic modulus relates stress increments and strain increments (unlike the continuum elastoplastic modulus, which relates stress rates and strain rates). This relationship is expressed as:

$$E^{\mathrm{ep}}_{c,n+1} = \frac{\partial \sigma_{n+1}}{\partial \varepsilon_{n+1}} = \frac{\partial \Delta\sigma_{n+1}}{\partial \Delta\varepsilon_{n+1}} \tag{6.37}$$

In Eq. (6.37), the subscript c stands for consistent. To derive the consistent elastoplastic modulus, we take the total derivative of Eq. (6.32), which gives:

$$\begin{bmatrix} \frac{1}{E} d\sigma_{n+1} - d\varepsilon^{p}_{n+1} \\ d\varepsilon^{p}_{n+1} - d\Delta\gamma \mathrm{sign}\,(\xi_{n+1}) \\ d\alpha_{n+1} - d\Delta\gamma \\ dq_{n+1} - H \cdot d\Delta\gamma \mathrm{sign}\,(\xi_{n+1}) \\ d\sigma_{n+1}\mathrm{sign}\,(\xi_{n+1}) - \sigma'_{y}\,(\alpha_{n+1})\, d\alpha_{n+1} - dq_{n+1}\mathrm{sign}\,(\xi_{n+1}) \end{bmatrix} = \begin{bmatrix} d\varepsilon_{n+1} \\ 0 \\ 0 \\ 0 \\ 0 \end{bmatrix} \tag{6.38}$$

Rearranging Eq. (6.38), we obtain:

$$\begin{bmatrix} 1/E & -1 & 0 & 0 & 0 \\ 0 & 1 & 0 & 0 & -\mathrm{sign}\,(\xi_{n+1}) \\ 0 & 0 & 1 & 0 & -1 \\ 0 & 0 & 0 & 1/H & -\mathrm{sign}\,(\xi_{n+1}) \\ \mathrm{sign} = (\xi_{n+1}) & 0 & -\frac{\partial \sigma_y}{\partial \alpha} & -\mathrm{sign}\,(\xi_{n+1}) & 0 \end{bmatrix} \begin{bmatrix} d\sigma_{n+1} \\ d\varepsilon^{p}_{n+1} \\ d\alpha_{n+1} \\ dq_{n+1} \\ d\Delta\gamma \end{bmatrix} = \begin{bmatrix} d\varepsilon_{n+1} \\ 0 \\ 0 \\ 0 \\ 0 \end{bmatrix} \tag{6.39}$$

In Eq. (6.39), all the variables are the final state variables at the end of the step, i.e. the converged values from the stress iteration process. Taking the inverse of Eq. (6.39), we get:

$$\begin{bmatrix} d\sigma_{n+1} \\ d\varepsilon^{p}_{n+1} \\ d\alpha_{n+1} \\ dq_{n+1} \\ d\Delta\gamma \end{bmatrix} = \begin{bmatrix} A_{11} & A_{12} & A_{13} & A_{14} & A_{15} \\ A_{21} & A_{22} & A_{23} & A_{24} & A_{25} \\ A_{31} & A_{32} & A_{33} & A_{34} & A_{35} \\ A_{41} & A_{42} & A_{43} & A_{44} & A_{45} \\ A_{51} & A_{52} & A_{53} & A_{54} & A_{55} \end{bmatrix} \begin{bmatrix} d\varepsilon_{n+1} \\ 0 \\ 0 \\ 0 \\ 0 \end{bmatrix} \tag{6.40}$$

In Eq. (6.40), $[\mathbf{A}]^{-1} = \begin{bmatrix} 1/E & -1 & 0 & 0 & 0 \\ 0 & 1 & 0 & 0 & -\mathrm{sign}\,(\xi_{n+1}) \\ 0 & 0 & 1 & 0 & -1 \\ 0 & 0 & 0 & 1/H & -\mathrm{sign}\,(\xi_{n+1}) \\ \mathrm{sign}\,(\xi_{n+1}) & 0 & -\frac{\partial \sigma_y}{\partial \alpha} & -\mathrm{sign}\,(\xi_{n+1}) & 0 \end{bmatrix}$

From Eq. (6.37), we can define the consistent elastoplastic tangent modulus as:

$$E_{c,n+1}^{\mathrm{ep}} = \frac{\partial \sigma_{n+1}}{\partial \varepsilon_{n+1}} = A_{11} \tag{6.41}$$

Additionally,

$$\frac{d\alpha_{n+1}}{d\varepsilon_{n+1}} = A_{31} \quad \frac{dq_{n+1}}{d\varepsilon_{n+1}} = A_{41} \tag{6.42}$$

In finite element analysis, the consistent elastoplastic modulus ensures that the quadratic convergence characteristic of the Newton–Raphson method is preserved near the solution of the system of equations. However, as long as the computation converges, the consistent elastoplastic modulus does not affect the final accuracy of the result.

By comparing Eq. (6.39) with Eq. (6.35), we observe that the left side of the equation corresponds to the Jacobian matrix [**J**]. Therefore, in practical computations, the inverse of the converged Jacobian matrix [**J**] can be used directly. The submatrices of this inverse matrix provide the required consistent elastoplastic modulus. During the program debugging phase, the consistent elastoplastic modulus can initially be approximated by the elastic modulus. Once the stress integration algorithm has been confirmed to converge, the submatrices of the inverse Jacobian matrix can be assigned to the consistent elastoplastic modulus for final calculations.

6.2 Three-dimensional Elastoplastic Model and Numerical Solution Framework

6.2.1 Model Overview

The framework of the three-dimensional elastoplastic model is fundamentally consistent with the one-dimensional elastoplastic model. Therefore, only the key components are outlined here:

Under the assumption of small strain, according to the principle of superposition, the total strain tensor $\boldsymbol{\varepsilon}$ is the sum of the elastic strain tensor $\boldsymbol{\varepsilon}^e$ and the plastic strain tensor $\boldsymbol{\varepsilon}^p$:

$$\boldsymbol{\varepsilon} = \boldsymbol{\varepsilon}^e + \boldsymbol{\varepsilon}^p \quad \varepsilon_{ij} = \varepsilon_{ij}^e + \varepsilon_{ij}^p \quad i, j = 1, 2, 3 \tag{6.43}$$

The relationship between the stress tensor $\boldsymbol{\sigma}$ and the elastic strain tensor $\boldsymbol{\varepsilon}^e$ is determined by Hooke's law, which is expressed as:

$$\boldsymbol{\sigma} = \mathbb{C} : \boldsymbol{\varepsilon}^e = \mathbb{C} : (\boldsymbol{\varepsilon} - \boldsymbol{\varepsilon}^p) \quad \sigma_{ij} = \mathrm{C}_{ijkl} : \left(\varepsilon_{kl} - \varepsilon_{kl}^p\right) \tag{6.44}$$

where $\mathbb{C}$ is the fourth-order elasticity tensor.

The internal variable vector $\boldsymbol{\alpha} = \{\alpha_1, \alpha_2, \alpha_3, \cdots \alpha_m\}$ includes quantities that represent the material's hardening behavior, similar to the internal variables and back stress tensor in the one-dimensional elastoplastic model. These internal variables are used to track the evolution of plastic deformation and the material's response during hardening. The yield function $f(\boldsymbol{\sigma}, \boldsymbol{\alpha})$ defines the boundary between the elastic and plastic regions in stress space. In elastoplastic models, the stress state cannot lie outside the yield surface. The set of admissible stress states is defined as

$$\mathbb{E}_\sigma = \{(\sigma, \boldsymbol{\alpha}) \in \mathbb{S} \times \mathbb{R}^m | \ f(\sigma, \boldsymbol{\alpha}) \leq 0\} \tag{6.45}$$

The elastic region int $(\mathbb{E}_\sigma)$ is defined as

$$\text{int}(\mathbb{E}_\sigma) = \{(\sigma, \boldsymbol{\alpha}) \in \mathbb{S} \times \mathbb{R}^m | \ f(\sigma, \boldsymbol{\alpha}) < 0\} \tag{6.46}$$

The yield surface in stress space $\partial(\mathbb{E}_\sigma)$ is defined as

$$\partial(\mathbb{E}_\sigma) = \{(\sigma, \boldsymbol{\alpha}) \in \mathbb{S} \times \mathbb{R}^m | \ f(\sigma, \boldsymbol{\alpha}) = 0\} \tag{6.47}$$

The flow rule and hardening rule are as follows:

$$\begin{cases} \dot{\varepsilon}^p = \dot{\gamma}\mathbf{r}(\sigma, \boldsymbol{\alpha}) \\ \dot{\boldsymbol{\alpha}} = -\dot{\gamma}\mathbf{h}(\sigma, \boldsymbol{\alpha}) \end{cases} \tag{6.48}$$

where **r** and **h** are custom functions typically determined through experiments, when using an associated flow rule, $\mathbf{r}(\sigma, \boldsymbol{\alpha}) = \partial f / \partial \sigma$. γ is a nonnegative multiplier determined by the Kuhn–Tucker complementary conditions (also known as loading/unloading conditions) and consistency conditions:

$$\begin{cases} \dot{\gamma} \geq 0 \quad f(\sigma, \boldsymbol{\alpha}) \leq 0 \quad \dot{\gamma} f(\sigma, \boldsymbol{\alpha}) = 0 \\ \dot{\gamma}\dot{f}(\sigma, \boldsymbol{\alpha}) = 0 \end{cases} \tag{6.49}$$

Referencing the preceding content, if the stress state is within the elastic region $\{\sigma, \boldsymbol{\alpha}\} \in \text{int}(\mathbb{E}_\sigma)$, then $f(\sigma, \boldsymbol{\alpha}) < 0$, and by the Kuhn–Tucker complementary conditions $\dot{\gamma} = 0$. If the stress state is on the yield surface $\{\sigma, \boldsymbol{\alpha}\} \in \partial(\mathbb{E}_\sigma), f(\sigma, \boldsymbol{\alpha}) = 0$, then γ is determined by the consistency condition:

a. If $\dot{f}(\sigma, \boldsymbol{\alpha}) < 0 \Rightarrow \dot{\gamma} = 0$. At this point, the material transitions from a yielding state to an elastic state, suggesting the material has undergone elastic unloading.
b. If $\dot{f}(\sigma, \boldsymbol{\alpha}) = 0$, the consistency condition is automatically satisfied.

Since $\dot{\gamma}$ is a nonnegative multiplier, it corresponds to the following cases: 1) $\dot{\gamma} > 0$, the material transitions from the current plastic state to a new plastic state, known as plastic loading; 2) $\dot{\gamma} = 0$, the stress state of the material moves from its current position on the yield surface to another position on the same yield surface, a process known as neutral loading.

The equations related to the three-dimensional elastoplastic model are as follows:

- Elastic Stress–Strain Relationship

$$\sigma = \mathbb{C} : \varepsilon^e = \mathbb{C} : (\varepsilon - \varepsilon^p)$$

- Elastic Domain

$$\text{int}(\mathbb{E}_\sigma) = \{(\sigma, \alpha) \in \mathbb{S} \times \mathbb{R}^m | \ f(\sigma, \alpha) < 0\}$$

- Flow Rule and Hardening Rule

$$\begin{cases} \dot{\varepsilon}^p = \dot{\gamma} r(\sigma, \alpha) \\ \dot{\alpha} = -\dot{\gamma} h(\sigma, \alpha) \end{cases}$$

- Kuhn–Tucker Conditions and Consistency Conditions

$$\begin{cases} \dot{\gamma} \geq 0 \quad f(\sigma, \alpha) \leq 0 \quad \dot{\gamma} f(\sigma, \alpha) = 0 \\ \dot{\gamma} \dot{f}(\sigma, \alpha) = 0 \end{cases}$$

6.2.2 Stress–Strain Rate Relations

Under elastic conditions, the relationship between the stress rate and strain rate is given by:

$$\dot{\sigma} = \mathbb{C} : \dot{\varepsilon} \tag{6.50}$$

In the state of plastic loading, the relationship between stress rate and strain rate satisfies:

$$\dot{\sigma} = \mathbb{C} : (\dot{\varepsilon} - \dot{\varepsilon}^p) \tag{6.51}$$

From the consistency conditions, it follows that:

$$\begin{aligned} \dot{f}(\boldsymbol{\sigma}, \boldsymbol{\alpha}) &= \frac{\partial f}{\partial \boldsymbol{\sigma}} \dot{\boldsymbol{\sigma}} + \frac{\partial f}{\partial \boldsymbol{\alpha}} \dot{\boldsymbol{\alpha}} = \frac{\partial f}{\partial \boldsymbol{\sigma}} : \mathbb{C} : (\dot{\varepsilon} - \dot{\varepsilon}^p) + \frac{\partial f}{\partial \boldsymbol{\alpha}} \dot{\boldsymbol{\alpha}} \\ &= \frac{\partial f}{\partial \boldsymbol{\sigma}} : \mathbb{C} : \dot{\varepsilon} - \dot{\gamma} \left(\frac{\partial f}{\partial \boldsymbol{\sigma}} : \mathbb{C} : \mathbf{r} + \frac{\partial f}{\partial \boldsymbol{\alpha}} \cdot \mathbf{h} \right) = 0 \end{aligned} \tag{6.52}$$

Thus, the plastic multiplier is expressed as:

$$\dot{\gamma} = \frac{\frac{\partial f}{\partial \sigma} : \mathbb{C} : \dot{\varepsilon}}{\left(\frac{\partial f}{\partial \sigma} : \mathbb{C} : \mathbf{r} + \frac{\partial f}{\partial \alpha} \cdot \mathbf{h} \right)} \tag{6.53}$$

Substituting Eqs. (6.53) and (6.48) into Eq. (6.51) yields the relationship between the stress rate and the strain rate.

$$\dot{\sigma} = \mathbb{C}^{ep} : \dot{\varepsilon} \tag{6.54}$$

where $\mathbb{C}^{ep}$ is the consistent elastoplastic modulus:

$$\mathbb{C}^{ep} = \mathbb{C} - \frac{\mathbb{C} : \mathbf{r} \otimes \mathbb{C} : \frac{\partial f}{\partial \sigma}}{\frac{\partial f}{\partial \sigma} : \mathbb{C} : \mathbf{r} + \frac{\partial f}{\partial \alpha} \cdot \mathbf{h}} \tag{6.55}$$

$\mathbb{C}^{ep}$ is symmetric if and only if $\mathbf{r}(\boldsymbol{\sigma}, \boldsymbol{\alpha}) = \partial_{\sigma} f(\boldsymbol{\sigma}, \boldsymbol{\alpha})$ is symmetric, which means that the plastic potential function and the yield function are identical (associated flow rule).

6.2.3 Elastic Predictor/Plastic Corrector Solution Algorithm

The equations related to the three-dimensional elastoplastic model are as follows:

$$\begin{cases} \dot{\sigma} = \mathbb{C} : \left(\dot{\varepsilon} - \dot{\varepsilon}^{p}\right) \\ \dot{\varepsilon}^{p} = \dot{\gamma}\mathbf{r}\left(\sigma, \alpha\right) \\ \dot{\alpha} = -\dot{\gamma}\mathbf{h}\left(\sigma, \alpha\right) \\ f\left(\sigma, \alpha\right) \leq 0 \\ \dot{\gamma} > 0 \quad f = 0 \quad \dot{\gamma} f = 0 \end{cases} \tag{6.56}$$

Following the Euler backward integration method, discretize Eq. (6.56) as:

$$\begin{cases} \sigma_{n+1} = \mathbb{C} : \left(\varepsilon_{n+1} - \varepsilon^{p}_{n+1}\right) \\ \varepsilon^{p}_{n+1} = \varepsilon^{p}_{n} + \Delta\gamma\mathbf{r}\left(\sigma_{n+1}, \alpha_{n+1}\right) = \varepsilon^{p}_{n} + \Delta\gamma\mathbf{r}_{n+1} \\ \alpha_{n+1} = \alpha_{n} - \Delta\gamma\mathbf{h}\left(\sigma_{n+1}, \alpha_{n+1}\right) = \alpha_{n} - \Delta\gamma\mathbf{h}_{n+1} \\ f\left(\sigma_{n+1}, \alpha_{n+1}\right) = f_{n+1} \leq 0 \\ \Delta\gamma > 0 \quad f_{n+1} = 0 \quad \Delta\gamma f_{n+1} = 0 \end{cases} \tag{6.57}$$

Applying the elastic predictor and plastic corrector algorithm shown in Figure 6.5 to Eq. (6.57) under plastic loading conditions, the system of equations to be solved is:

$$\begin{cases} \mathbb{C}^{-1} : \sigma_{n+1} - \varepsilon_{n+1} + \varepsilon^{p}_{n+1} = 0 \\ \varepsilon^{p}_{n+1} = \varepsilon^{p}_{n} + \Delta\gamma\mathbf{r}\left(\sigma_{n+1}, \alpha_{n+1}\right) = \varepsilon^{p}_{n} + \Delta\gamma\mathbf{r}_{n+1} \\ \alpha_{n+1} = \alpha_{n} - \Delta\gamma\mathbf{h}\left(\sigma_{n+1}, \alpha_{n+1}\right) = \alpha_{n} - \Delta\gamma\mathbf{h}_{n+1} \\ f\left(\sigma_{n+1}, \alpha_{n+1}\right) = f_{n+1} = 0 \end{cases} \tag{6.58}$$

Employing the Newton method to solve the above nonlinear system of equations. Rewrite Eq. (6.59) as:

$$\boldsymbol{g}\left(\boldsymbol{x}_{n+1}\right) = \begin{cases} \boldsymbol{g}_1\left(\sigma_{n+1}, \varepsilon^{p}_{n+1}, \alpha_{n+1}, \Delta\gamma\right) = \mathbb{C}^{-1} : \sigma_{n+1} - \varepsilon_{n+1} + \varepsilon^{p}_{n+1} = \mathbf{0} \\ \boldsymbol{g}_2\left(\sigma_{n+1}, \varepsilon^{p}_{n+1}, \alpha_{n+1}, \Delta\gamma\right) = \varepsilon^{p}_{n+1} - \varepsilon^{p}_{n} - \Delta\gamma\mathbf{r}_{n+1} = \mathbf{0} \\ \boldsymbol{g}_3\left(\sigma_{n+1}, \varepsilon^{p}_{n+1}, \alpha_{n+1}, \Delta\gamma\right) = \alpha_{n+1} - \alpha_{n} + \Delta\gamma\mathbf{h}_{n+1} = \mathbf{0} \\ g_4\left(\sigma_{n+1}, \varepsilon^{p}_{n+1}, \alpha_{n+1}, \Delta\gamma\right) = f_{n+1} = 0 \end{cases} \tag{6.59}$$

where $\boldsymbol{g}_1, \boldsymbol{g}_2, \boldsymbol{g}_3$, and g_4 represent 6, 6, $\boldsymbol{m}$, and one independent nonlinear equations, respectively. Rewrite the equations in a more compact form as follows

$$\boldsymbol{g}\left(\boldsymbol{x}\right) = 0 \tag{6.60}$$

where

$$\boldsymbol{g}\left(\boldsymbol{x}\right) = \left[\boldsymbol{g}_1\left(\boldsymbol{x}\right), \boldsymbol{g}_2\left(\boldsymbol{x}\right), \boldsymbol{g}_3\left(\boldsymbol{x}\right), g_4\left(\boldsymbol{x}\right)\right]^T$$

$$\boldsymbol{x} = \left[\boldsymbol{x}_1, \boldsymbol{x}_2, \boldsymbol{x}_3, x_4\right]^T = \left[\sigma, \varepsilon^{p}, \alpha, \Delta\gamma\right]^T$$

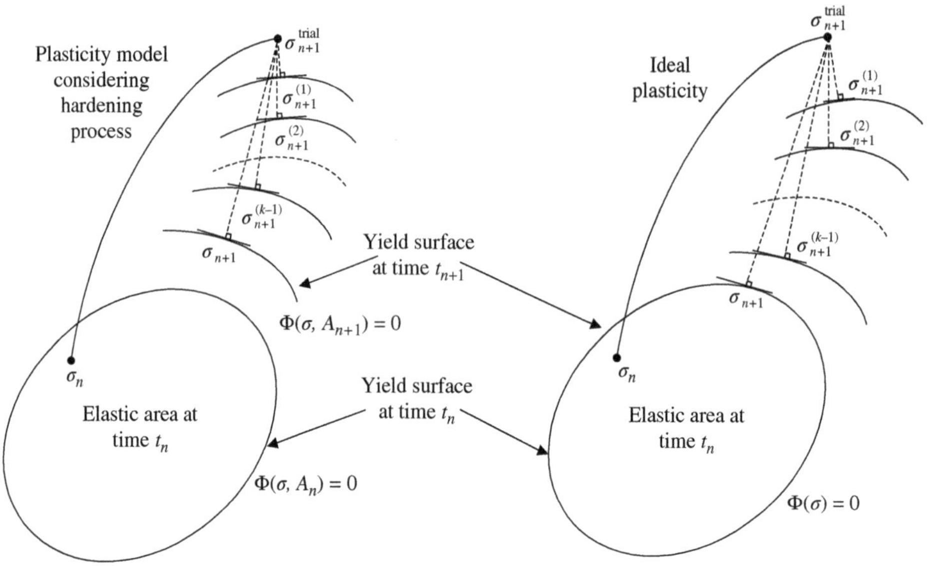

Figure 6.5 Three-dimensional elastic predictor/plastic corrector algorithm.

The unknown vector $\boldsymbol{x}$ also contains $\boldsymbol{m}$+13 unknown variables, making the system of equations closed. For applying the Newton–Raphson method, the corresponding Jacobian matrix [**J**] is

$$[\mathbf{J}(\boldsymbol{x})] = \begin{bmatrix} \frac{\partial \boldsymbol{g}_1(\boldsymbol{x})}{\partial \boldsymbol{x}_1} & \frac{\partial \boldsymbol{g}_1(\boldsymbol{x})}{\partial \boldsymbol{x}_2} & \frac{\partial \boldsymbol{g}_1(\boldsymbol{x})}{\partial \boldsymbol{x}_3} & \frac{\partial \boldsymbol{g}_1(\boldsymbol{x})}{\partial x_4} \\ \frac{\partial \boldsymbol{g}_2(\boldsymbol{x})}{\partial \boldsymbol{x}_1} & \frac{\partial \boldsymbol{g}_2(\boldsymbol{x})}{\partial \boldsymbol{x}_2} & \frac{\partial \boldsymbol{g}_2(\boldsymbol{x})}{\partial \boldsymbol{x}_3} & \frac{\partial \boldsymbol{g}_2(\boldsymbol{x})}{\partial x_4} \\ \frac{\partial \boldsymbol{g}_3(\boldsymbol{x})}{\partial \boldsymbol{x}_1} & \frac{\partial \boldsymbol{g}_3(\boldsymbol{x})}{\partial \boldsymbol{x}_2} & \frac{\partial \boldsymbol{g}_3(\boldsymbol{x})}{\partial \boldsymbol{x}_3} & \frac{\partial \boldsymbol{g}_3(\boldsymbol{x})}{\partial x_4} \\ \frac{\partial g_4(\boldsymbol{x})}{\partial \boldsymbol{x}_3} & \frac{\partial g_4(\boldsymbol{x})}{\partial \boldsymbol{x}_2} & \frac{\partial g_4(\boldsymbol{x})}{\partial \boldsymbol{x}_3} & \frac{\partial g_4(\boldsymbol{x})}{\partial x_4} \end{bmatrix} \tag{6.61}$$

$$= \begin{bmatrix} \mathbf{A}_1 & \mathbf{A}_2 & \mathbf{A}_3 & \mathbf{A}_4 \\ \mathbf{B}_1 & \mathbf{B}_2 & \mathbf{B}_3 & \mathbf{B}_4 \\ \mathbf{C}_1 & \mathbf{C}_2 & \mathbf{C}_3 & \mathbf{C}_4 \\ \mathbf{D}_1 & \mathbf{D}_2 & \mathbf{D}_3 & \mathbf{D}_4 \end{bmatrix}$$

where

$$\mathbf{A}_1 = [\mathbf{C}^{-1}]_{6\times6}, \mathbf{A}_2 = [\mathbf{I}^S]_{6\times6}, \mathbf{A}_3 = [\mathbf{0}]_{6\times m}, \mathbf{A}_4 = 0$$

$$\mathbf{B}_1 = \left[-\frac{\partial \mathbf{r}}{\partial \boldsymbol{\sigma}}\right]_{6\times6}, \mathbf{B}_2 = [\mathbf{I}^S]_{6\times6}, \mathbf{B}_3 = \left[-\frac{\partial \mathbf{r}}{\partial \boldsymbol{\alpha}}\right]_{6\times m}, \mathbf{B}_4 = [-\mathbf{r}]_{6\times1}$$

$$\mathbf{C}_1 = \left[-\frac{\partial \mathbf{h}}{\partial \sigma}\right]_{m\times 6}, \mathbf{C}_2 = [\mathbf{0}]_{m\times 6}, \mathbf{C}_3 = [\mathbf{I}]_{m\times m}, \mathbf{C}_4 = [\mathbf{h}]_{m\times 1}$$

$$\mathbf{D}_1 = \left[\frac{\partial f}{\partial \sigma}\right]_{1\times 6}, \mathbf{D}_2 = [\mathbf{0}]_{1\times 6}, \mathbf{D}_3 = \left[\frac{\partial f}{\partial \alpha}\right]_{1\times m}, \mathbf{D}_4 = 0$$

In Eq. (6.61), the 6*6 matrix represents the Voigt form of the fourth-order tensor, while the 1*6 matrix represents the Voigt form of the second-order tensor. $\mathbf{I}^s$ and $\mathbf{I}$ are the Voigt forms of the fourth-order symmetric tensor and the fourth-order tensor, respectively.

The initial value for the Newton–Raphson method is $\boldsymbol{x}_{n+1}^{(0)} = \left[\boldsymbol{\sigma}_{n+1}^{(0)}, \boldsymbol{\varepsilon}_{n+1}^{p,(0)}, \boldsymbol{\alpha}_{n+1}^{(0)}, \Delta\boldsymbol{\gamma}_{n+1}^{(0)}\right] = \left[\boldsymbol{\sigma}_{n+1}^{\text{trial}}, \boldsymbol{\varepsilon}_n^p, \boldsymbol{\alpha}_n, 0\right]$. For the $k+1$-th iteration step, the corresponding solution is:

$$\boldsymbol{x}_{n+1}^{(k+1)} = \boldsymbol{x}_{n+1}^{(k)} - \left[\mathbf{J}\left(\boldsymbol{x}_{n+1}^{(k)}\right)\right]^{-1} \boldsymbol{g}\left(\boldsymbol{x}_{n+1}^{(k)}\right) \tag{6.62}$$

If $\left\|\boldsymbol{g}\left(\boldsymbol{x}_{n+1}^{(k+1)}\right)\right\| \geq \text{Tol}$ and Tol are the error tolerance values, the algorithm proceeds with the iterations until $\left\|\boldsymbol{g}\left(\boldsymbol{x}_{n+1}^{(k+1)}\right)\right\| < \text{Tol}$ is satisfied. The corresponding algorithmic flow is as follows.

I. Elastic Predictor

$$\varepsilon_{n+1}^{\text{trial}} = \varepsilon_n + \Delta\varepsilon_{n+1} \quad \varepsilon_{n+1}^{p,\text{trial}} = \varepsilon_n^p$$

$$\sigma_{n+1}^{\text{trial}} = \mathbb{C} : \left(\varepsilon_{n+1}^{\text{trial}} - \varepsilon_{n+1}^{\text{p,trial}}\right) \quad \alpha_{n+1}^{\text{trial}} = \alpha_n$$

II. Check the Stress State

If $f_{n+1}^{\text{trial}} = f\left(\sigma_{n+1}^{\text{trial}}, \alpha_{n+1}^{\text{trial}}\right) < 0$ then

$$(\cdot)_{n+1} = (\cdot)_{n+1}^{\text{trial}}$$

Go to step IV)
Else
Go to step III)
End

III. Plastic Corrector (Stress Iteration)

Do while $k < k^{\max}$ (where $k^{\max}$ is the maximum iteration value)
Compute the residual on the left-hand side of Eq. (6.60).
If $\|g(x)\| < \text{Tol}$ then

$$(\cdot)_{n+1} = (\cdot)_{n+1}^{(k)}$$

Go to step IV)
End
Determine the correction vector using the Jacobian matrix and residuals.
Update the state variables according to Eq. (6.62).
End

IV. Solve the consistency elastoplastic model and return to the main program.

6.2.4 Consistency Elastoplastic Modulus

This section will derive the three-dimensional consistency elastoplastic modulus, as defined below

$$\mathbb{C}_{c,n+1}^{\mathrm{ep}} = \frac{\partial \sigma_{n+1}}{\partial \varepsilon_{n+1}} = \frac{\partial \Delta \sigma_{n+1}}{\partial \Delta \varepsilon_{n+1}} \tag{6.63}$$

taking the total differential of Eq. (6.58) yields

$$[\mathbf{J}_{n+1}(\boldsymbol{x})] \begin{bmatrix} d\boldsymbol{\sigma}_{n+1} \\ d\boldsymbol{\varepsilon}_{n+1}^{p} \\ d\boldsymbol{\alpha}_{n+1} \\ d\Delta\gamma \end{bmatrix} = \begin{bmatrix} d\boldsymbol{\varepsilon}_{n+1} \\ \mathbf{0} \\ \mathbf{0} \\ 0 \end{bmatrix} \tag{6.64}$$

Let the elastoplastic consistency tangent modulus correspond to the submatrix of the inverse of the convergent Jacobian matrix [**J**], where

$$[\mathbf{J}_{n+1}(\boldsymbol{x})]^{-1} = \begin{bmatrix} \mathbf{A}_{1,n+1} & \mathbf{A}_{2,n+1} & \mathbf{A}_{3,n+1} & \mathbf{A}_{4,n+1} \\ \mathbf{B}_{1,n+1} & \mathbf{B}_{2,n+1} & \mathbf{B}_{3,n+1} & \mathbf{B}_{4,n+1} \\ \mathbf{C}_{1,n+1} & \mathbf{C}_{2,n+1} & \mathbf{C}_{3,n+1} & \mathbf{C}_{4,n+1} \\ \mathbf{D}_{1,n+1} & \mathbf{D}_{2,n+1} & \mathbf{D}_{3,n+1} & \mathbf{D}_{4,n+1} \end{bmatrix} \tag{6.65}$$

According to the definition of the elastoplastic consistency tangent modulus

$$\mathbb{C}_{c,n+1}^{\mathrm{ep}} = \frac{\partial \sigma_{n+1}}{\partial \varepsilon_{n+1}} = \frac{\partial \Delta \sigma_{n+1}}{\partial \Delta \varepsilon_{n+1}} = \left[\mathbf{A}_{1,n+1}\right]_{6\times 6} \tag{6.66}$$

6.3 Three-dimensional Elastoviscoplastic Model with Yield Surface and Solution Framework

6.3.1 Model Overview

The main content of the three-dimensional elastoviscoplastic model is essentially consistent with that of the elastoplastic model. The total strain tensor,$\boldsymbol{\varepsilon}$, is the sum of the elastic strain tensor, $\boldsymbol{\varepsilon}^e$, and the viscoplastic strain tensor, $\boldsymbol{\varepsilon}^{\mathrm{vp}}$:

$$\boldsymbol{\varepsilon} = \boldsymbol{\varepsilon}^e + \boldsymbol{\varepsilon}^{\mathrm{vp}} \quad \varepsilon_{ij} = \varepsilon_{ij}^e + \varepsilon_{ij}^{\mathrm{vp}} \quad i, j = 1, 2, 3 \tag{6.67}$$

The relationship between the stress tensor $\boldsymbol{\sigma}$ and the elastic strain tensor $\boldsymbol{\varepsilon}^e$ is still governed by Hooke's law:

$$\boldsymbol{\sigma} = \mathbb{C} : \boldsymbol{\varepsilon}^e = \mathbb{C} : (\boldsymbol{\varepsilon} - \boldsymbol{\varepsilon}^{\mathrm{vp}}) \quad \sigma_{ij} = \mathbb{C}_{ijkl} : \left(\varepsilon_{kl} - \varepsilon_{kl}^{\mathrm{vp}}\right) \tag{6.68}$$

where $\mathbb{C}$ is the fourth-order elastic tensor.

The internal variable vector $\boldsymbol{\alpha} = \{\alpha_1, \alpha_2, \alpha_3, \cdots \alpha_m\}$. $f(\boldsymbol{\sigma}, \boldsymbol{\alpha})$ is the yield function. For the elastoviscoplastic model, the stress state can be located outside the yield surface, defining $\mathrm{int}(\mathbb{E}_\sigma)$ as the elastic region:

$$\mathrm{int}(\mathbb{E}_\sigma) = \{(\boldsymbol{\sigma}, \boldsymbol{\alpha}) \in \mathbb{S} \times \mathbb{R}^m \,|\, f(\boldsymbol{\sigma}, \boldsymbol{\alpha}) < 0\} \tag{6.69}$$

$\partial(\mathbb{E}_\sigma)$ represents the yield surface in stress space:

$$\partial(\mathbb{E}_\sigma) = \{(\sigma, \alpha) \in \mathbb{S} \times \mathbb{R}^m | f(\sigma, \alpha) = 0\} \tag{6.70}$$

Flow Rule and Hardening Rule

$$\begin{cases} \dot{\varepsilon}^{\text{vp}} = \dot{\gamma}\mathbf{r}(\sigma, \alpha) \\ \dot{\alpha} = -\dot{\gamma}\mathbf{h}(\sigma, \alpha) \end{cases} \tag{6.71}$$

If the associated flow rule is applied, $\mathbf{r}(\sigma, \alpha) = \partial f / \partial \sigma$, unlike in elastoplastic models, the viscoelastic-plastic model does not possess a continuous tangent modulus, meaning there is no direct relationship between the stress rate and strain rate as follows:

$$\dot{\sigma} \neq \mathbb{C}^{\text{evp}} : \dot{\varepsilon} \tag{6.72}$$

6.3.2 The Definition of the Plastic Multiplier

Another significant distinction between viscoplastic models and elastoplastic models lies in the definition of the plastic multiplier. In elastoplastic models, the stress state cannot lie outside the yield surface, and the plastic multiplier is determined by the Kuhn–Tucker complementary conditions and the consistency condition. In contrast, in viscoplastic models, the plastic multiplier is an explicit function of the stress state, determined by time-dependent deformation mechanisms [3], such as:

$$\dot{\gamma} = \dot{\gamma}(\sigma, \alpha) \tag{6.73}$$

Combining with Eq. (6.71), the flow rule at this point is:

$$\begin{cases} \dot{\varepsilon}^{p} = \dot{\gamma}\mathbf{r}(\sigma, \alpha) \\ \dot{\alpha} = -\dot{\gamma}\mathbf{h}(\sigma, \alpha) \end{cases} \tag{6.74}$$

$\dot{\gamma}(\sigma, \alpha)$ is zero within the elastic region or on the yield surface, and only becomes nonzero outside the elastic region. Therefore, the evolution of ε^p and α can only occur under the state defined by $f(\sigma, \alpha) > 0$.

Essentially, the explicit function of the plastic multiplier reflects how the viscoplastic strain rate varies with the stress level. The literature provides several explicit expressions for the plastic multiplier: for instance, in the Bingham model, the plastic multiplier is a linear function of the von Mises effective stress. In the Perzyna model, the plastic multiplier is a power function of the von Mises effective stress. Notably, when the power index of the Perzyna model approaches its limit, the viscoplastic model degenerates into an elastoplastic model. Given its flexibility, the Perzyna model is widely applied.

The equations related to the three-dimensional elastoplastic model are as follows:

- Elastic stress–strain relationship:

  $$\sigma = \mathbb{C} : \varepsilon^e = \mathbb{C} : (\varepsilon - \varepsilon^p)$$

- Elastic domain

  $$\text{int}(\mathbb{E}_\sigma) = \{(\sigma, \alpha) \in \mathbb{S} \times \mathbb{R}^m | f(\sigma, \alpha) < 0\}$$

- Flow Rule and Hardening Rule

$$\begin{cases} \dot{\varepsilon}^p = \dot{\gamma} r(\sigma, \alpha) \\ \dot{\alpha} = -\dot{\gamma} h(\sigma, \alpha) \end{cases}$$

where the plastic multiplier is defined as follows:

$$\dot{\gamma} = \dot{\gamma}(\sigma, \alpha)$$

6.3.3 Elastic Predictor/Plastic Corrector Solution Algorithm

Given the definition of the plastic multiplier, the numerical solution process for the viscoplastic model can also employ the elastic predictor/plastic corrector algorithm. The main steps are as follows: 1) elastic prediction. Apply a strain increment $\Delta\boldsymbol{\varepsilon}_{n+1}$, assuming the material remains in an elastic state, resulting in $\Delta\gamma = 0$ $\left(\Delta\boldsymbol{\varepsilon}^{\text{vp}}_{n+1} = 0\right)$. The state variables at this point are $\boldsymbol{\sigma}^{\text{trial}}_{n+1}$, $\boldsymbol{\varepsilon}^{\text{vp,trial}}_{n+1}$, $\boldsymbol{\alpha}^{\text{trial}}_{n+1}$. If $f^{\text{trial}}_{n+1} = \left(\boldsymbol{\sigma}^{\text{trial}}_{n+1}, \boldsymbol{\alpha}^{\text{trial}}_{n+1}\right) < 0$, then the elastic predicted state is the material's true state, i.e. $\boldsymbol{\sigma}_{n+1} = \boldsymbol{\sigma}^{\text{trial}}_{n+1}$, $\boldsymbol{\varepsilon}^{\text{vp}}_{n+1} = \boldsymbol{\varepsilon}^{p,\text{trial}}_{n+1}$, $\boldsymbol{\alpha}_{n+1} = \boldsymbol{\alpha}^{\text{trial}}_{n+1}$; 2) $f^{\text{trial}}_{n+1} = \left(\boldsymbol{\sigma}^{\text{trial}}_{n+1}, \boldsymbol{\alpha}^{\text{trial}}_{n+1}\right) > 0$, the material undergoes viscoplastic flow, and the evolution of $\boldsymbol{\varepsilon}^{\text{vp}}_{n+1}$ and $\boldsymbol{\alpha}_{n+1}$ is determined using the backward Euler method.

In the state of plastic flow, the equations related to the three-dimensional elastoviscoplastic model are as follows:

$$\begin{cases} \dot{\boldsymbol{\sigma}} = \mathbb{C} : (\dot{\boldsymbol{\varepsilon}} - \dot{\boldsymbol{\varepsilon}}^p) \\ \dot{\boldsymbol{\varepsilon}}^p = \dot{\gamma}\mathbf{r}(\boldsymbol{\sigma}, \boldsymbol{\alpha}) \\ \dot{\boldsymbol{\alpha}} = -\dot{\gamma}\mathbf{h}(\boldsymbol{\sigma}, \boldsymbol{\alpha}) \\ \dot{\gamma} = \dot{\gamma}(\boldsymbol{\sigma}, \boldsymbol{\alpha}) \end{cases} \tag{6.75}$$

Using the backward Euler method, Eq. (6.75) is discretized as follows:

$$\begin{cases} \mathbb{C}^{-1} : \boldsymbol{\sigma}_{n+1} - \boldsymbol{\varepsilon}_{n+1} + \boldsymbol{\varepsilon}^{\text{vp}}_{n+1} = 0 \\ \boldsymbol{\varepsilon}^{\text{vp}}_{n+1} - \boldsymbol{\varepsilon}^{\text{vp}}_{n} + \Delta\gamma\mathbf{r}\left(\boldsymbol{\sigma}_{n+1}, \boldsymbol{\alpha}_{n+1}\right) = \boldsymbol{\varepsilon}^{\text{vp}}_{n} + \Delta\gamma\mathbf{r}_{n+1} \\ \boldsymbol{\alpha}_{n+1} = \boldsymbol{\alpha}_n - \Delta\gamma\mathbf{h}\left(\boldsymbol{\sigma}_{n+1}, \boldsymbol{\alpha}_{n+1}\right) = \boldsymbol{\alpha}_n - \Delta\gamma\mathbf{h}_{n+1} \\ \Delta\gamma = \dot{\gamma}\left(\boldsymbol{\sigma}_{n+1}, \boldsymbol{\alpha}_{n+1}\right)\Delta t \end{cases} \tag{6.76}$$

Equation (6.76) is also a nonlinear system of equations and requires solving using the Newton–Raphson method. Following the approach used in Section 6.2 for the elastoplastic model, Eq. (6.76) can be rewritten as follows:

$$\boldsymbol{g}\left(\boldsymbol{x}_{n+1}\right) = \begin{cases} \boldsymbol{g}_1\left(\boldsymbol{\sigma}_{n+1}, \boldsymbol{\varepsilon}^{\text{vp}}_{n+1}, \boldsymbol{\alpha}_{n+1}, \Delta\gamma\right) = \mathbb{C}^{-1} : \boldsymbol{\sigma}_{n+1} - \boldsymbol{\varepsilon}_{n+1} + \boldsymbol{\varepsilon}^{p}_{n+1} = \mathbf{0} \\ \boldsymbol{g}_2\left(\boldsymbol{\sigma}_{n+1}, \boldsymbol{\varepsilon}^{\text{vp}}_{n+1}, \boldsymbol{\alpha}_{n+1}, \Delta\gamma\right) = \boldsymbol{\varepsilon}^{\text{vp}}_{n+1} - \boldsymbol{\varepsilon}^{\text{vp}}_{n} - \Delta\gamma\mathbf{r}_{n+1} = \mathbf{0} \\ \boldsymbol{g}_3\left(\boldsymbol{\sigma}_{n+1}, \boldsymbol{\varepsilon}^{\text{vp}}_{n+1}, \boldsymbol{\alpha}_{n+1}, \Delta\gamma\right) = \boldsymbol{\alpha}_{n+1} - \boldsymbol{\alpha}_n + \Delta\gamma\mathbf{h}_{n+1} = \mathbf{0} \\ g_4\left(\boldsymbol{\sigma}_{n+1}, \boldsymbol{\varepsilon}^{\text{vp}}_{n+1}, \boldsymbol{\alpha}_{n+1}, \Delta\gamma\right) = \Delta\gamma - \dot{\gamma}\left(\boldsymbol{\sigma}_{n+1}, \boldsymbol{\alpha}_{n+1}\right)\Delta t = 0 \end{cases} \tag{6.77}$$

where $\boldsymbol{g}_1, \boldsymbol{g}_2, \boldsymbol{g}_3$, and g_4 represent 6, 6, $\boldsymbol{m}$, and one independent nonlinear equation, respectively. Rewrite the equations in a more compact form as follows

$$\boldsymbol{g}(\boldsymbol{x}) = 0 \tag{6.78}$$

where

$$\boldsymbol{g}(\boldsymbol{x}) = [\boldsymbol{g}_1(\boldsymbol{x}), \boldsymbol{g}_2(\boldsymbol{x}), \boldsymbol{g}_3(\boldsymbol{x}), g_4(\boldsymbol{x})]^T$$

$$\boldsymbol{x} = [\boldsymbol{x}_1, \boldsymbol{x}_2, \boldsymbol{x}_3, x_4]^T = [\boldsymbol{\sigma}, \boldsymbol{\varepsilon}^{\mathrm{vp}}, \boldsymbol{\alpha}, \Delta\gamma]^T$$

The unknown vector $\boldsymbol{x}$ also contains $\boldsymbol{m}+13$ unknown variables, making the system of equations closed. For applying the Newton–Raphson method, the corresponding Jacobian matrix [**J**] is

$$[\mathbf{J}(\boldsymbol{x})] = \begin{bmatrix} \frac{\partial \boldsymbol{g}_1(\boldsymbol{x})}{\partial \boldsymbol{x}_1} & \frac{\partial \boldsymbol{g}_1(\boldsymbol{x})}{\partial \boldsymbol{x}_2} & \frac{\partial \boldsymbol{g}_1(\boldsymbol{x})}{\partial \boldsymbol{x}_3} & \frac{\partial \boldsymbol{g}_1(\boldsymbol{x})}{\partial x_4} \\ \frac{\partial \boldsymbol{g}_2(\boldsymbol{x})}{\partial \boldsymbol{x}_1} & \frac{\partial \boldsymbol{g}_2(\boldsymbol{x})}{\partial \boldsymbol{x}_2} & \frac{\partial \boldsymbol{g}_2(\boldsymbol{x})}{\partial \boldsymbol{x}_3} & \frac{\partial \boldsymbol{g}_2(\boldsymbol{x})}{\partial x_4} \\ \frac{\partial \boldsymbol{g}_3(\boldsymbol{x})}{\partial \boldsymbol{x}_1} & \frac{\partial \boldsymbol{g}_3(\boldsymbol{x})}{\partial \boldsymbol{x}_2} & \frac{\partial \boldsymbol{g}_3(\boldsymbol{x})}{\partial \boldsymbol{x}_3} & \frac{\partial \boldsymbol{g}_3(\boldsymbol{x})}{\partial x_4} \\ \frac{\partial g_4(\boldsymbol{x})}{\partial \boldsymbol{x}_3} & \frac{\partial g_4(\boldsymbol{x})}{\partial \boldsymbol{x}_2} & \frac{\partial g_4(\boldsymbol{x})}{\partial \boldsymbol{x}_3} & \frac{\partial g_4(\boldsymbol{x})}{\partial x_4} \end{bmatrix}$$

$$= \begin{bmatrix} \mathbf{A}_1 & \mathbf{A}_2 & \mathbf{A}_3 & \mathbf{A}_4 \\ \mathbf{B}_1 & \mathbf{B}_2 & \mathbf{B}_3 & \mathbf{B}_4 \\ \mathbf{C}_1 & \mathbf{C}_2 & \mathbf{C}_3 & \mathbf{C}_4 \\ \mathbf{D}_1 & \mathbf{D}_2 & \mathbf{D}_3 & \mathbf{D}_4 \end{bmatrix} \tag{6.79}$$

where

$$\mathbf{A}_1 = [\mathbf{C}^{-1}]_{6\times6}, \mathbf{A}_2 = [\mathbf{I}^s]_{6\times6}, \mathbf{A}_3 - [\mathbf{0}]_{6\times m}, \mathbf{A}_4 - 0$$

$$\mathbf{B}_1 = \left[-\frac{\partial \mathbf{r}}{\partial \boldsymbol{\sigma}}\right]_{6\times6}, \mathbf{B}_2 = [\mathbf{I}^s]_{6\times6}, \mathbf{B}_3 = \left[-\frac{\partial \mathbf{r}}{\partial \boldsymbol{\alpha}}\right]_{6\times m}, \mathbf{B}_4 = [-\mathbf{r}]_{6\times1}$$

$$\mathbf{C}_1 = \left[-\frac{\partial \mathbf{h}}{\partial \boldsymbol{\sigma}}\right]_{m\times6}, \mathbf{C}_2 = [\mathbf{0}]_{m\times6}, \mathbf{C}_3 = [\mathbf{I}]_{m\times m}, \mathbf{C}_4 = [\mathbf{h}]_{m\times1}$$

$$\mathbf{D}_1 = \left[\frac{\partial f}{\partial \boldsymbol{\sigma}}\right]_{1\times6}, \mathbf{D}_2 = [\mathbf{0}]_{1\times6}, \mathbf{D}_3 = \left[\frac{\partial f}{\partial \boldsymbol{\alpha}}\right]_{1\times m}, \mathbf{D}_4 = 0$$

the subscripts have the same meaning as in the corresponding parts of the elastoplastic model.

The initial value for the Newton–Raphson method is $\boldsymbol{x}_{n+1}^{(0)} = \left[\boldsymbol{\sigma}_{n+1}^{(0)}, \boldsymbol{\varepsilon}_{n+1}^{p,(0)}, \boldsymbol{\alpha}_{n+1}^{(0)}, \Delta\gamma_{n+1}^{(0)}\right] = \left[\boldsymbol{\sigma}_{n+1}^{\mathrm{trial}}, \boldsymbol{\varepsilon}_n^p, \boldsymbol{\alpha}_n, 0\right]$. For the $k+1$-th iteration step, the corresponding solution is:

$$\boldsymbol{x}_{n+1}^{(k+1)} = \boldsymbol{x}_{n+1}^{(k)} - \left[\mathbf{J}\left(\boldsymbol{x}_{n+1}^{(k)}\right)\right]^{-1} \boldsymbol{g}\left(\boldsymbol{x}_{n+1}^{(k)}\right) \tag{6.80}$$

If $\left\| \boldsymbol{g}\left(\boldsymbol{x}_{n+1}^{(k+1)}\right) \right\| \geq \text{Tol}$ and Tol are the error tolerance values, the algorithm proceeds with the iterations until $\left\| \boldsymbol{g}\left(\boldsymbol{x}_{n+1}^{(k+1)}\right) \right\| < \text{Tol}$ is satisfied. The corresponding algorithmic flow is as follows.

I. Elastic predictor

$$\varepsilon_{n+1}^{\text{trial}} = \varepsilon_n + \Delta\varepsilon_{n+1} \quad \varepsilon_{n+1}^{\text{vp,trial}} = \varepsilon_n^{\text{vp}}$$

$$\sigma_{n+1}^{\text{trial}} = \mathbb{C} : \left(\varepsilon_{n+1}^{\text{trial}} - \varepsilon_{n+1}^{\text{vp,trial}}\right) \quad \alpha_{n+1}^{\text{trial}} = \alpha_n$$

II. Check the stress state

If $f_{n+1}^{\text{trial}} = f\left(\sigma_{n+1}^{\text{trial}}, \alpha_{n+1}^{\text{trial}}\right) < 0$ then

$$(\cdot)_{n+1} = (\cdot)_{n+1}^{\text{trial}}$$

Go to step IV)
Else
Go to step III)
End

III. Plastic corrector (stress iteration)

Do while $k < k^{\max}$ (where $k^{\max}$ is the maximum iteration value)
Compute the residual on the left-hand side of Eq. (6.60)
If $\|g(x)\| < \text{Tol}$ then

$$(\cdot)_{n+1} = (\cdot)_{n+1}^{(k)}$$

Go to step IV)
End
Determine the correction vector using the Jacobian matrix and residuals.
Update the state variables according to Eq. (6.80).
End

IV. Solve the consistency elastoplastic model and return to the main program.

The consistent tangent modulus for the three-dimensional elastoviscoplastic model is defined as follows:

$$\mathbb{C}_{c,n+1}^{\text{evp}} = \frac{\partial \sigma_{n+1}}{\partial \varepsilon_{n+1}} = \frac{\partial \Delta\sigma_{n+1}}{\partial \Delta\varepsilon_{n+1}} \tag{6.81}$$

According to the conclusion, the elastoplastic consistent tangent modulus corresponds to a submatrix of the inverse of the converged Jacobian matrix [**J**], denoted as:

$$[\mathbf{J}_{n+1}(\boldsymbol{x})]^{-1} = \begin{bmatrix} \mathbf{A}_{1,n+1} & \mathbf{A}_{2,n+1} & \mathbf{A}_{3,n+1} & \mathbf{A}_{4,n+1} \\ \mathbf{B}_{1,n+1} & \mathbf{B}_{2,n+1} & \mathbf{B}_{3,n+1} & \mathbf{B}_{4,n+1} \\ \mathbf{C}_{1,n+1} & \mathbf{C}_{2,n+1} & \mathbf{C}_{3,n+1} & \mathbf{C}_{4,n+1} \\ \mathbf{D}_{1,n+1} & \mathbf{D}_{2,n+1} & \mathbf{D}_{3,n+1} & \mathbf{D}_{4,n+1} \end{bmatrix} \tag{6.82}$$

From the definition of the elastoviscoplastic consistent tangent modulus, it can be inferred that:

$$\mathbb{C}^{\text{evp}}_{c,n+1} = \frac{\partial \boldsymbol{\sigma}_{n+1}}{\partial \boldsymbol{\varepsilon}_{n+1}} = \frac{\partial \Delta \boldsymbol{\sigma}_{n+1}}{\partial \Delta \boldsymbol{\varepsilon}_{n+1}} = \left[\mathbf{A}_{1,n+1}\right]_{6\times 6} \tag{6.83}$$

6.4 Three-dimensional Elastoplastic Damage Model and Solution Framework

The elastoplastic damage constitutive model discussed in this section is based on the concept of effective stress, as well as the assumptions of strain equivalence and damage being coupled only with elasticity. The state variables introduced in the constitutive equations are used to describe the evolution of internal damage, as well as nonlinear isotropic hardening and kinematic hardening.

6.4.1 Model Overview

In the case of small strains, according to the superposition principle, the total strain tensor $\boldsymbol{\varepsilon}$ is the sum of the elastic strain tensor $\boldsymbol{\varepsilon}^e$ and the plastic strain tensor $\boldsymbol{\varepsilon}^p$:

$$\boldsymbol{\varepsilon} = \boldsymbol{\varepsilon}^e + \boldsymbol{\varepsilon}^p \quad \varepsilon_{ij} = \varepsilon^e_{ij} + \varepsilon^p_{ij} \quad i, j = 1, 2, 3 \tag{6.84}$$

The relationship between the stress tensor $\boldsymbol{\sigma}$ and the elastic strain tensor $\boldsymbol{\varepsilon}^e$ is governed by Hooke's law, expressed as follows:

$$\boldsymbol{\sigma} = (1 - D)\,\mathbb{C} : \boldsymbol{\varepsilon}^e = (1 - D)\,\mathbb{C} : (\boldsymbol{\varepsilon} - \boldsymbol{\varepsilon}^p) \quad \sigma_{ij} = (1 - D)\,\mathrm{C}_{ijkl} : \left(\varepsilon_{kl} - \varepsilon^p_{kl}\right) \tag{6.85}$$

In Eq. (6.85), $\mathbb{C}$ is the fourth-order elastic tensor, and D is a variable describing material isotropic damage. The definition of the effective stress tensor $\boldsymbol{\sigma}^{\text{eff}}$ is as follows:

$$\boldsymbol{\sigma}^{\text{eff}} = \mathbb{C} : \boldsymbol{\varepsilon}^e = \mathbb{C} : (\boldsymbol{\varepsilon} - \boldsymbol{\varepsilon}^p) \quad \sigma^{\text{eff}}_{ij} = \mathrm{C}_{ijkl} : \left(\varepsilon_{kl} - \varepsilon^p_{kl}\right) \tag{6.86}$$

Therefore, the relationship between the stress tensor $\boldsymbol{\sigma}$ and the effective stress tensor $\boldsymbol{\sigma}^{\text{eff}}$ is given by:

$$\boldsymbol{\sigma} = (1 - D)\,\boldsymbol{\sigma}^{\text{eff}} \quad \text{or} \quad \boldsymbol{\sigma}^{\text{eff}} = \boldsymbol{\sigma} / (1 - D) \tag{6.87}$$

The internal variable vector is $\boldsymbol{\alpha} = \{D, \alpha_1, \alpha_2\}$, where α_1 and α_2 are used to characterize isotropic hardening and kinematic hardening, respectively. $f(\boldsymbol{\sigma}, \boldsymbol{\alpha})$ represents the yield function. For the elastoplastic damage model, the set of permissible stress states is defined as follows:

$$\mathbb{E}_\sigma = \left\{(\boldsymbol{\sigma}, \boldsymbol{\alpha}) \in \mathbb{S} \times \mathbb{R}^3 \middle| f(\boldsymbol{\sigma}, \boldsymbol{\alpha}) \leq 0\right\} \tag{6.88}$$

$\operatorname{int}(\mathbb{E}_\sigma)$ represents the elastic region:

$$\operatorname{int}(\mathbb{E}_\sigma) = \left\{(\boldsymbol{\sigma}, \boldsymbol{\alpha}) \in \mathbb{S} \times \mathbb{R}^3 \middle| f(\boldsymbol{\sigma}, \boldsymbol{\alpha}) < 0\right\} \tag{6.89}$$

$\partial(\mathbb{E}_\sigma)$ represents the yield surface in stress space:

$$\partial(\mathbb{E}_\sigma) = \left\{(\boldsymbol{\sigma}, \boldsymbol{\alpha}) \in \mathbb{S} \times \mathbb{R}^2 \middle| f(\boldsymbol{\sigma}, \boldsymbol{\alpha}) = 0\right\} \tag{6.90}$$

The flow rule, damage evolution rule, and hardening rule are as follows:

$$\begin{cases} \dot{\varepsilon}^p = \dot{\gamma}\mathbf{r}(\sigma, \alpha) \\ \dot{\alpha} = -\dot{\gamma}\mathbf{h}(\sigma, \alpha) \end{cases} \tag{6.91}$$

where r and h are user-defined functions. γ is a nonnegative multiplier, determined by the Kuhn–Tucker complementary conditions (also known as the loading/unloading conditions) and the consistency condition:

$$\begin{cases} \dot{\gamma} \geq 0 \quad f(\sigma, \alpha) \leq 0 \quad \dot{\gamma} f(\sigma, \alpha) = 0 \\ \dot{\gamma} \dot{f}(\sigma, \alpha) = 0 \end{cases} \tag{6.92}$$

Damage Threshold. When the accumulated plastic strain is relatively low, the reduction in the elastic modulus is minimal. Therefore, it is assumed that damage evolution begins only when the accumulated plastic strain (denoted as $\bar{\varepsilon}^p$) exceeds a critical value, known as the damage threshold, denoted as $\bar{\varepsilon}^p_D$.

6.4.2 Elastic Prediction/Plastic Correction-based Solution Algorithm

The elastoplastic damage model discussed in this section can be viewed as a special case of the elastoplastic model. Therefore, the algorithm mentioned in Section 6.4.1 can be applied to the elastoplastic damage model with only minor modifications.

Assuming all state variables are known at time t_n, and the displacement increment ΔU_{n+1} and corresponding strain increment $\Delta\varepsilon_{n+1}$ due to external loads are given, the numerical algorithm proceeds as follows:

If $\bar{\varepsilon}^p_n < \bar{\varepsilon}^p_D$ is given, initially assume that under the current strain increment $\Delta\varepsilon_{n+1}$, the material does not experience damage, i.e.

$$D_{n+1} = D_n = 0 \tag{6.93}$$

In this case, Eq. (6.93) representing damage evolution will not be included in the iteration process. Referring to Section 6.3, the nonlinear equations to be solved are:

$$\begin{cases} \mathbb{C}^{-1} : \boldsymbol{\sigma}_{n+1} : -\boldsymbol{\varepsilon}_{n+1} + \boldsymbol{\varepsilon}^p_{n+1} = 0 \\ \boldsymbol{\varepsilon}^p_{n+1} = \boldsymbol{\varepsilon}^p_n + \Delta\gamma\mathbf{r}\left(\boldsymbol{\sigma}_{n+1}, \alpha_{1,n+1}, \alpha_{2,n+1}\right) = \boldsymbol{\varepsilon}^p_n + \Delta\gamma\mathbf{r}_{n+1} \\ \alpha_{1,n+1} = \alpha_{1,n} - \Delta\gamma h_2\left(\boldsymbol{\sigma}_{n+1}, \alpha_{1,n+1}, \alpha_{2,n+1}\right) = \alpha_{1,n} - \Delta\gamma h_2 \\ \alpha_{2,n+1} = \alpha_{2,n} - \Delta\gamma h_3\left(\boldsymbol{\sigma}_{n+1}, \alpha_{1,n+1}, \alpha_{2,n+1}\right) = \alpha_{2,n} - \Delta\gamma h_3 \\ f\left(\boldsymbol{\sigma}_{n+1}, \alpha_{1,n+1}, \alpha_{2,n+1}\right) = f_{n+1} = 0 \end{cases} \tag{6.94}$$

Calculate $\bar{\varepsilon}^p_{n+1}$ after the stress iteration has converged. To verify the previous assumption, the following additional checks are required:

- If $\bar{\varepsilon}^p_{n+1} < \bar{\varepsilon}^p_D$ is reasonable, assume the converged state variables from the iteration are taken as the final variables for the ***n*+1** step.

- If $\bar{\varepsilon}^p_{n+1} \geq \bar{\varepsilon}^p_D$ is unreasonable, it is assumed that the material has already experienced damage under the current strain increment. The equation representing damage evolution needs to be included in the stress iteration, and the iteration process must be recalculated.

$$\begin{cases} \mathbb{C}^{-1} : \boldsymbol{\sigma}_{n+1} - (1-D)\,\boldsymbol{\varepsilon}_{n+1} + (1-D)\,\boldsymbol{\varepsilon}^p_{n+1} = 0 \\ \boldsymbol{\varepsilon}^p_{n+1} = \boldsymbol{\varepsilon}^p_n + \Delta\gamma \mathbf{r}\left(\boldsymbol{\sigma}_{n+1}, D, \alpha_{1,n+1}, \alpha_{2,n+1}\right) = \boldsymbol{\varepsilon}^p_n + \Delta\gamma \mathbf{r}_{n+1} \\ D_{n+1} = D_n - \Delta\gamma h_1\left(\boldsymbol{\sigma}_{n+1}, D, \alpha_{1,n+1}, \alpha_{2,n+1}\right) = D_n - \Delta\gamma h_1 \\ \alpha_{1,n+1} = \alpha_{1,n} - \Delta\gamma h_2\left(\boldsymbol{\sigma}_{n+1}, D, \alpha_{1,n+1}, \alpha_{2,n+1}\right) = \alpha_{1,n} - \Delta\gamma h_2 \\ \alpha_{2,n+1} = \alpha_{2,n} - \Delta\gamma h_3\left(\boldsymbol{\sigma}_{n+1}, D, \alpha_{1,n+1}, \alpha_{2,n+1}\right) = \alpha_{2,n} - \Delta\gamma h_3 \\ f\left(\boldsymbol{\sigma}_{n+1}, D, \alpha_{1,n+1}, \alpha_{2,n+1}\right) = f_{n+1} = 0 \end{cases} \tag{6.95}$$

If damage has already occurred in the previous step for $\bar{\varepsilon}^p_n \geq \bar{\varepsilon}^p_D$, this book does not consider the recovery of damage, so the damage will further increase under the current strain increment. The nonlinear Eq. (6.95) is solved to obtain the final variables at step $\boldsymbol{n+1}$.

6.4.3 Numerical Issues

Compared to elastoplastic models, the evolution of damage affects the accuracy of numerical algorithms on one hand, and on the other, it also impacts the convergence of the Newton–Raphson method. The reasons are as follows:

The plastic multiplier may converge to a negative value during stress iteration, which contradicts the definition of the plastic multiplier. As damage increases, the convergence region of the Newton–Raphson method narrows.

For cases of highly damaged materials, the non-convergence of stress iteration at a certain Gauss integration point can cause the main program to restart the iterative calculation, significantly increasing computational costs, especially in the analysis of large structures.

To address these issues, relevant scholars have proposed a series of solutions:

Introduce the line-search method into the standard Newton-Raphson iterative method:

Select a reasonable initial value. If $\boldsymbol{x}^{(0)}_{n+1} = \left[\boldsymbol{\sigma}^{\text{trial}}_{n+1}, \boldsymbol{\varepsilon}^p_n, \boldsymbol{\alpha}_n, 0\right]$ does not converge, set the initial value to $\boldsymbol{x}^{(0)}_{n+1} = \left[\boldsymbol{\sigma}^{\text{proj}}_{n+1}, \boldsymbol{\varepsilon}^p_n, \boldsymbol{\alpha}_n, 0\right]$, where $\boldsymbol{\sigma}^{\text{proj}}_{n+1}$ is the root of the following equation:

$$f\left(\boldsymbol{\sigma}^{\text{proj}}_{n+1}, \boldsymbol{\alpha}_n\right) = f\left(\boldsymbol{\sigma}^{\text{proj}}_{n+1}, D_n, \alpha_{1,n}, \alpha_{2,n}\right) = 0 \tag{6.96}$$

The geometric interpretation of Eq. (6.96) is as follows: without considering the evolution of the yield surface, find the projection point $\boldsymbol{\sigma}^{\text{trial}}_{n+1}$ of the trial stress $\boldsymbol{\sigma}^{\text{proj}}_{n+1}$ on the yield surface.

The sub-increment method, which involves subdividing the current strain increment $\Delta\varepsilon_{n+1}$ into several sub-increments, is as follows:

$$\Delta\boldsymbol{\varepsilon}_{n+1} = \Delta\boldsymbol{\varepsilon}_{n+1,1} + \Delta\boldsymbol{\varepsilon}_{n+1,2} + \Delta\boldsymbol{\varepsilon}_{n+1,3} + \cdots + \Delta\boldsymbol{\varepsilon}_{n+1,m} \tag{6.97}$$

Then, incrementally apply the sub-step loads until reaching the current strain increment. Details of the above solutions can be found in the relevant literature. The derivation of the consistent elastoplastic damage modulus is as follows:

If $f(\boldsymbol{\sigma}_{n+1}, \boldsymbol{\alpha}_{n+1}) < 0$, the material is in the elastic region or in an elastic unloading state, then:

$$\boldsymbol{\sigma}_{n+1} = (1 - D_{n+1})\,\mathbb{C} : \boldsymbol{\varepsilon}^{e}_{n+1} \Rightarrow \mathbb{C}^{\mathrm{edp}}_{c,n+1} = (1 - D_{n+1})\,\mathbb{C} \tag{6.98}$$

If $f(\boldsymbol{\sigma}_{n+1}, \boldsymbol{\alpha}_{n+1}) = 0$, the material is in the plastic loading phase. Following the approach of Section 6.3, after stress iteration convergence, the full derivative of the corresponding nonlinear equation system is computed and simplified to obtain:

$$[\mathrm{J}_{n+1}(x)] \begin{bmatrix} d\sigma_{n+1} \\ d\varepsilon^{p}_{n+1} \\ dD_{n+1} \\ d\alpha_{1,n+1} \\ d\alpha_{2,n+1} \\ d\Delta\gamma \end{bmatrix} = \begin{bmatrix} (1 - D_{n+1})\, d\varepsilon_{n+1} \\ 0 \\ 0 \\ 0 \\ 0 \\ 0 \end{bmatrix} \Rightarrow \begin{bmatrix} d\sigma_{n+1} \\ d\varepsilon^{p}_{n+1} \\ dD_{n+1} \\ d\alpha_{1,n+1} \\ d\alpha_{2,n+1} \\ d\Delta\gamma \end{bmatrix} = [\mathrm{J}_{n+1}(x)]^{-1} \begin{bmatrix} (1 - D_{n+1})\, d\varepsilon_{n+1} \\ 0 \\ 0 \\ 0 \\ 0 \\ 0 \end{bmatrix} \tag{6.99}$$

According to the definition of the consistent tangent modulus:

$$\mathbb{C}^{\mathrm{edp}}_{c,n+1} = \frac{\partial \sigma_{n+1}}{\partial \varepsilon_{n+1}} = \frac{\partial \Delta\sigma_{n+1}}{\partial \Delta\varepsilon_{n+1}} = \left[\mathbf{A}_{1,n+1}\right]_{6\times 6} \tag{6.100}$$

In Eq. (6.100), $\left[\mathbf{A}_{1,n+1}\right]_{6\times 6}$ is a submatrix of the inverse of the Jacobian matrix.

6.5 Custom Constitutive Models Based on ABAQUS

ABAQUS is a general-purpose finite element analysis software used for mechanical, civil, electronics, fluid, acoustics, and optical analyses. ABAQUS was originally a product of HKS Inc. in the United States and was sold to Dassault Systèmes in 2005, thus the software is also known as Dassault Systèmes SIMULIA [4].

ABAQUS features a user-friendly pre- and post-processing interface, a powerful nonlinear solver, a variety of functional modules, and extensive interface extension capabilities. The content of this chapter primarily relates to the UMAT module (User-defined Material Constitutive Model) of ABAQUS. Through the UMAT interface, users can write custom programs based on their own material constitutive theories and assign these custom material models to existing elements in the program, leveraging the robust nonlinear solver of the general-purpose analysis software to achieve specific analysis objectives.

The general procedure for running the UMAT module in ABAQUS is as follows:

The main program solves the corresponding governing equations, such as momentum conservation, mass conservation, and energy conservation equations, and then passes the solutions for the primary unknowns (such as displacements, temperature, pressure, etc.) and state variables (primarily for nonlinear problems) to the UMAT program.

Based on the specific constitutive model and corresponding numerical integration algorithm, update the stress and state variables at each integration point under the current iteration's strain increment, and return these values to the main program.

Given that the main program generally employs an incremental iterative method to solve nonlinear equations, UMAT also needs to provide the consistent tangent modulus for the current iteration step. Although the consistent tangent modulus will only accelerate the convergence rate and will not affect the convergence accuracy when the program is convergent.

Figure 6.6 illustrates the basic format of the UMAT user-defined material constitutive interface. The variables that need to be updated in the program mainly include:

STRESS(N): The Cauchy stress tensor, represented as an $\boldsymbol{N}$-dimensional column vector according to the Voigt notation, includes $\boldsymbol{N1}$ normal stress components and $\boldsymbol{N2}$ shear stress components. At the beginning of the iteration step, STRESS(N) contains the values from the previous increment. The stress values for the current step need to be computed using the stress update algorithm discussed in this chapter and returned to the main program.

DDSDDE (N, N): The Jacobian matrix related to the integration algorithm. Since the Jacobian matrix does not affect the accuracy of the program's calculations, it is recommended to assign the elastic matrix to DDSDDE during the early stages of program development. Once the stress integration algorithm has been confirmed to converge, the submatrix of the inverse Jacobian matrix should be assigned to DDSDDE.

STATEV(m): An mmm-dimensional column vector that stores state variables, such as the vector $\boldsymbol{\alpha}$ mentioned in the integration algorithm. At the beginning of the iteration step, STATEV(m) contains the values from the previous increment. The values for the current step need to be computed using the stress update algorithm discussed in the previous sections and returned to the main program.

```
      SUBROUTINE UMAT(STRESS,STATEV,DDSDDE,SSE,SPD,SCD,
     1 RPL,DDSDDT,DRPLDE,DRPLDT,
     2 STRAN,DSTRAN,TIME,DTIME,TEMP,DTEMP,PREDEF,DPRED,CMNAME,
     3 NDI,NSHR,NTENS,NSTATV,PROPS,NPROPS,COORDS,DROT,PNEWDT,
     4 CELENT,DFGRD0,DFGRD1,NOEL,NPT,LAYER,KSPT,KSTEP,KINC)
C
      INCLUDE 'ABA PARAM.INC'
C
      CHARACTER*80 CMNAME
      DIMENSION STRESS(NTENS),STATEV(NSTATV),
     1 DDSDDE(NTENS,NTENS),DDSDDT(NTENS),DRPLDE(NTENS),
     2 STRAN(NTENS),DSTRAN(NTENS),TIME(2),PREDEF(1),DPRED(1),
     3 PROPS(NPROPS),COORDS(3),DROT(3,3),DFGRD0(3,3),DFGRD1(3,3)
     4 JSTEP(4)

      User coding to define: DDSDDE, STRESS, STATEV, SSE, SPD, SCD
      And if necessary: RPL, DDSDDT, DRPLDE, DRPLDT,PNEWDT

      RETURN
      END
```

Figure 6.6 UMAT user-defined material constitutive interface.

The specific meanings of other related physical quantities can be found in the detailed descriptions in the ABAQUS manual.

6.6 Conclusion

ABAQUS and its subroutines provide a wide range of constitutive models that can be used to simulate the mechanical behavior of different materials. By selecting an appropriate constitutive model and inputting material parameters, numerical simulation of material behavior can be achieved, allowing for the prediction of material responses under various loading conditions. By comparing the constitutive model with experimental data, it is possible to determine the accuracy of the material parameters used in the model and to optimize the model for better prediction of real-world behavior. Using ABAQUS and its subroutines, engineers can evaluate various material properties such as strength, stiffness, plastic behavior, and fatigue life. This helps in assessing the reliability of materials and optimizing designs to enhance the safety and durability of structures. The application of ABAQUS provides a rapid and efficient method for evaluating material behavior and structural performance in engineering practice. This approach can save a significant amount of time and cost while improving engineering efficiency and quality.

In summary, ABAQUS and its subroutines not only enable numerical simulation of material behavior but also allow for the verification of model accuracy, assessment of material reliability, and improvement of engineering efficiency.

References

1 Li, Q.Y., Wang, N.C., and Yi, D.Y. (2008). *Numerical Analysis*. Beijing: Tsinghua University Press.
2 Simo, J.C. and Hughes, T.J.R. (1998). *Computational Inelasticity*. New York: Springer.
3 Neto, E.A.S., Perić, D., and Owen, D.R.J. (2008). *Computational Methods for Plasticity: Theory and Applications*. United Kingdom: Wiley.
4 *Abaqus User Subroutines Reference Guide, Version 6.14*. Vélizy-Villacoublay, France: Dassault Systemes Simulia, Inc, 2014.

7

Artificial Intelligence in Constitutive Research

As the core technology of the ongoing industrial revolution, artificial intelligence (AI) plays an important role in the current multidisciplinary and cross-disciplinary research. As a black box tool, AI has advantages in decision-making and multi-parameter optimization. By combining with AI, traditional industries can be empowered from many angles, thereby achieving quality improvement and increased efficiency in traditional industries [1, 2].

7.1 Tensor Bases of Machine Learning

As a mathematical concept, the tensor was widely used in physics, engineering, and now in machine learning (ML) and deep learning. In the broadest sense, a tensor can be viewed as a multidimensional array that represents linear relationships in some basis. They are a generalization of the concepts of scalars, vectors, and matrices that can be used to describe higher-dimensional data. Therefore, the tensor was always treated as a mathematical tool and a set of operational rules in many ML algorithms.

7.1.1 Function of Tensor in Machine Learning Algorithms

In a traditional linear algorithm, the multidimensional feature of the tensor can precisely describe the dynamic feature of a complex system. For example, scalars are zero-order tensors (such as mass and temperature), vectors are first-order tensors (such as velocity and force), and second-order tensors can describe physical phenomena (such as stress or strain). Taking advantage of information-carrying capacity of the tensor, the tensor was set as a standard data type in many frameworks of ML. Meanwhile, many deep learning tools and libraries, including PyTorch and TensorFlow, are designed around tensors.

Primarily, as a basic mathematical concept, the data-storing strategy in AI shares same idea as the storing strategy in physics and math. Similarly, like the counterpart in physics, the dimension of data structure in AI can reflect the dimension of the decision matrix, which is similar to that of a vector in physics. Both of them reflect how a single sample or vector is characterized by the different quantities in a tensor. Meanwhile, for tasks of classification and regressive analysis, it requires an encoded tensor to stand for the output or the

Constitutive Models of Solid Materials: From Mechanical Principles to Engineering Applications, First Edition.

label of the series samples, such as the compressive strength of the material and different material processing techniques.

$$[v] = \begin{bmatrix} v_x & v_y & v_z \end{bmatrix} \tag{7.1}$$

$$[\text{Characteristics of sample}] = [x_1\ x_2\ x_3\ x_4 \cdots x_n] \tag{7.2}$$

Then, with the increasing complexity of the analyzed data, the tensor also extended the capability of describing the data with temporal characteristics and higher dimensions. To be specific, while AI models pay more attention to the language understanding and classification of images, the data of the computer vision and large language models (LLMs), including sentences and images, can be structured into the form of a tensor. For example, as a typical algorithm of computer vision, convolutional neural networks (CNN) are meant to classify the type of the image, and the image should be digitized into a tensor to be operated. The way of digitalizing the image data can be illustrated as follows:

Although the given illustrations have shown the capability of a tensor to structure different data, the tensor cannot be empowered with such capability without the help of a unique operation. Therefore, some unique operations of the tensor in AI applications will be introduced in Section 7.1.2 to illustrate.

7.1.2 Key Operational Rule of Tensor Application in Machine Learning

According to the introduction in Section 7.1.1, the data-storing function of the tensor in AI algorithms has been introduced. By neatly using the features of the tensor with different ranks and shapes, the tensor empowered AI algorithms with the capability of processing data types with complex structures, such as image data, word embeddings, and temporal data. However, to process the source data into the required tensor-formed data, a unique process should be introduced. Therefore, this section will introduce some operations that empower the tensor with the mentioned function in the AI application.

Convolution

As a central concept within deep learning frameworks (DFMs), the operations of the tensors are conceptually highly similar to multidimensional arrays. However, the convolution of tensors provided key functionality and advantages compared to traditional multidimensional arrays. These advantages render tensors far more effective and efficient when handling complex computations and extensive datasets.

For datasets that have extensive dimensions and volume in their raw data, it usually requires an operation to minimize the storage size of the dataset, and such an operation is also helpful in increasing the operational speed. Based on the above requirements of AI algorithms, researchers find that the operation of the convolutional functional analysis meets the requirements of data operations in AI.

Conceptually, the convolution is a mathematical operation that generates a third function from two functions. Its essence is a special integral transformation. By flipping and translating the overlapping parts of two functions, the integral of the two functions can be generated. If we apply the idea of convolution to the tensor operation, the convolution is helpful in minimizing the size of the original size of the non-processed data. The convolution of a 1D continuous convolution can be represented as following equations:

$$(f * g)(t) = \int_{-\infty}^{\infty} f(\tau)\, g(t-\tau)\, d\tau \tag{7.3}$$

where $f(t)$ and $g(t)$ are the two integrable functions; t is the axial position of the output; τ is the variable of the integrating operation, and it is the unit length of the $g(t)$ translating and flipping along the axis of t; $g(t-\tau)$ represent the value of $g(t)$ translate and flip along the axis of t by the length of τ.

Overall, the key action of the convolution can be interpreted as the value of the $g(t)$ multiply by the $f(t)$. Then, the $g(t)$ translates the alone the axis of the by τ.

According to the definition of the 1D continuous convolution, the 2D discrete convolution of two matrices can be illustrated as follows:

$$(I * K)[i, j] = \sum_{m=-\infty}^{\infty} \sum_{m=-\infty}^{\infty} I[m, n] \cdot K[i-m, j-n] \tag{7.4}$$

where I and K are the input matrices and the convolution kernel; $[i, j]$ is the position that integration or product happens; $[m, n]$ is the position of the convolution kernel. Therefore, the output size is determined by the size of the input matrix and the kernel size:

$$\text{Output} = (W - K_w + 2P + 1) \times (H - K_H + 2P + 1) \tag{7.5}$$

where $(W \times H)$ is the size of the input matrix; $(K_w \times K_h)$ is the size of the convolutional kernel; P is the padding size.

To have a better understanding of convolution, the example of the image data will be picked. According to the introduction in Section 7.1.1, the convolution operation of an image can be represented as a 2D sensor; the operation of the convolution can minimize the data size of an image into a manageable size of matrix.

Tensors serve as the mathematical bedrock of modern ML, bridging abstract theory and computational practice. As multidimensional generalizations of scalars, vectors, and matrices, they provide a unified framework for representing complex, high-dimensional data across diverse domains—from physics-inspired structural modeling to cutting-edge AI applications. Their capacity to encode dynamic relationships (e.g. Eq. (7.1) for vectors, Eq. (7.2) for feature spaces) enables precise characterization of intricate systems, whether describing material stress or digital images. Furthermore, the adoption of tensors as core data structures in frameworks like PyTorch and TensorFlow underscores their operational necessity. Crucially, raw data complexity (images, language, temporal sequences) is tamed through tensor encapsulation, as seen in Figure 7.1. However, tensors realize their full potential only through specialized operations—most notably convolution. This transformative process (Eqs. 7.3 and 7.4) compresses data dimensionality while extracting semantically rich features, as demonstrated by convolutional neural networks (CNNs) in computer vision. By sliding kernels across tensor-structured inputs (Eq. 7.5), convolution balances efficiency with expressiveness, enabling scalable processing of massive datasets. Ultimately, tensors are not mere containers but dynamic engines for AI innovation. Their synergy with operations like convolution has propelled breakthroughs from real-time image

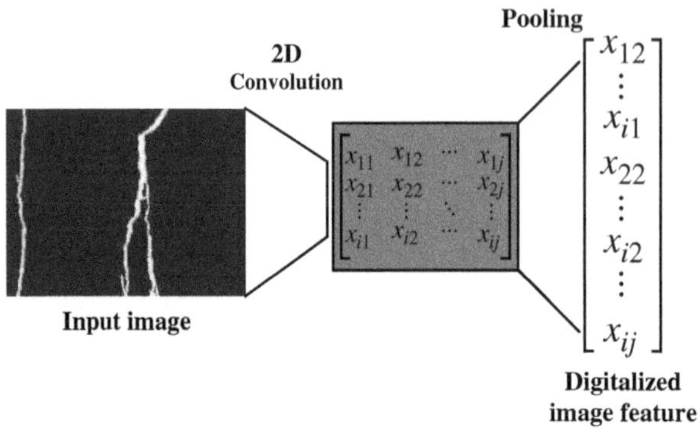

Figure 7.1 Digitalized image data.

classification to LLMs, proving indispensable in translating mathematical abstraction into intelligent systems. As ML continues to evolve, tensors will remain central to unlocking higher dimensions of understanding—one axis at a time.

7.2 Basic Mathematical Principles of Artificial Intelligence

Drawing upon the current landscape of ML applications across diverse fields over the past three decades [3, 4] and considering the precision requirements for numerical prediction within constitutive research, recent findings demonstrate that artificial neural networks (ANNs) generally meet the accuracy demands for evaluating numerical predictions in material constitutive studies [2]. Concurrently, DFMs built by combining ANNs with distinct functionalities can be leveraged to deduce complex material and component mechanical laws [5, 6]. Consequently, ANNs have gained widespread acceptance and application in related research on material properties.

However, with the advancement of ML algorithms, recent studies also indicate that Classification and Regression Tree (CART)-based eXtreme Gradient Boosting (XGBoost), as a tool for both numerical prediction and classification, exhibits significant advantages in both precision and accuracy for numerical prediction tasks when compared to a range of models including ANNs [7, 8]. In summary, XGBoost holds considerable potential within the domains of material development and component data prediction, and has consequently been selected as one of the prevalent analytical methods for material mechanical data.

As two representative ML algorithms, ANNs and XGBoost exhibit fundamental differences in their mathematical principles. Nevertheless, since the predictive performance of both satisfies the accuracy requirements for mechanical numerical predictions, we will provide a brief introduction to their underlying mathematical principles based on these two methods.

7.2.1 MLP-based Machine Learning Algorithms

As a quintessential ML model based on ANNs, the fundamental mathematical principle of the Multi-Layer Perceptron (MLP) involves minimizing an objective function by applying the theory of gradient descent. This process utilizes the concept of linear combinations to derive an optimized weight matrix "w" and bias vector "b" capable of making predictions [9].

As an ML algorithm widely accepted in contemporary research, the ANN model leverages diverse types of hidden layers, which endow it with the capability to handle various data processing tasks, such as numerical prediction, classification tasks, and image processing [2, 10].

Subsequently, a numerical prediction neural network featuring two hidden layers will serve as an example to illustrate its basic mathematical principles. Its architecture can be represented schematically, as shown in Figure 7.2.

Based on the neural network architecture depicted in Figure 7.2 and the principles of multilinear programming, the unprocessed output ($o'_{i,j}$) of each neuron can be calculated as follows:

$$o'_{ijk} = \sum_{k=1}^{k=n} w_{ij-1k} \cdot o_{ij-1k} + b_{ij-1} \tag{7.6}$$

where o'_{ijk} represents the unprocessed output value of the kth neuron in the jth layer before activation function application; w_{ijk} denotes the weight associated with the kth neuron in the jth layer; o_{i-1jk} signifies the activated output value (post-activation function) of the kth neuron in the jth layer from the $(i-1)$th training iteration; b_{ij} is the bias value for the jth layer during the ith training iteration. Here, "I" corresponds to the training iteration index, "j" indicates the neural network layer index, and "k" specifies the neuron index within layer j.

Drawing upon the current landscape of ML applications across diverse fields over the past three decades [3, 9] and considering the precision requirements for numerical prediction within constitutive research, recent findings demonstrate that ANNs generally meet the accuracy demands for evaluating numerical predictions in material constitutive studies [2].

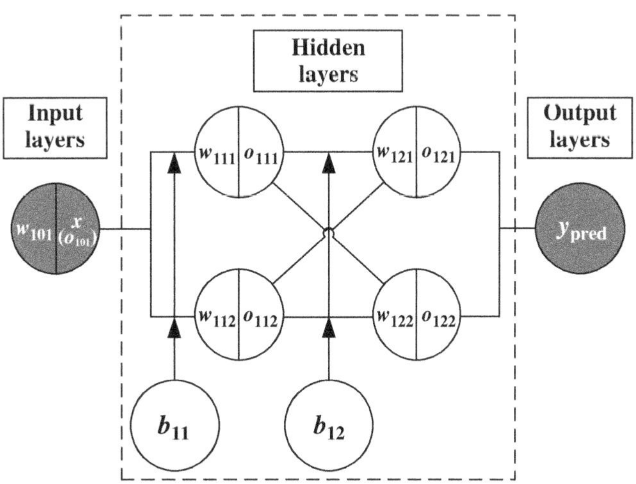

Figure 7.2 ANN neural network with two hidden layers.

Concurrently, DFMs built by combining ANNs with distinct functionalities can be leveraged to deduce complex material and component mechanical laws [6, 11]. Consequently, ANNs have gained widespread acceptance and application in related research on material properties.

$$\text{"ReLU" function: } f(x) = \begin{cases} x & x > 0 \\ 0 & x \leq 0 \end{cases} \tag{7.7}$$

$$\text{"Sigmoid" function: } f(x) = \frac{1}{1 - e^{-x}} \tag{7.8}$$

where $x = o'_{ijk}$ denotes the unprocessed output value of the corresponding neuron before the application of the activation function, and the commonly used activation function was illustrated in Figure 7.3.

Given the significant impact of activation function selection on ML model performance, the chosen function should fundamentally satisfy the following requirements:

1. The range of the activation function should align closely with the value domain of the output layer parameters.
2. The selection should correspond to the requirements of the mechanical parameter study and the specific objective task of the ML model.
3. Due to the application of gradient descent principles within ANN models, activation functions primarily designed for classification (e.g. "Sigmoid" or "tanh") should be selected cautiously to avoid the detrimental effects of exploding or vanishing gradients on model performance.

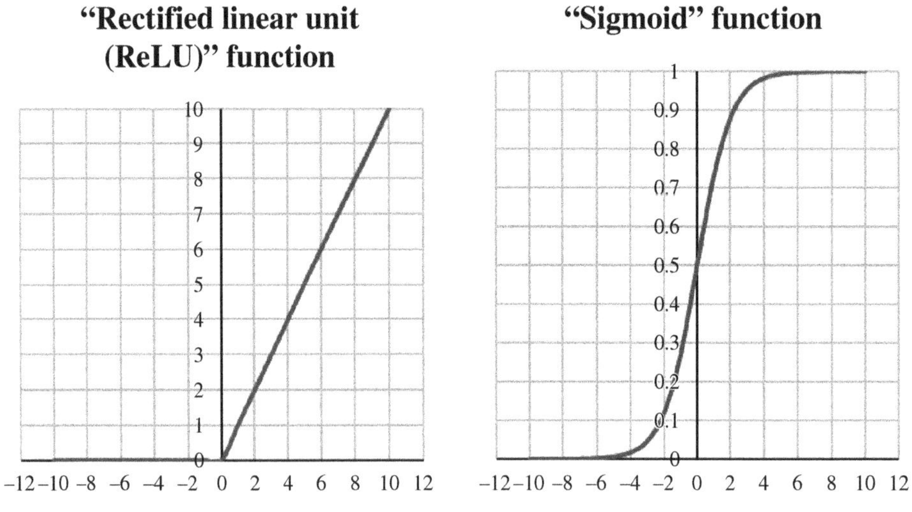

Figure 7.3 "ReLU" function and "Sigmoid" function.

Table 7.1 Loss function selection for different tasks.

Type of tasks	Name	Loss function	
Numerical prediction	Mean square error (MSE)	$\text{MSE} = \frac{1}{N}\sum_{i=1}^{N}\left(y_{i,\text{pred}} - y_{i,\text{true}}\right)^2$	(7.9)
	Mean absolute percentage error (MAPE)	$\text{MAPE} = \frac{1}{N}\sum_{i=1}^{N}\left\lvert\frac{y_{i,\text{pred}} - y_{i,\text{true}}}{y_{i,\text{true}}}\right\rvert * 100\%$	(7.10)
Classification and recognition tasks	Categorical cross entropy (CCE)	$\text{CCE} = -\frac{1}{N}\sum_{i=1}^{N} y_{i,\text{true}} * \log y_{i,\text{pred}}$	(7.11)
	Binary cross entropy (BCE)	$\text{BCE} = -\frac{1}{N} \times \sum_{i=1}^{N}\left(y_{i,\text{true}}\log y_{i,\text{pred}} + \left(1 - y_{i,\text{true}}\right)\log\left(1 - y_{i,\text{pred}}\right)\right)$	(7.12)

Subsequently, when the neural network's forward propagation computation reaches the output layer, this marks the completion of one forward pass (or training iteration). Concurrently, the loss function is computed for each iteration to quantify the discrepancy between the model's prediction $\left(y_{\text{pred}}\right)$ and the true value (y_{true}).

The choice of loss function itself varies depending on the nature of the ANN task:

1. Numerical prediction models typically favor the mean squared error (MSE) or mean absolute percentage error (MAPE).
2. Classification and image recognition tasks are more inclined toward categorical cross entropy (CCE) or binary cross entropy (BCE).

Based on this loss evaluation, the ANN can then be optimized using techniques such as backpropagation.

Back-propagation (BP), as a methodology for enhancing ANN models, aims to update the weights of individual neurons and the biases of each layer through the reverse application of gradient descent theory. Consequently, optimizers and back-propagation are introduced into ANNs, building upon forward propagation, to further improve computational efficiency.

Since the back-propagation process in ANNs utilizes the chain rule to compute the gradients of the output layer's loss function with respect to the weights of individual neurons, the selected loss function must be differentiable (see Table 7.1). The specific mathematical expression for this chain rule computation is detailed in Eq. (7.13).

$$\frac{\partial L_{\text{total}}}{\partial w_{ijk}} = \sum_{k=1}^{k=n} \frac{\partial L_{ijk}}{\partial o_{ijk}} \times \frac{\partial o_{ijk}}{\partial o'_{ijk}} \times \frac{\partial o'_{ijk}}{\partial w_{ijk}} \tag{7.13}$$

where L_{ijk} denotes the contribution of the kth neuron in the jth layer to the loss function during the ith training iteration.

Therefore, based on the gradient descent algorithm and the gradients of the output layer's loss function with respect to the weights of individual neurons, the updated weights can be computed using a linear optimizer (Eq. 7.14). However, since weight updates influence model convergence and generalization capability, research into optimizers' impact on ML model predictive performance has intensified alongside advancements in mechanical science and ML algorithm development. Considering recent progress in optimizer research, both within mechanical applications and ML algorithm development, Adaptive Moment Estimation (Adam) (Eq. 7.15) has gained significant traction in both application and research domains [4]. This preference stems from Adam's advantages of rapid gradient descent convergence and its tendency to oscillate effectively near optimal solutions.

$$\text{Linear optimizer: } w_{i+1jk} = w_{ijk} - \eta \frac{\partial L_{\text{total}}}{\partial w_{ijk}} \tag{7.14}$$

$$\text{Adaptive moment estimation: } w_{i+1jk} = w_{ijk} - \frac{\eta}{\sqrt{\widehat{v}_i} + \epsilon} \widehat{m}_i \tag{7.15}$$

where η denotes the learning rate, i.e. the descent rate per unit gradient weight. $\widehat{m}_t = \frac{m_t}{1-\beta_1^i}$ represents the bias-corrected first moment estimate. $\widehat{v}_i = \frac{v_t}{1-\beta_2^i}$ represents the bias-corrected second moment estimate. $m_i = \beta_1 m_{i-1} + (1-\beta_1)\, g_i$ is the first moment estimate (momentum-based) of the gradient at ith iteration. $v_i = \beta_2 v_{i-1} + (1-\beta_2)\, g_i^2$ is the second moment estimate (momentum-based) of the gradient at ith iteration. $g_i = \frac{\partial L_{\text{total}}}{\partial w_{ijk}}$ is the gradient of the output layer loss function with respect to the neuron weights. β_1, β_2, and ϵ are hyperparameters for correcting momentum bias, and most ML algorithm implementations set their default values to 0.9, 0.999, and 10–810–8, respectively.

In summary, as a ML algorithm fundamentally centered on gradient descent theory, ANNs exhibit significant advantages in model performance and prediction accuracy for evaluating mechanical properties of materials when compared with conventional statistical methods. This has led to a substantial body of research leveraging ML approaches to predict material mechanical properties.

7.2.2 CART-based Machine Learning Algorithms

Initially, the decision tree was introduced as a concept of statistics, and it was applied in aspects of decision-making [12]. While the ML algorithm was rapidly developing in recent decades, the idea of the decision tree was borrowed to conduct regression and classification analysis. In 1984, Breiman proposed a new supervised learning algorithm, which formed the basic concept of CART [13]. Considering the application present of the ML algorithm, despite the wide application of the NN-based algorithm, the CART-based algorithms are another popular choice for data analysis. The categories of the CART-based algorithms can be identified based on the number of applied decision trees, which include single-tree algorithms and ensemble CART algorithms.

Many data-analysis-oriented studies have proved that single-tree algorithms and ensemble CART algorithms can generate predictions with higher accuracy and robustness [14]. However, many comparative studies between different CART-based algorithms also proved that ensemble algorithms of CART can have superior performance over the single-tree algorithms. Furthermore, with even more rapid development of the ensemble algorithms of CART, XGBoost, and RF were widely applied in many aspects of material mechanics, because of the satisfactory accuracy and the convenience of the deployment [14, 15]. Therefore, in this section, the basic idea of the CART and the ensemble CART-based algorithm will be introduced.

7.2.2.1 Classification and Regression Tree

Regarding the general structures of the CART, the binary tree is the unit structure of the CART-based algorithms. Thus, borrowing the basic concept of the binary tree in statistics, the node selection and splitting of the feature are two significant parts of generating the CART. Consequently, the leaf nodes of the CART are the output of each split, which forms the prediction of the CART. In this case, the curtail processes of the CART follow the listed principle:

1. The input feature of a sample that would be determined.
2. The splitting point of the input feature should be determined.

When selecting an input feature for splitting, the decision-making process varies from the standard CART approach depending on the task. In classification tasks, the splitting feature at a node is chosen by examining each feature in the dataset and identifying the smallest subset of samples that can still be split without decreasing the purity of the resulting groups. The split point for that feature is then determined by computing the impurity of this smallest subset. In this study, Gini impurity is used as the metric for measuring the impurity of the subset, as expressed in Eq. (7.16) [16].

$$\text{Gini impurity} = 1 - \sum_{i=1}^{C} p_i^2 \tag{7.16}$$

where C is the number of classes or categories; p_i is the proportion of samples in the split group that belong to class "i."

Meanwhile, because of the application of the information theory in the CART-based algorithm, the concept of the information entropy (Eq. (7.17)) was applied to evaluate the impurity of the smallest split group [16].

$$\text{Entropy} = \sum_{i=1}^{C} p_i \log_2 p_i \tag{7.17}$$

A lower Gini impurity or entropy value signifies a cleaner split, where the majority of samples belong to the target class [16]. In regression tasks, however, the goal shifts toward minimizing the loss function of the predicted values, making the selection criterion for a split point the minimization of this loss.

For regression in CART, the feature's optimization domain is continuous, and the model seeks to identify the split that achieves the smallest loss across these continuous ranges. The primary objective is to reduce prediction error and enhance the accuracy of regression results, with MSE commonly used as the loss metric [16]. The splitting process is applied recursively until either the tree reaches its maximum predefined depth or the number of samples in a node drops below the minimum threshold.

In summary, CART serves as a versatile ML algorithm capable of handling both regression and classification tasks. When multiple CART models are built independently and their outputs aggregated, an ensemble CART-based method is formed, which can significantly enhance prediction accuracy compared to a single tree.

7.2.2.2 Ensemble CART-based Algorithm

Owing to the independence and scalability of CART, numerous ensemble techniques have been developed based on it, with prominent examples including Random Forest (RF), LightGBM (LGBM), and XGBoost. Among these, XGBoost and RF have been extensively applied to both regression and classification tasks in ML-assisted studies of material mechanics [17–19]. Consequently, the following sections will focus on discussing the unique characteristics of RF as a representative CART-based ensemble algorithm.

Random Forest

Following Leo Breiman's establishment of the CART framework in 1984, the concept of Bootstrap Aggregating (Bagging) was integrated into CART, leading to the creation of RF as an ML algorithm. In 2001, Breiman formally introduced RF, where multiple CART models are trained and combined to enhance both the accuracy and robustness of predictions [15].

RF was initially developed as a classification tool, mapping feature combinations to specific classes. A key aspect of its training involves selecting a subset of "m" features from the total available in the dataset. The method for forming these subsets plays a central role in the RF training process [15]. In each iteration, a random subset of features is drawn, and a CART is built using only those features. Simultaneously, bagging is applied to generate training datasets: bootstrap samples are created by sampling with replacement from the original data, and each CART is trained on a different bootstrap sample. While these samples may overlap, they are never identical, ensuring greater independence between the trees. Once the predefined number of CART models has been trained, RF makes its final prediction by aggregating the outputs, typically through majority voting for classification tasks [15].

Moreover, RF can compute feature importance scores, which quantify how much each feature contributes to the model's performance. These scores are derived by evaluating the feature's role in reducing impurity at decision nodes throughout the forest. The formula for calculating feature importance is provided below [15]:

$$\mathrm{FI}\left(X_j\right) = \frac{\sum_{t=1}^{T} \text{Impurity decrease}\left(X_j, t\right)}{T} \tag{7.18}$$

where T is a set number of established CARTs; Impurity decrease $\left(X_j, t\right)$ is the decrease in impurity by splitting on the feature X_j in the tth CART, which can be calculated by the Gini impurity or entropy.

Although feature importance does not directly quantify correlations between features, it can highlight the most informative variables and provide indirect insights into potential interrelationships.

For regression tasks, the RF training process parallels that of classification, with splitting and prediction as the central operations. As in CART, the split point for a chosen feature is determined by minimizing a loss function, which serves as the splitting criterion in RF's

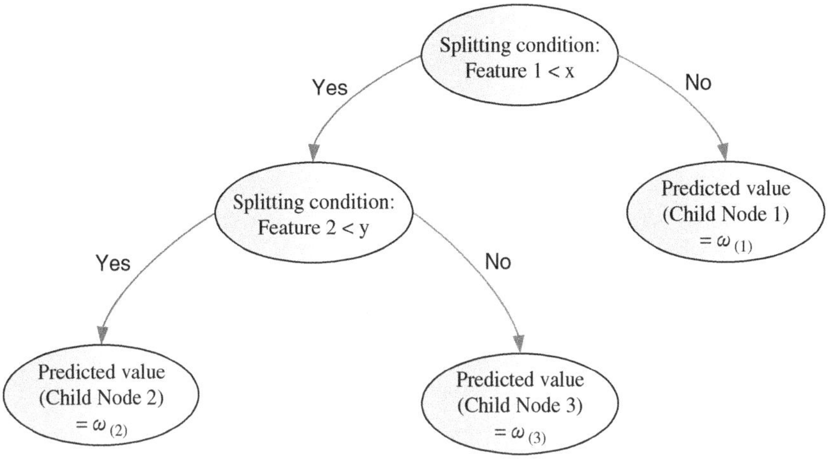

$$\text{Prediction made by single CART } (f_i) = \sum_{q=1}^{q=n} \omega_q$$

$$\text{Prediction made by XGBoost } (F_j) = \sum_{j=1}^{j=n} \eta * f_i$$

Figure 7.4 One of the CART in the Ensembled XGBoost Forest.

tree construction. The final regression output of RF is obtained by averaging the predictions from all constituent CART models. Owing to its versatility in both classification and regression, RF has been widely adopted in ML-assisted studies of material mechanics as a preferred choice for classification and regression analysis.

Building on the core principles of RF, additional ensemble CART-based algorithms have been developed, offering improved accuracy and robustness. Notable examples include XGBoost and Light Gradient Boosting Machine (LightGBM). The process of CART construction in XGBoost is illustrated in Figure 7.4.

7.3 Current State of Machine Learning in Research of Constative Models

In both material development and engineering applications, it is a prevalent methodology that employs fitting formulas derived from experimental data to establish relationships between physicochemical properties and macroscopic performance [20, 21]. Concurrently, within multi-field coupling constitutive frameworks, the increasing complexity of associated field parameters leads to the frequent adoption of fitted or semi-empirical parameters in constitutive theories to represent the role of material characteristics [22]. Consequently, linear programming and fitting techniques grounded in statistics and operations research are commonly applied for post-experimental data processing and pattern extraction in material property characterization [23, 24].

As an emerging data optimization technology, ML shares the same fundamental mathematical concepts and statistical evaluation parameters as conventional fitting approaches. However, with advancements in ML algorithms enhancing numerical prediction accuracy and inductive capacity for nonlinear patterns [3], traditional statistics-based fitting and prediction methods have gradually lost their edge in prediction precision. Furthermore, driven by the engineering and material development sectors' escalating demand for higher-precision data processing methodologies, ML has emerged as one of the key tools for performing numerical prediction tasks.

7.3.1 Data-driven Research of the Material Constitutive Models

The prediction of material mechanical properties and the induction of material mechanical laws constitute one of the primary applications of ML in mechanics research. While ML models have been widely adopted for composite material design and elucidating macro-micro constitutive relationships, continuous advancements in AI algorithms are simultaneously unveiling the potential of newer ML algorithms within the mechanics domain, as explored by an increasing number of relevant studies.

Driven by considerations of model accuracy and prediction correlation, comparative studies evaluating different ML algorithms for material property prediction are being extensively conducted [2, 18, 25]. These studies aim to explore the application potential of various ML algorithms in mechanics research.

Among these algorithms, ANNs [26, 27], support vector machines (SVMs) [28, 29], and decision trees (DTs) [30] represent three distinct ML approaches based on fundamentally different computational principles. Each exhibits unique strengths for different mechanical pattern recognition tasks. Consequently, they are frequently employed as control groups in ML usability studies to benchmark and validate the applicability of novel ML algorithms to specific mechanical problems.

With the ongoing evolution of computational science, newer ML models demonstrating high predictive accuracy are continually being developed and applied to relevant mechanics problems. Algorithms such as XGBoost [7], Light Gradient Boosting Machine (LGBM) [31], and Adaptive Boosting (AdaBoost) [32, 33] have been extensively introduced into mechanics research to address tasks including classification and numerical prediction.

Statistical analysis of comparative studies, examining both the ML models used and the specific mechanical problems solved, reveals that XGBoost—a model based on CART—holds a distinct advantage over other models in terms of accuracy for both numerical prediction and classification tasks [14]. Table 7.2 summarizes a series of these comparative studies focused on ML model evaluation and selection.

In classification and numerical prediction tasks within mechanics research, while the accuracy of various ML models can vary depending on factors such as data quality and model architecture, XGBoost and LGBM are widely employed for numerical prediction of material physicochemical properties due to their advantages in usability and predictive accuracy [7, 8].

Concurrently, it is noteworthy that when addressing mechanics problems involving complex computational demands, deep neural networks (DNNs)—built upon ANN principles and serving as a prevalent ML model—demonstrate considerable utility. This stems not

Table 7.2 Summary of ML research on material properties analysis.

Research topics	Applied algorithms	Reference
Bond strength prediction of concrete-encased steel structures	GA-ANN (genetic algorithm), PSO-ANN (particle swarm optimization)	Wang et al. [27]
Compressive strength of concrete	ANN, SVM, Boosted Tree, RF	Young et al. [28]
Compressive strength of cement-based mortar materials	ANN, Adaptive Neuro Fuzzy Inference System (ANFIS)	Armaghani and Asteris [26]
Compressive strength of geopolymer composites	ANN, SVM, AdaBoost, RF	Amin et al. [32]
Service Life prediction of metallic components under fatigue and creep-fatigue conditions	ANN, SVM, RF, Gaussian process regression	Zhang et al. [29]
Failure mode classification of reinforced concrete columns	ANN, SVM, AdaBoost, CART, RF	Feng et al. [33]
Creep behavior of concrete	RF, XGBoost, Light Gradient Boosting Machine, LGBM	Liang et al. [8]
Compressive strength prediction of ultra-high-performance concrete (UHPC) incorporating solid waste	DT, ANN, SVM, RF, MLR	Farooq et al. [30]
Influence of carbon nanotubes (CNTs) on compressive strength of cement-based materials	LR, SVM, RF, XGBoost	Li et al. [7]

only from the significant influence different intermediate layers exert on task performance but also from their inherent adaptability [11, 34–36].

7.3.2 Machine-learning Based Prediction of Life Cycle Material Properties

In the context of increasingly sophisticated constitutive research, engineering practice, and maintenance, the significance of material mechanical performance data particularly that related to loading history during both the manufacturing and service life of materials and components—is becoming ever more pronounced within material mechanics research. Examples include the creep performance of materials and components, and the monitoring of material properties throughout their entire lifecycle [34, 37, 38].

Such lifecycle monitoring data and loading-history-related data constitute typical time-series data characterized by:

1. Large Volume: Due to the high acquisition frequency of mechanical sensors, time-series data in mechanical studies is typically massive.
2. Strong Temporal Correlation: The loading history in mechanical studies often exhibits significant periodicity (e.g. cyclic loading, temperature hold times), resulting in a strong correlation between the data and time or loading sequence.

Firstly, ML algorithms for analyzing this time-series data must be capable of leveraging its structural features to capture correlations between parameters and identify mechanical features linked to the loading history. Fueled by rapid advancements in natural language processing (NLP) and time-series prediction algorithms [39], mechanical time-series data, sharing commonalities with general time-series data, holds potential for processing by these specialized algorithms, with relevant research already demonstrating progress.

Taking material hysteresis loop data as an example—a quintessential representation of material mechanical behavior tied to component loading history—time-series-aware ML models exhibit unique advantages in capturing the characteristics of hysteresis loops throughout the loading sequence [40]. Algorithms employed for such studies include long short-term memory networks (LSTM) [5] and cyclical generative adversarial networks (cGAN) [37]. DFMs, conceptualized as algorithm frameworks built upon ANN principles, can integrate ANNs with distinct functionalities in series to analyze complex data. This approach has already been applied to the analysis of material hysteresis behavior [6].

Secondly, engineering practice frequently encounters changes in mechanical properties related to loading history during the fabrication and maintenance of materials and components. These changes can significantly impact the performance of the final products. Furthermore, alterations in macro- and micro-mechanical properties can introduce damage, detrimentally affecting the evolution of material mechanical performance [41]. Therefore, monitoring parameters during this evolution enables the application of targeted maintenance strategies to repair or mitigate the loss of mechanical properties, ultimately improving the yield rate of manufactured materials [42, 43]. Related research involves integrating advanced sensors during composite material fabrication to collect macro- and micro-scale parameters, subsequently modeling and predicting the evolution of composite strength [44].

Consequently, supported by mature experimental datasets, the aforementioned ML models can effectively capture the characteristics of time-series data in mechanics research and generate time-dependent mechanical behavior rules for materials under specific loading and environmental conditions. Utilizing these models enables efficient prediction of material fatigue behavior, substantially reducing the need for extensive physical testing in fatigue performance studies. This offers significant potential for minimizing both experimental and temporal costs in material research.

7.3.3 Physic-informed Research of the Solid Material

Leveraging the ML models discussed previously for applications in material mechanics, these models demonstrate the capability to induce material mechanical laws based on experimental data and computational mechanics results. Once the usability of an ML model for a specific mechanical problem is established through model evaluation and selection studies, more profound applications emerge for tackling complex mechanical challenges. A key example is solving partial differential equations (PDEs) governing fluid dynamics and material behavior.

Solving PDEs constitutes a core methodology in constitutive research and the motion analysis of materials and components. This intersection of ML and material mechanics has spurred the development of physics-informed neural networks (PINNs). PINNs operate within the framework of known constitutive and motion PDEs to perform ML-based prediction and optimization of material properties. This enables material optimization guided more strictly by the fundamental laws of material mechanics.

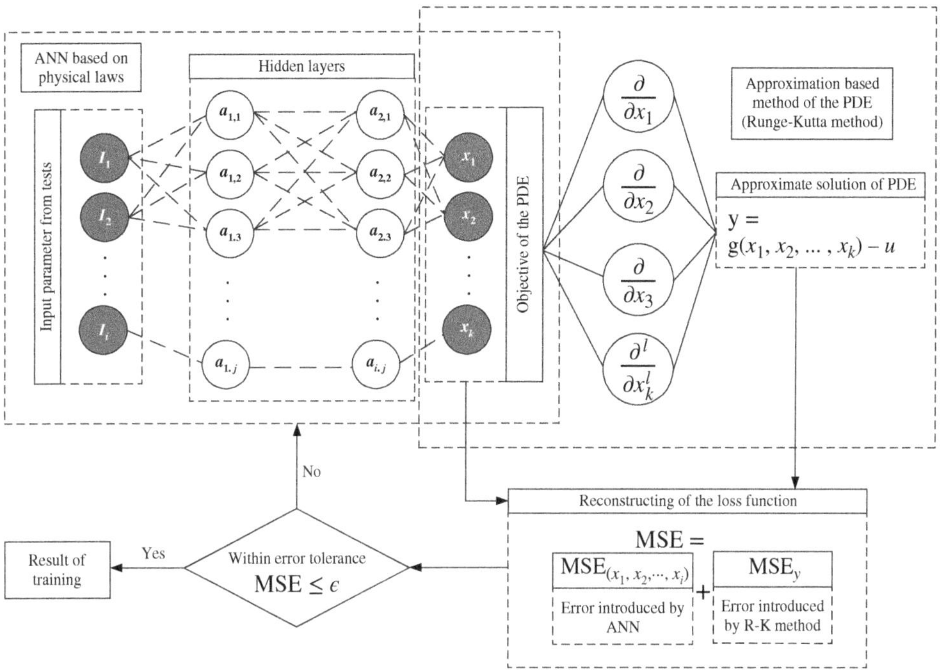

Figure 7.5 The basic idea of PINN for approximating solutions to partial differential equations.

A seminal study by Raissi, M. (2019) focused on solving the Navier-Stokes equations (N-S equations) for nonlinear fluid motion. This work employed a Deep Learning Framework (DFM) combined with Runge–Kutta methods. Utilizing an approximation approach, the study constructed a custom loss function within an ANN algorithm. This methodology achieved more accurate predictions of mechanical laws under the constraints of physical principles, as illustrated in Figure 7.5 [11].

Related concepts and methods have also been applied to optimize the prediction of material workability. For instance, Tianjie Zhang et al. (2023) utilized the PINN approach to investigate the rheological properties of cement-based materials. They validated the usability of their specialized ANN architecture, "RheologyNet," in mechanics research by benchmarking its predictions against finite element models [45].

7.3.4 Practices to Improve the Applicability of Machine Learning Models

As AI has emerged as a widely discussed scientific optimization methodology within both academia and industry, its rapid development in recent years has spurred the convergence of AI algorithms—particularly ML—with various traditional disciplines. With the deepening of such interdisciplinary research, extensive applications of ML models have identified a subset generally well-suited for mechanics research. Examples include CNNs [10, 25] based on ANNs, and XGBoost [7], rooted in CART.

While the application outcomes of these models in solving mechanics problems are broadly accepted, their use in material design and mechanics problem-solving has also revealed certain limitations within the mechanics research context:

1. High Sensitivity to Data: Optimization and prediction results are highly sensitive to data volume, quality, and deficiencies (e.g. missing data).
2. Limited Interpretability: ML models exhibit poor interpretability regarding the underlying mechanical laws they capture.

Consequently, addressing these limitations and enhancing the applicability of ML models in mechanics research remains crucial for improving their future prospects in this field. Related research focuses on mitigating these issues by employing other algorithms or computational mechanics methods to interpolate mechanical test data. This aims to improve the ML models' ability to induce mechanical laws during application.

For instance, Gideon A. Lyngdoh et al. (2022) utilized the K-nearest neighbors (KNN) algorithm to interpolate experimental data. Submitting this processed data to an XGBoost model for training yielded significantly improved prediction results [18]. Similarly, Xingyue Sun et al. (2022) employed a cyclical generative adversarial network (cGAN) approach for data augmentation of experimental data. Utilizing this augmented data during the ML model training phase substantially enhanced the prediction accuracy across various models [37].

In summary, ML algorithms, particularly those focused on numerical prediction and pattern recognition, offer novel approaches for contemporary constitutive research. This stems from their superior predictive accuracy compared to traditional statistical methods. Furthermore, their strong inductive capability for uncovering underlying laws provides unique advantages in new material development. However, due to the stringent requirements ML algorithms impose on both the quality and quantity of experimental data, coupled with their inherent weakness in explicitly describing material mechanical laws, most scholars maintain a cautiously optimistic view regarding their application prospects within constitutive studies.

7.4 Example: Predicting Constitutive Parameters of Concrete

Subsequently, this section will present a numerical example where an Artificial Neural Network (ANN) predicts various mechanical property parameters of concrete. This serves to illustrate the fundamental procedures and rationale involved in establishing an ML model. Concurrently, the feasibility of ML models for numerical prediction tasks will be validated by contrasting the ANN's results with those obtained from statistical models. The ML model for this case study will be developed using Python programming language and its associated ML libraries, with core algorithm implementation relying on Keras and Scikit-learn (Sklearn).

As ML model efficacy exhibits significant data dependency, establishing a reliable dataset represents the foundational requirement for valid analytical outcomes. Dataset construction necessitates adherence to three core principles:

1. Population Robustness: Ensure statistically sufficient sampling to minimize overfitting susceptibility inherent in sparse datasets.

2. Orthogonality Integrity: Curate non-redundant experimental data with verified orthogonality and physical significance.
3. Feature Interdependence: Maintain correlated feature spaces to preserve model expressiveness regarding physical phenomena.

Subsequent data preprocessing–encompassing cleaning, integration, and transformation–optimizes model training efficiency. Within physics-informed ML (PIML) paradigms, where empirical data incorporates physical constraints, data cleaning and transformation emerge as particularly vital preprocessing operations for feature-space regularization.

Data from material physicochemical and mechanical tests inevitably contain experimental and environmental errors, leading to significant deviations from theoretical or true values. Consequently, data cleaning rectifies anomalies by imputing missing values, flagging outliers, and eliminating noisy data. This process enhances model accuracy in learning underlying physical laws. As a data processing technique for quality improvement, data cleaning plays a vital role in leveraging ML models' strengths for inducing nonlinear physical relationships. Thus, developing physically consistent data imputation and denoising methodologies constitutes a critical research direction in mechanics-AI interdisciplinary studies [18].

Secondly, data transformation—another essential preprocessing technique—converts raw samples into machine-learning-appropriate formats through standardization. Data normalization, a prevalent transformation method, scales sample features according to their statistical characteristics. This accelerates convergence during gradient descent optimization, thereby improving computational efficiency and indirectly enhancing model accuracy. Common normalization techniques include max-min scaling and Z-score standardization.

$$x_{i-\text{normalized}} = \frac{x_i - x_{\min}}{x_{\max} - x_{\min}} \tag{7.19}$$

$$x_{\text{std}} = \frac{x_i - \mu}{\sigma_{\text{std}}} \tag{7.20}$$

where x_i denotes the raw sample data point; $x_{i-\text{normalized}}$ represents the normalized value corresponding to x_i; $x_{\min}$ signifies the minimum value of the feature across the dataset; $x_{\max}$ indicates the maximum value of the feature across the dataset; x_{std} is the standardized score of the sample; μ refers to the mean value of the feature in the dataset; σ_{std} corresponds to the standard deviation of the feature in the dataset.

Standard practice in ML model development involves partitioning the available dataset into three mutually exclusive subsets:

1. Training Set: Directly used to fit the model's learnable parameters (weights).
2. Validation Set: Reserved for iterative model evaluation during training to guide hyperparameter optimization and mitigate overfitting.
3. Testing Set: Used exclusively for the final, unbiased assessment of the model's generalization performance to unseen data. Under no circumstances should the test set inform feature engineering, parameter tuning, or model selection.

To maximize the effective use of limited data and obtain more robust performance estimates, cross-validation is commonly implemented. This methodology systematically rotates data through the training/validation roles. The configuration of data splits relative to training epochs is illustrated schematically in Figure 7.6 prior to model training and evaluation.

Consistent with the ML algorithm data prerequisites detailed in the preceding sections, this case study utilizes experimental data published by Ehsan Ghafari et al. (2015) [23]. Their study examined the compressive strength of the steel fiber-reinforced ultra-high-performance concrete (UHPC).

According to the mathematical principles governing ANN training, the efficiency and performance of ANNs are primarily determined by two factors:

1. The network architecture comprising the neural network during training.
2. The selection of functions and parameters constituting this architecture.

Key functions within the network—such as the loss function, activation function, and optimizer—can be determined based on the network's intended objectives. Furthermore, based on the mathematical principles of ANN-based predictive fitting, the number of training iterations and the learning rate are parameters directly influencing model performance. Consequently, by defining the correlation between the trained model and the data as the performance metric, optimal parameters can be derived through orthogonal optimization.

The network architecture itself is also a critical factor affecting ANN performance. Following the same optimization logic, architectural hyperparameters should likewise be optimizable via orthogonal methods to yield the most suitable configuration for the material system. Specifically, the number of hidden layers and the number of neurons per hidden layer—central components of ANN architecture—can be optimized through orthogonal experimentation to identify the optimal structure for the prediction task.

As a reference for ANN development, parametric optimization studies from related mechanics research can be leveraged. For instance, Ghafari and Bandarabadi employed parametric variation analysis in their ANN study on concrete compressive strength, identifying an optimal network structure for this specific prediction task [2].

This case study employs orthogonal optimization to determine the ANN architecture. The primary parameters optimized are:

1. The number of hidden layers
2. The number of neurons per hidden layer

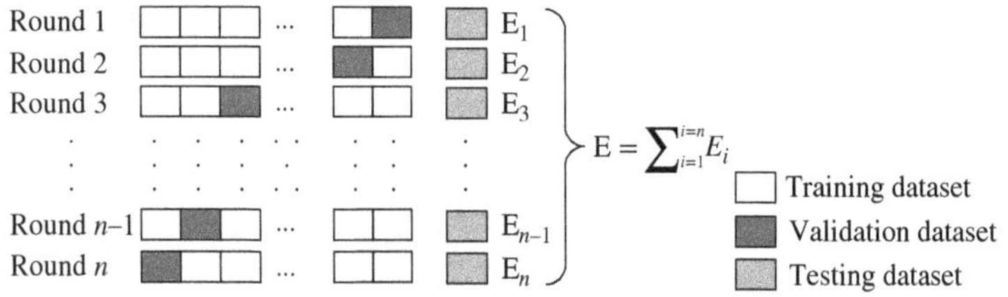

Figure 7.6 The relationship between the cross-validation dataset and the number of folds.

The optimization procedure is as follows:

1. Optimize Hidden Layer Count: Evaluate sub-models with 2, 5, 10, and 15 hidden layers. Identify the layer count yielding the highest model-data correlation.
2. Optimize Neuron Count: Using the optimal layer count from Step 1, evaluate sub-models with 3, 5, 10, and 15 neurons per layer.
3. Final Selection: Select the sub-model (layer count + neuron count) demonstrating the highest correlation coefficient as the final ANN architecture.

Correlation coefficients for all sub-models were obtained by averaging results from 10 independent training trials. The orthogonal optimization process is detailed in Table 7.3.

This case study employs the coefficient of determination (R^2) to quantify the relationship between the model predictions and the observed data. The coefficient of determination (R^2), which measures the goodness-of-fit between model predictions and true data values, ranges between 0 and 1. Values approaching 1 indicate a near-perfect positive linear relationship, while values near 0 signify the absence of a linear correlation.

$$r^2 = \frac{\text{ESS}}{\text{TSS}} = 1 - \frac{\text{RSS}}{\text{TSS}} = \frac{\sum_{i=1}^{n} (\hat{y}_i - \bar{y})^2}{\sum_{i=1}^{n} (y_i - \bar{y})^2} = 1 - \frac{\sum_{i=1}^{n} (y_i - \hat{y}_i)^2}{\sum_{i=1}^{n} (y_i - \bar{y})^2} \tag{7.21}$$

where $\hat{y}_i$ denotes the predicted output from the fitted model; y_i represents the true observed value corresponding to the input data; $\bar{y}$ signifies the mean of the true observed values; ESS (Explained Sum of Squares) quantifies the variation explained by the model; TSS (Total Sum of Squares) measures the total variation in the data; RSS (Residual Sum of Squares) captures the unexplained variation.

Both the coefficient of determination (r^2) and RSS serve as metrics for evaluating model performance. A lower r^2 value corresponds to a higher RSS, indicating poorer model fit. Consequently, an r^2 value approaching 1 signifies optimal predictive performance.

The parametric analysis reveals that, under identical training iterations and optimizer settings, sub-models with fewer hidden layers (two or five) and neurons (three or five) can regress nonlinear relationships within the data. However, they exhibit inferior performance compared to models with more layers (10 or 15) and neurons (10 or 15) regarding prediction correlation, training stability across multiple runs, and numerical prediction accuracy. Notably, the 15-layer sub-model demonstrated reduced prediction correlation relative to the 10-layer variant. This performance degradation stems from increased model complexity

Table 7.3 Orthogonal optimization parameters for the ANN model network structure ("•" indicates the network structure of the sub-model; "*" indicates the network structure of the output model).

Optimized parameters		No. of hidden layers			
		2	5	10	15
No. of neuron	3			•	
	5			•	
	10			•	
	15	•	•	*	•

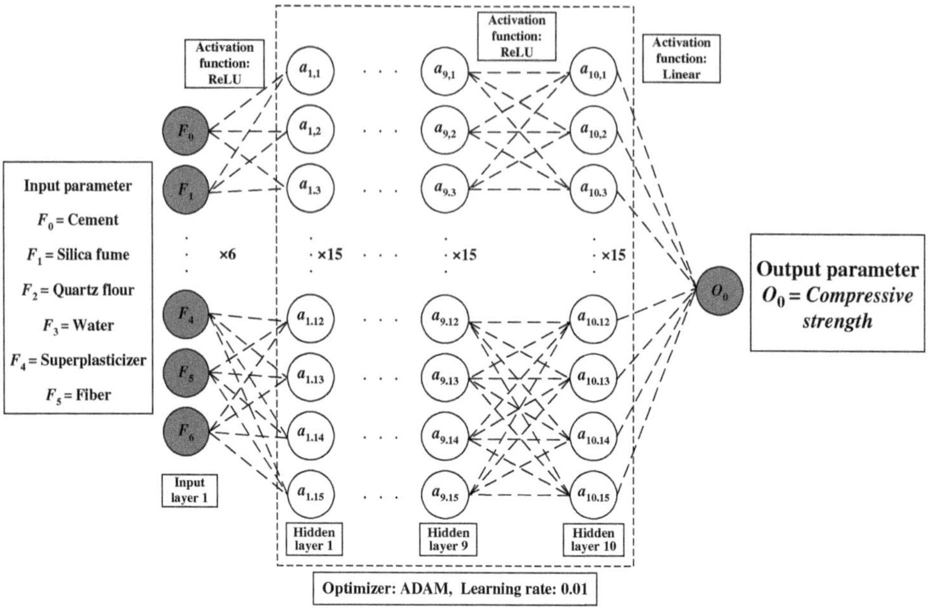

Figure 7.7 Artificial neural network architecture and parameters.

associated with excessive layers, which heightens sensitivity to training data idiosyncrasies. Consequently, the ANN developed overfitting—increasing model complexity while simultaneously diminishing predictive accuracy.

Synthesizing the optimization process, the case study yields the optimal ANN architecture and parameters depicted in Figure 7.7: a numerical prediction network featuring 10 hidden layers with 15 neurons per layer.

The loss function for the numerical prediction neural network in this case study is MSE, and the optimizer is adaptive moment estimation (Adam) with a learning rate of 0.01.

Comparing the predictive outcomes of the ANN model and the statistical model for concrete mechanical properties—evaluated through prediction correlation and stability—yields the following conclusions:

1. Superior Correlation: The ANN demonstrates significantly higher prediction correlation with actual data than the statistical method.
2. Enhanced Result Concentration: While both methods yield predictions within a 10% error margin, the coefficient of determination (R^2) accounts for the dispersion of predictions relative to actual values. An R^2 of 0.84 approaches 1 for the trained ANN, which indicates more concentrated (less scattered) predictions.

Predictive results and correlation analyses for both numerical models are presented in Figure 7.8.

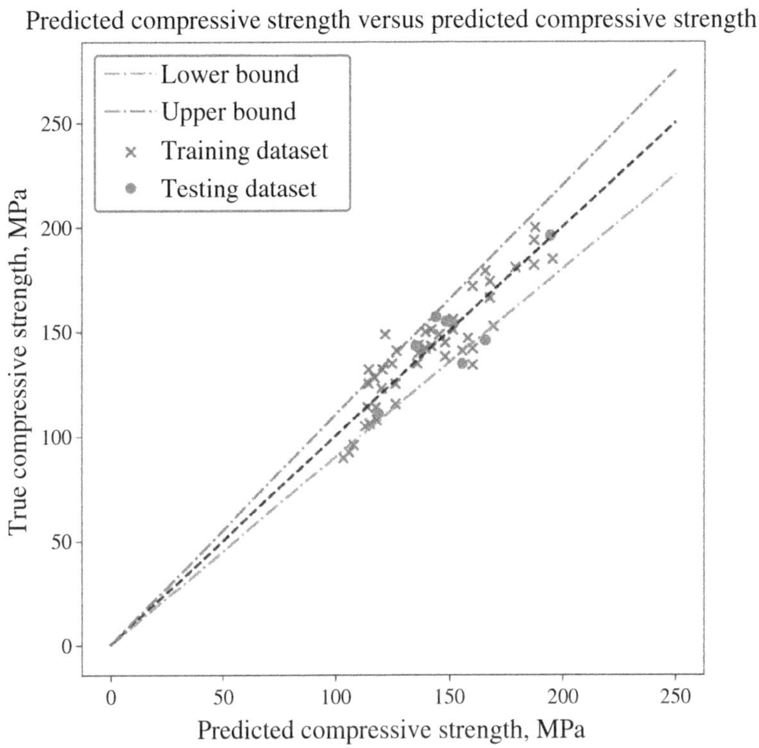

Figure 7.8 Prediction result trained ANN model.

7.5 Conclusion

In summary, AI, as a core technology driving the ongoing industrial revolution, is already playing a pivotal role in interdisciplinary research. By integrating AI, traditional industries gain multifaceted empowerment, enabling enhanced quality and efficiency. Within these sectors, improving material properties represents a crucial upgrade pathway, where research and development (R&D) and fabrication often center on understanding constitutive relationships.

However, as constitutive research delves deeper, traditional statistical methods increasingly struggle to meet the demands of handling complex parameter systems and achieving high fitting accuracy. Fortunately, AI excels in pattern recognition and optimizing complex nonlinear hyperparameters. By accelerating the iterative process of fitting material mechanical parameters and deriving material mechanical laws, AI enhances both constitutive research and computational efficiency.

Although AI-assisted material performance research currently lacks groundbreaking breakthroughs, advancements in AI algorithms and computational power are reigniting interest among material developers. Consequently, AI is progressively becoming a prominent research methodology in material performance studies and fabrication. ML (ML) methods, in particular, play a vital role in data-driven constitutive research, hybrid constitutive modeling, model selection and evaluation, and adaptive modeling.

Ultimately, the application of ML in constitutive research has exerted a profound influence across various research scenarios. With the continuous advancement of AI, we can anticipate the emergence of further innovation and pioneering breakthroughs in the future of materials science.

References

1 Lu, J.-X., Shen, P., Ali, H.A., and Poon, C.S. (2022). Mix design and performance of lightweight ultra high-performance concrete. *Materials & Design* 216: 110553. https://doi.org/10.1016/j.matdes.2022.110553

2 Ghafari, E., Bandarabadi, M., Costa, H., and Júlio, E. (2015). Prediction of fresh and hardened state properties of UHPC: comparative study of statistical mixture design and an artificial neural network model. *Journal of Materials in Civil Engineering* 27 (11): 04015017. https://doi.org/10.1061/(asce)mt.1943-5533.0001270

3 Gorr, W.L., Nagin, D., and Szczypula, J. (1994). Comparative study of artificial neural network and statistical models for predicting student grade point averages. *International Journal of Forecasting* 10 (1): 17–34.

4 Kingma, D.P. and Ba, J. (2014). Adam: a method for stochastic optimization. *arXiv preprint 1412.6980*. https://doi.org/10.48550/arXiv.1412.6980

5 Bartošák, M. (2022). Using machine learning to predict lifetime under isothermal low-cycle fatigue and thermo-mechanical fatigue loading. *International Journal of Fatigue* 163: 107067. https://doi.org/10.1016/j.ijfatigue.2022.107067

6 Wang, C., Xu, L.-y., and Fan, J.-s. (2020). A general deep learning framework for history-dependent response prediction based on UA-Seq2Seq model. *Computer Methods in Applied Mechanics and Engineering* 372: 113357. https://doi.org/10.1016/j.cma.2020.113357

7 Li, Y., Li, H., jin, C., and Shen, J. (2022). The study of effect of carbon nanotubes on the compressive strength of cement-based materials based on machine learning. *Construction and Building Materials* 358: 129435. https://doi.org/10.1016/j.conbuildmat.2022.129435

8 Liang, M., Chang, Z., Wan, Z. et al. (2022). Interpretable Ensemble-Machine-Learning models for predicting creep behavior of concrete. *Cement and Concrete Composites* 125: 104295. https://doi.org/10.1016/j.cemconcomp.2021.104295

9 Jain, A.K., Mao, J., and Mohiuddin, K.M. (1996). Artificial neural networks: a tutorial. *Computer* 29 (3): 31–44. https://doi.org/10.1109/2.485891

10 Gao, Y., Berger, M., and Duddu, R. (2023). CNN-based surrogate for the phase field damage model: generalization across microstructure parameters for composite materials. *Journal of Engineering Mechanics* 149 (6): 04023025. https://doi.org/doi:10.1061/JENMDT.EMENG-6936

11 Raissi, M., Perdikaris, P., and Karniadakis, G.E. (2019). Physics-informed neural networks: a deep learning framework for solving forward and inverse problems involving nonlinear partial differential equations. *Journal of Computational Physics* 378: 686–707. https://doi.org/10.1016/j.jcp.2018.10.045

12 Hastie, T., Tibshirani, R., and Friedman, J.H. (2009). *The Elements of Statistical Learning: Data Mining, Inference, and Prediction*. New York, NY: Springer. https://doi.org/10.1007/978-0-387-84858-7

13 Breiman, L., Friedman, J., Olshen, R., and Stone, C. (1984). *Classification and Regression Trees*, 1e. New York, NY: Taylor & Francis Group. https://doi.org/10.1201/9781315139470

14 Chen, T. and Guestrin, C. (2016). XGBoost: a scalable tree boosting system. *Proceedings of the 22nd ACM SIGKDD International Conference on Knowledge Discovery and Data Mining*, pp. 785–794. San Francisco, CA, USA. https://doi.org/10.1145/2939672.2939785

15 Breiman, L. (2001). Random forests. *Machine Learning* 45: 5–32. https://doi.org/10.1023/A:1010933404324

16 Grajski, K.A., Breiman, L., Di Prisco, G.V., and Freeman, W. (1986). Classification of EEG spatial patterns with a tree-structured methodology: CART. *IEEE Transactions on Biomedical Engineering* 33 (12): 1076–1086. https://doi.org/10.1109/TBME.1986.325684

17 Zhang, F., Wang, C., Liu, J. et al. (2023). Prediction of FRP-concrete interfacial bond strength based on machine learning. *Engineering Structures* 274: 115156. https://doi.org/10.1016/j.engstruct.2022.115156

18 Lyngdoh, G.A., Zaki, M., Krishnan, N.M.A., and Das, S. (2022). Prediction of concrete strengths enabled by missing data imputation and interpretable machine learning. *Cement and Concrete Composites* 128: 104414. https://doi.org/10.1016/j.cemconcomp.2022.104414

19 DeRousseau, M.A., Laftchiev, E., Kasprzyk, J.R. et al. (2019). A comparison of machine learning methods for predicting the compressive strength of field-placed concrete. *Construction and Building Materials* 228: 116661. https://doi.org/10.1016/j.conbuildmat.2019.08.042

20 Tyagi, L., Butola, R., Kem, L., and Singari, R.M. (2021). Comparative analysis of response surface methodology and artificial neural network on the wear properties of surface composite fabricated by friction stir processing. *Journal of Bio- and Tribo-Corrosion* 7 (2): 36. https://doi.org/10.1007/s40735-020-00469-1

21 Sun, X., Liu, Z., Wang, X., and Chen, X. (2022). Determination of ductile fracture properties of 16MND5 steels under varying constraint levels using machine learning methods. *International Journal of Mechanical Sciences* 224: 107331. https://doi.org/10.1016/j.ijmecsci.2022.107331

22 Yao, Y., Fang, H., and Guo, H. (2022). Unified damage constitutive model for fiber-reinforced concrete at high temperature. *Journal of Engineering Mechanics* 148 (1): 04021132. https://doi.org/10.1061/(asce)em.1943-7889.0002057

23 Ghafari, E., Costa, H., and Júlio, E. (2015). Statistical mixture design approach for eco-efficient UHPC. *Cement and Concrete Composites* 55: 17–25. https://doi.org/10.1016/j.cemconcomp.2014.07.016

24 Zhang, Y., Sun, X., Zhu, X., and Xiao, J. (2022). Multi-criteria optimization of concrete mixes incorporating cenosphere waste and multi-minerals. *Journal of Cleaner Production* 367: 133102. https://doi.org/10.1016/j.jclepro.2022.133102

25 Sun, X., Zhou, T., Song, K., and Chen, X. (2023). An image recognition based multiaxial low-cycle fatigue life prediction method with CNN model. *International Journal of Fatigue* 167: 107324. https://doi.org/10.1016/j.ijfatigue.2022.107324

26 Armaghani, D.J. and Asteris, P.G. (2020). A comparative study of ANN and ANFIS models for the prediction of cement-based mortar materials compressive strength. *Neural Computing and Applications* 33 (9): 4501–4532. https://doi.org/10.1007/s00521-020-05244-4

27 Wang, X., Liu, Y., and Xin, H. (2021). Bond strength prediction of concrete-encased steel structures using hybrid machine learning method. *Structures* 32: 2279–2292. https://doi.org/10.1016/j.istruc.2021.04.018

28 Young, B.A., Hall, A., Pilon, L. et al. (2019). Can the compressive strength of concrete be estimated from knowledge of the mixture proportions?: New insights from statistical analysis and machine learning methods. *Cement and Concrete Research* 115: 379–388. https://doi.org/10.1016/j.cemconres.2018.09.006

29 Zhang, X.-C., Gong, J.-G., and Xuan, F.-Z. (2021). A deep learning based life prediction method for components under creep, fatigue and creep-fatigue conditions. *International Journal of Fatigue* 148: 106236. https://doi.org/10.1016/j.ijfatigue.2021.106236

30 Farooq, F., Ahmed, W., Akbar, A. et al. (2021). Predictive modeling for sustainable high-performance concrete from industrial wastes: a comparison and optimization of models using ensemble learners. *Journal of Cleaner Production* 292: 126032. https://doi.org/10.1016/j.jclepro.2021.126032

31 Lashari, N., Ganat, T., Otchere, D. et al. (2021). Navigating viscosity of GO-SiO_2/HPAM composite using response surface methodology and supervised machine learning models. *Journal of Petroleum Science and Engineering* 205: 108800. https://doi.org/10.1016/j.petrol.2021.108800

32 Amin, M.N., Khan, K., Ahmad, W. et al. (2022). Compressive strength estimation of geopolymer composites through novel computational approaches. *Polymers (Basel)* 14 (10): 2128. https://doi.org/10.3390/polym14102128

33 Feng, D.-C., Liu, Z.-T., Wang, X.-D. et al. (2020). Failure mode classification and bearing capacity prediction for reinforced concrete columns based on ensemble machine learning algorithm. *Advanced Engineering Informatics* 45: 101126. https://doi.org/10.1016/j.aei.2020.101126

34 Gao, M.-Y., Zhang, N., Shen, S.-L., and Zhou, A. (2020). Real-time dynamic earth-pressure regulation model for shield tunneling by integrating GRU deep learning method with GA optimization. *IEEE Access* 8: 64310–64323. https://doi.org/10.1109/access.2020.2984515

35 Zhang, X., Wu, X., and Huang, X. (2022). Smart real-time forecast of transient tunnel fires by a dual-agent deep learning model. *Tunnelling and Underground Space Technology* 129: 104631. https://doi.org/10.1016/j.tust.2022.104631

36 Sun, C., Gu, D., and Lu, X. (2023). Three-dimensional structural displacement measurement using monocular vision and deep learning based pose estimation. *Mechanical Systems and Signal Processing* 190: 110141. https://doi.org/10.1016/j.ymssp.2023.110141

37 Sun, X., Zhou, K., Shi, S. et al. (2022). A new cyclical generative adversarial network based data augmentation method for multiaxial fatigue life prediction. *International Journal of Fatigue* 162: 106996. https://doi.org/10.1016/j.ijfatigue.2022.106996

38 Marchand, B., Chamoin, L., and Rey, C. (2016). Real-time updating of structural mechanics models using Kalman filtering, modified constitutive relation error, and proper generalized decomposition. *International Journal for Numerical Methods in Engineering* 107 (9): 786–810. https://doi.org/10.1002/nme.5197

39 Li, N., Liu, S., Liu, Y. et al. (2019). Neural speech synthesis with transformer network. *Proceedings of the AAAI Conference on Artificial Intelligence*, Vol. 33, pp. 6706–6713. https://doi.org/10.1609/aaai.v33i01.33016706

40 Wang, J.-J., Wang, C., Fan, J.-S., and Mo, Y.L. (2022). A deep learning framework for constitutive modeling based on temporal convolutional network. *Journal of Computational Physics* 449: 110784. https://doi.org/10.1016/j.jcp.2021.110784

41 Ghafoori, N., Najimi, M., Sobhani, J., and Aqel, M.A. (2013). Predicting rapid chloride permeability of self-consolidating concrete: a comparative study on statistical and neural network models. *Construction and Building Materials* 44: 381–390. https://doi.org/10.1016/j.conbuildmat.2013.03.039

42 Andrushia, A.D., Anand, N., Neebha, T.M. et al. (2022). Autonomous detection of concrete damage under fire conditions. *Automation in Construction* 140: 104364. https://doi.org/10.1016/j.autcon.2022.104364

43 Li, G., Luo, M., Huang, J., and Li, W. (2023). Early-age concrete strength monitoring using smart aggregate based on electromechanical impedance and machine learning. *Mechanical Systems and Signal Processing* 186: 109865. https://doi.org/10.1016/j.ymssp.2022.109865

44 Almasaeid, H.H., Suleiman, A., Alawneh, R. (2022). Assessment of high-temperature damaged concrete using non-destructive tests and artificial neural network modelling. *Case Studies in Construction Materials* 16: e01080. https://doi.org/10.1016/j.cscm.2022.e01080

45 Zhang, T., Wang, D., and Lu, Y. (2023). RheologyNet: a physics-informed neural network solution to evaluate the thixotropic properties of cementitious materials. *Cement and Concrete Research* 168: 107157. https://doi.org/10.1016/j.cemconres.2023.107157

8

Tensile Creep Failure Mechanism and Theoretical Model of Sintered Nano-silver

8.1 Introduction

Nano-silver materials fabricated through sintering have demonstrated excellent adaptability to the harsh service conditions associated with third-generation semiconductor applications [1], making them promising candidates for high-temperature packaging in advanced electronics such as electric vehicles and aerospace systems. Prior to the sintering process, the material typically consists of silver nanoparticles dispersed in organic solvents [2, 3]. Owing to the high surface energy inherent to nanoscale silver particles, sintering can be achieved at comparatively low temperatures, resulting in the formation of a porous metallic network [2, 4]. During this transformation, organic components are entirely volatilized, leaving behind a conductive solid where the interparticle necks enable electrical signal transmission [3]. Simultaneously, as shown in Figure 8.1, the evaporation of solvents leads to the development of numerous microvoids within the sintered structure. These pores help accommodate internal stress induced by thermal expansion mismatches among packaging components. However, their presence may also impair the electrical and mechanical performance of the packaging, as has been demonstrated in prior studies [5, 6]. To address this, efforts have been directed toward minimizing porosity by refining sintering parameters and processes [7–9]. Despite various advancements, a fully effective strategy to eliminate the detrimental effects of pores in sintered nano-silver remains elusive. While ensuring adequate electrical and thermal performance, the mechanical behavior of sintered nano-silver plays a critical role in defining the overall reliability of electronic packaging systems. Thermal expansion mismatches among packaging materials often induce stress concentrations, leading to typical failure modes such as compressive fracture [10, 11], tensile failure [12–15], and shear failure [2, 16, 17], as shown in Figure 8.2, all of which have garnered significant research attention. In response to these challenges, several theoretical frameworks have been proposed to describe and predict such failure mechanisms [14, 18, 19]. Furthermore, prior investigations have identified creep deformation as a key factor compromising structural integrity. Both tensile and shear creep behaviors of sintered nano-silver have been explored through experimental approaches [4, 20, 21].

Nevertheless, existing creep constitutive models for sintered nano-silver are primarily empirical in nature, often relying on fitting equations that lack a well-established physical foundation. Moreover, environmental temperature significantly affects atomic diffusion

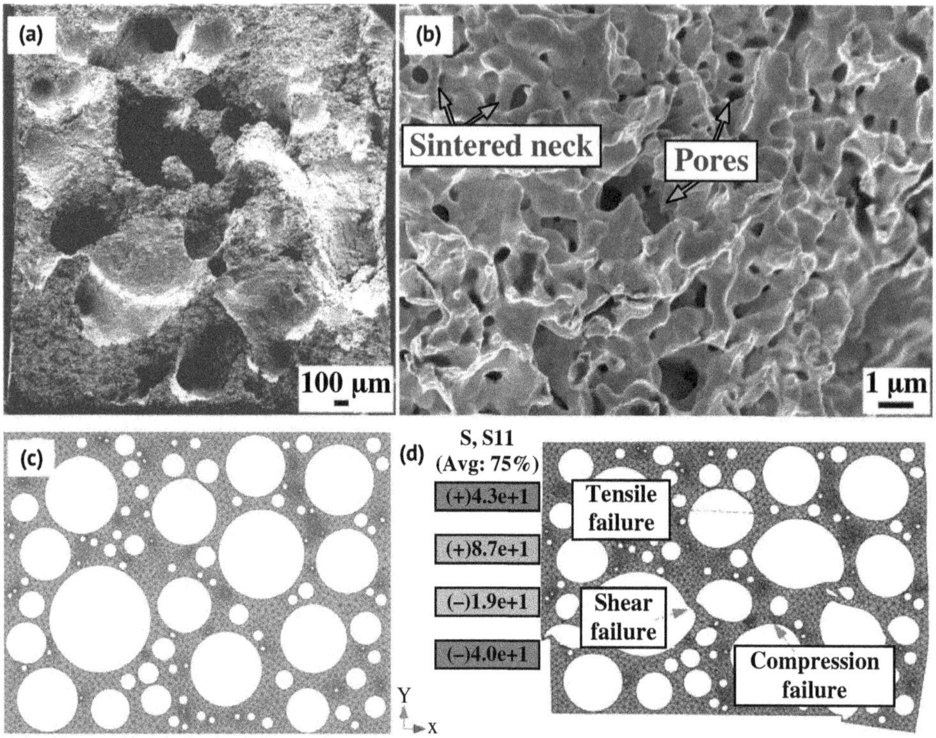

Figure 8.1 Microstructural features and finite element representation of sintered nano-silver: (a) macroscopic fracture surface captured using scanning electron microscopy (SEM); (b) SEM image showing the internal microstructure of sintered nano-silver; (c) finite element mesh constructed to simulate the mechanical behavior of sintered nano-silver; and (d) observed failure pattern under mechanical loading.

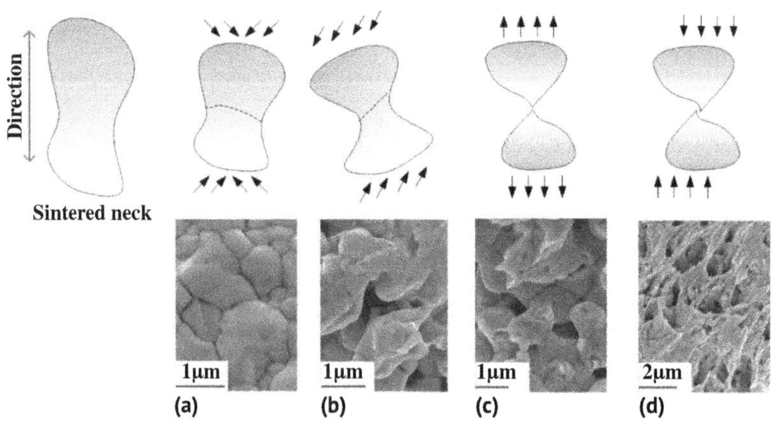

Figure 8.2 Failure morphology of sintered neck: (a) interfacial damage (*Source:* Adapted from [2, 10]); (b) compression-shear failure; (c) tensile failure; and (d) shear failure.

behavior and the stress exponent, both of which are critical parameters in creep deformation [1]. Given its application as a high-temperature interconnect material, the temperature dependence of the mechanical response of sintered nano-silver warrants careful consideration [1, 22]. As a highly promising candidate for power electronics packaging, both experimental validation and theoretical modeling of its creep behavior are essential to ensure structural reliability. Despite ongoing efforts, a physically grounded theoretical framework for modeling the creep of sintered nano-silver remains underdeveloped. To address this, a constitutive creep model incorporating damage mechanics was proposed to describe the deformation and failure processes of sintered nano-silver. The robustness and generality of the model were demonstrated through comparisons with experimental results under compressive, tensile, and shear creep conditions. Additionally, a porosity evolution model was formulated for tensile creep, based on the damage progression observed in the sintered structure. Complementary compression creep tests and theoretical analyses were also carried out to enrich the experimental dataset and further validate the model.

8.2 Molecular Dynamics Model of Sintered Neck

For isotropic damage, the damage variable D can be defined as [23, 24]:

$$D = \frac{S_d}{S_0} \tag{8.1}$$

where S_0 represents the area of the normal cross-section of the volume unit, and S_d represents the defect area on the cross-section. Here, $D = 0$ indicates undamaged material, $D = 1$ indicates complete destruction [24]. Therefore:

$$\sigma_T = \frac{\sigma_N}{1 - D} \tag{8.2}$$

where σ_T is the actual stress on the sintered neck at the microscopic level, and σ_N is the nominal stress in the local area of sintered nano-silver.

Although the nominal stress at the macro level remains constant, due to the positive correlation between damage and creep time [25], the true stress of the sintered neck gradually increases at the micro level according to Eq. (8.1). First, we extract a tiny solder joint structure, as shown in Figure 8.3a. By magnifying the position of the red circle, we can see the necking section as shown in Figure 8.3b. As shown in Figure 8.3c. As damage increases, the sintered neck gradually fractures under the influence of true stress. Therefore, during the creep process, the effective bearing area of the sintered neck decreases after fracture in the brittle region, as shown in Figure 8.3d. Subsequently, the adjacent sintered neck undergoes stress redistribution. It again undergoes the aforementioned fracture process until the sintered silver nanowires experience macroscopic fracture, which is consistent with experimental results (Figure 8.3e and f).

Figure 8.4 presents the dislocation evolution during the fracture process of a sintered neck, as revealed by molecular dynamics simulations. As the necked region approaches its peak stress, the increasing length of Shockley dislocation lines corresponds with a reduction in tensile strength. This correlation suggests that the initiation of damage and subsequent necking are primarily driven by the generation and accumulation of Shockley dislocations.

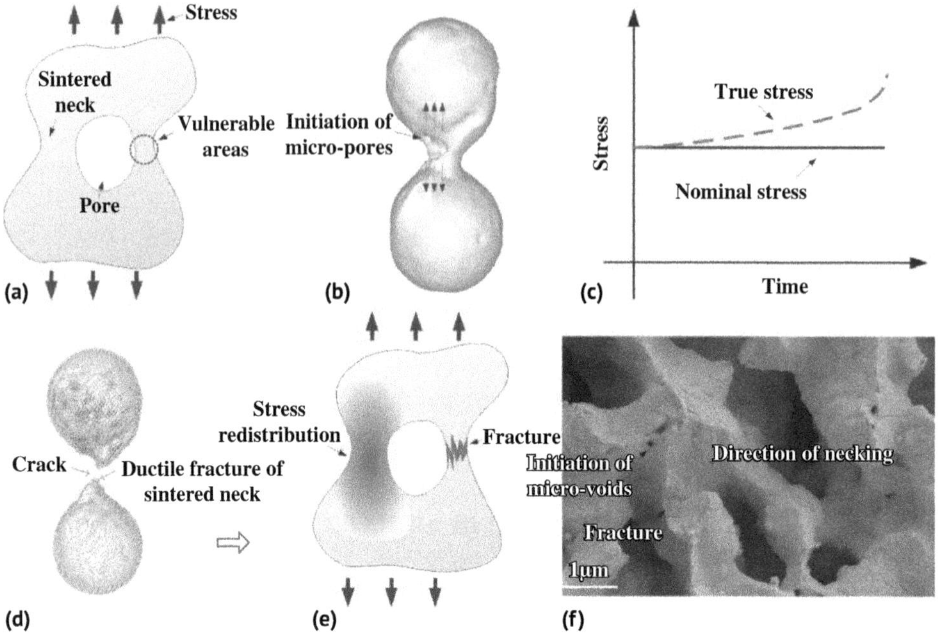

Figure 8.3 Illustration of failure mechanisms in the sintered neck region: (a) substructural features within sintered nano-silver; (b) molecular dynamics simulation model representing the sintered neck; (c) evolution of stress within the neck area during deformation; (d) fracture morphology of the sintered neck; (e) redistribution of local stress in the surrounding sintered nano-silver matrix; and (f) SEM image showing the fracture microstructure of sintered nano-silver.

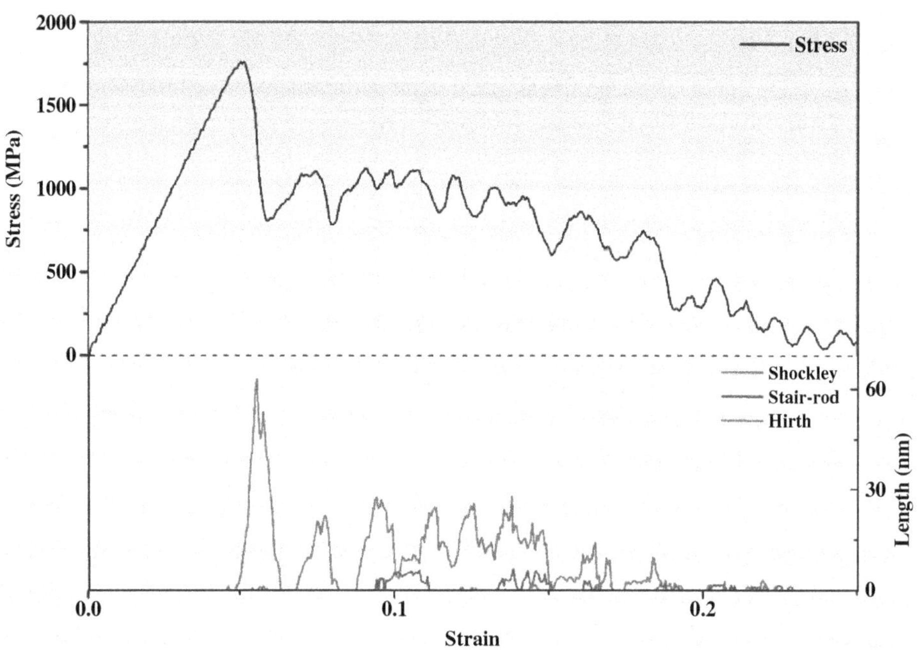

Figure 8.4 Stress–strain curve and dislocation evolution curve of the sintered neck at 100 °C obtained by molecular dynamics simulation.

During the transition from necking to final rupture, additional dislocation types such as Stair-rod and Hirth dislocations emerge. These dislocations partially restrict the mobility of Shockley dislocations, contributing to local strain hardening within the necked region. This hardening effect is beneficial in blunting crack tips and impeding their propagation. Upon reaching a critical damage threshold, the sintered neck ultimately fails, promoting the growth of internal cracks that extend into nearby nanopores and along the silver nanowire network [25]. The observed failure mode demonstrates significant ductility, attributed to the active motion of dislocations—a behavior supported by both molecular dynamics modeling and experimental evidence.

8.3 Damage Model

Experiments have confirmed that damage is controlled by plastic strain. According to thermodynamic methods, the damage rate $\dot{D}$ is defined as [26]:

$$\dot{D} = -\dot{\lambda}\frac{\partial F_D}{\partial Y} \tag{8.3}$$

where F_D is the damage potential equation, Y is the energy density release rate. $\dot{\lambda}$ is the plasticity multiplier, which can be obtained through the dissipation potential equation F, whose dissipation potential equation F is defined as [27]:

$$F = f + F_X + F_D \tag{8.4}$$

where f is the viscoplastic loading equation, which F_X is the follow-up hardening term. As a viscoplastic material [10, 28], according to von Mises' law, the viscoplastic loading equation of sintered nano-silver is [26]:

$$f = \left(\frac{\sigma}{1-D} - \boldsymbol{X}\right)_{\text{eq}} - R - \sigma_y = \sigma_v \tag{8.5}$$

where σ is the Cauchy stress, $\boldsymbol{X}$ is the back stress partial tensor, R is the isotropic hardening variable, σ_y is the yield stress, σ_v is the viscous excess stress, and $\dot{\lambda}$ is the plastic multiplier:

$$\dot{\lambda} = \dot{p}\,(1 - D) \tag{8.6}$$

where $\dot{p}$ is the cumulative plastic strain rate, which is defined as:

$$\dot{p} = \sqrt{\frac{2}{3}\dot{\varepsilon}^p\dot{\varepsilon}^p} \tag{8.7}$$

A damage dissipation potential function has been proposed for analyzing microstructural evolution during the failure process of metallic materials [23], and related work has confirmed that this damage dissipation potential can also be used for damage analysis of chip packaging materials [28, 29]. However, the damage rate obtained through this damage dissipation potential depends on the strain threshold. Therefore, it is difficult to apply it to creep damage.

According to the second law of thermodynamics, given the correlation between cumulative plastic strain and the evolution of internal microstructure, the damage dissipation potential also depends on cumulative plastic strain [30]. To apply it to the analysis of damage during creep, the damage dissipation potential is simplified to the following product form in this work:

$$F_D = Y^2\alpha\,(1-D)(D_{cr}-D)^{\frac{\alpha-1}{\alpha}}\,\vartheta p^{\vartheta-1} \tag{8.8}$$

where α is the damage index, which n is the hardening index of the material. ϑ is the plasticity index, which characterizes the relationship between damage and plastic deformation. Therefore:

$$\frac{\partial F_D}{\partial Y} = 2Y\alpha\,(1-D)(D_{\mathrm{cr}}-D)^{\frac{\alpha-1}{\alpha}}\,\vartheta p^{\vartheta-1} \tag{8.9}$$

where Y is the energy density release rate,

$$Y = -\frac{\sigma_{\mathrm{eq}}^2}{2E\,(1-D)^2} f\left(\frac{\sigma_H}{\sigma_{\mathrm{eq}}}\right) \tag{8.10}$$

Therefore:

$$\frac{\partial F_D}{\partial Y} = -\frac{{\sigma_{\mathrm{eq}}}^2}{E\,(1-D)} f\left(\frac{\sigma_H}{\sigma_{\mathrm{eq}}}\right)\alpha\,(D_{\mathrm{cr}}-D)^{\frac{\alpha-1}{\alpha}}\,\vartheta p^{\vartheta-1} \tag{8.11}$$

According to the Ramberg–Osgood relationship [30]:

$$\frac{\sigma_{\mathrm{eq}}}{1-D} = Kp^{\frac{1}{n}} \tag{8.12}$$

Then, combining the above formulas, we obtain:

$$\dot{D} = \frac{{\sigma_{\mathrm{eq}}}^2}{E} f\left(\frac{\sigma_H}{\sigma_{\mathrm{eq}}}\right)\alpha\,(D_{\mathrm{cr}}-D)^{\frac{\alpha-1}{\alpha}}\,\vartheta p^{\vartheta-1}\dot{p} \tag{8.13}$$

where σ_{eq} is the von Mises equivalent stress, s_{ij} is the stress vector, $\sigma_{\mathrm{eq}} = \left(\frac{3}{2}s_{ij}s_{ij}\right)^{1/2}$, $s_{ij} = \sigma_{ij} - \frac{1}{3}\sigma_{kk}\delta_{ij}$. σ_H is the hydrostatic stress, $\sigma_H = \frac{1}{3}\sigma_{kk}$.

$f\left(\frac{\sigma_H}{\sigma_{\mathrm{eq}}}\right)$ is a parameter related to the Poisson ratio:

$$f\left(\frac{\sigma_H}{\sigma_{\mathrm{eq}}}\right) = 2/3\,(1+\upsilon) + 3\,(1-2\upsilon)\left(\frac{\sigma_H}{\sigma_{\mathrm{eq}}}\right)^2 \tag{8.14}$$

For creep damage, at the initial state (the starting point of load application, when the creep strain is 0), the damage caused by creep is 0. At time t, the creep strain is ε, and the corresponding damage is D. Integrate Eq. (8.13):

$$\int_0^D \frac{1}{\alpha}\,(D_{\mathrm{cr}}-D)^{\frac{1}{\alpha}-1}\,dD = \int_0^{\varepsilon} \frac{{\sigma_{\mathrm{eq}}}^2}{E} f\left(\frac{\sigma_H}{\sigma_{\mathrm{eq}}}\right)\vartheta p^{\vartheta-1}dp \tag{8.15}$$

The damage evolution can be obtained as follows:

$$D = D_{\mathrm{cr}} - \left(D_{\mathrm{cr}}^{\frac{1}{\alpha}} - \frac{\sigma_{\mathrm{eq}}^2}{E} f\left(\frac{\sigma_H}{\sigma_{\mathrm{eq}}}\right)\varepsilon^{\vartheta}\right)^{\alpha} \tag{8.16}$$

Therefore, damage evolution equations based on the thermodynamic framework were obtained.

8.4 Creep Life Model

The relationship between damage and time is defined as [20, 31]:

$$D = 1 - \left(1 - \frac{t}{t_r}\right)^{\frac{1}{1+r}} \tag{8.17}$$

where r is a parameter related to creep damage.

Combining Eqs. (8.16) and (8.17), we obtain the creep evolution equation:

$$\varepsilon = \left(\frac{D_{\mathrm{cr}}^{1/\alpha} - \left(D_{\mathrm{cr}} - 1 + \left(1 - \frac{t}{t_r}\right)^{1/(1+r)} \right)^{1/\alpha}}{\frac{\sigma_{\mathrm{eq}}^2}{E} f\left(\frac{\sigma_H}{\sigma_{\mathrm{eq}}}\right)} \right)^{1/\vartheta} \tag{8.18}$$

8.5 Parameter Determination

All parameters involved in the model are D_{cr}, $f\left(\frac{\sigma_H}{\sigma_{\mathrm{eq}}}\right)$, t_r, E, r, α, ϑ. $f\left(\frac{\sigma_H}{\sigma_{\mathrm{eq}}}\right)$, t_r, E can be directly determined through experiments under uniaxial creep conditions, $f\left(\frac{\sigma_H}{\sigma_{\mathrm{eq}}}\right) = 1$. For the elastic modulus, according to the elastic modulus evolution law of sintered nano-silver given in Ref. [18], the elastic modulus corresponding to different temperatures of sintered nano-silver can be obtained through linear fitting, as shown in Table 8.1.

The parameters r can be obtained from the relationship between creep life and stress [21], where the relationship between damage and applied stress is [31]:

$$\frac{dD}{dt} = \left(\frac{\sigma}{b(1-D)}\right)^r \tag{8.19}$$

where b and r are parameters representing creep damage.

Integrating Eq. (8.19) yields [21]:

$$t = \frac{1-(1-D)^{1+r}}{1+r}\left(\frac{\sigma}{b}\right)^{-r} \tag{8.20}$$

Therefore, based on the above equation, the creep life can be obtained:

$$t_r = \frac{1}{1+r}\left(\frac{\sigma}{b}\right)^{-r} \tag{8.21}$$

Table 8.1 Elastic modulus of sintered nano-silver at different ambient temperatures.

Temperature/°C	100	125	150	175
Modulus of elasticity/GPa	3.45	2.45	1.45	0.45

Then, take the logarithm for both sides of Eq. (8.21) respectively:

$$lnt_r = rlnb + ln\frac{1}{1+r} - rln\sigma \tag{8.22}$$

Therefore, the parameters r can be obtained through the slope.

8.6 Damage Evolution Analysis of Sintered Nano-silver

The creep behavior data of sintered nano-silver were sourced from the study conducted by Chen et al. [21]. In their experiments, a nano-silver paste was sintered into thin films to evaluate the material's creep characteristics. The correlation between creep deformation and damage evolution under varying temperature conditions is presented in Figure 8.5a–d, based on the aforementioned calculations. The progression of damage in sintered nano-silver can be broadly divided into two distinct phases. During the initial and steady-state creep stages, the accumulation of damage is gradual, and the overall damage level remains relatively low. However, once the material enters the tertiary creep stage, a sharp increase in damage rate is observed, ultimately resulting in structural failure.

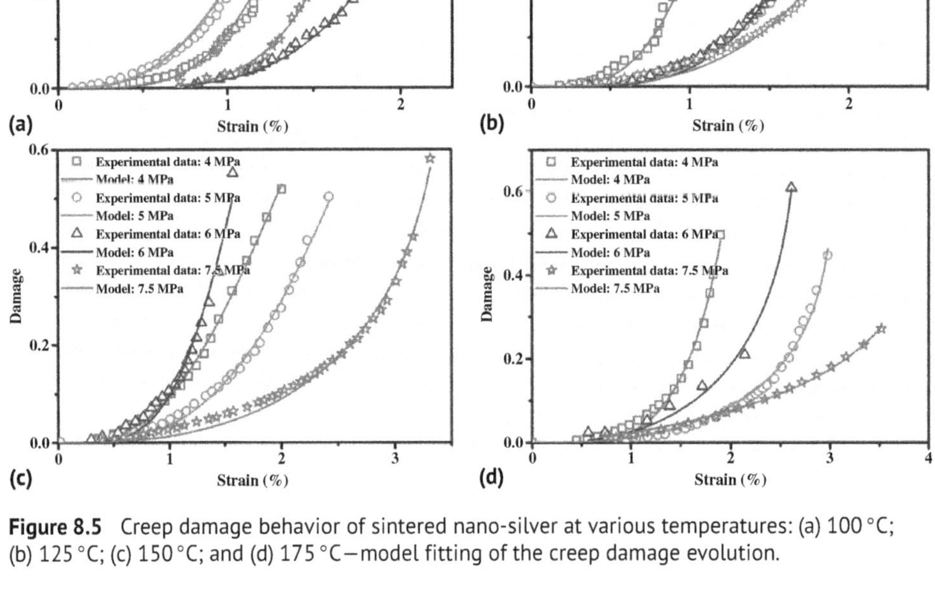

Figure 8.5 Creep damage behavior of sintered nano-silver at various temperatures: (a) 100 °C; (b) 125 °C; (c) 150 °C; and (d) 175 °C—model fitting of the creep damage evolution.

8.7 Theoretical Analysis of Sintered Nano-silver Creep

Theoretically, when the material is in a completely fractured state, $D_{cr} = 1$. The creep evolution formula can be simplified as:

$$\varepsilon = \left(\frac{1 - \left(1 - \frac{t}{t_r}\right)^{1/(\alpha + r\alpha)}}{\frac{\sigma_{eq}^2}{E}} \right)^{1/\vartheta} \tag{8.23}$$

By differentiating Eq. (8.23), the creep rate is obtained as:

$$\dot{\varepsilon} = \frac{E}{\vartheta \alpha \sigma_{eq}^2 t_r (1 - r)} \left(\frac{1 - \left(1 - \frac{t}{t_r}\right)^{1/(\alpha + r\alpha)}}{\frac{\sigma_{eq}^2}{E}} \right)^{\frac{1}{\vartheta} - 1} \left(1 - \frac{t}{t_r}\right)^{\frac{1}{\alpha + r\alpha} - 1} \tag{8.24}$$

The creep data obtained from the creep model established in this study for sintered nano-silver under different environmental temperatures and stresses are compared with experimental data as shown in Figure 8.6, with the experimental conditions for sintered nano-silver summarized in Table 8.2. The creep model constructed in this study matches the experimental data well, exhibiting three distinct stages of creep evolution. For the creep rate, both the experimental data and the creep model established in this study exhibit three stages corresponding to the creep evolution. When compared with damage evolution, at the first stage, the creep rate decreases as creep increases, as the hardening effect of sintered nano-silver is greater than the damage effect. At the second stage, both the damage rate and the creep rate remain constant. At the third stage, as damage increases, the creep rate of sintered silver nanoparticles gradually increases. The creep rate equation developed in this work can predict the evolution of creep rates at different stages of creep, unlike the power-law creep model [31].

8.8 Conclusion

In response to the creep of sintered nano-silver at high temperatures, the microscopic mechanism was studied through numerical simulation, and corresponding theoretical models were developed to describe the evolution of creep and creep rate based on the damage mechanism:

1. A random pore model was developed that matches the microstructure of sintered nano-silver, which was used to study the effect of pores on the creep of sintered nano-silver. It was found that the sintered neck is a vulnerable area for the creep failure of sintered nano-silver. The damage and fracture of the sintered neck lead to the initiation of microcracks, which then evolve into macrocracks, and this leads to the final failure of the sintered nano-silver.

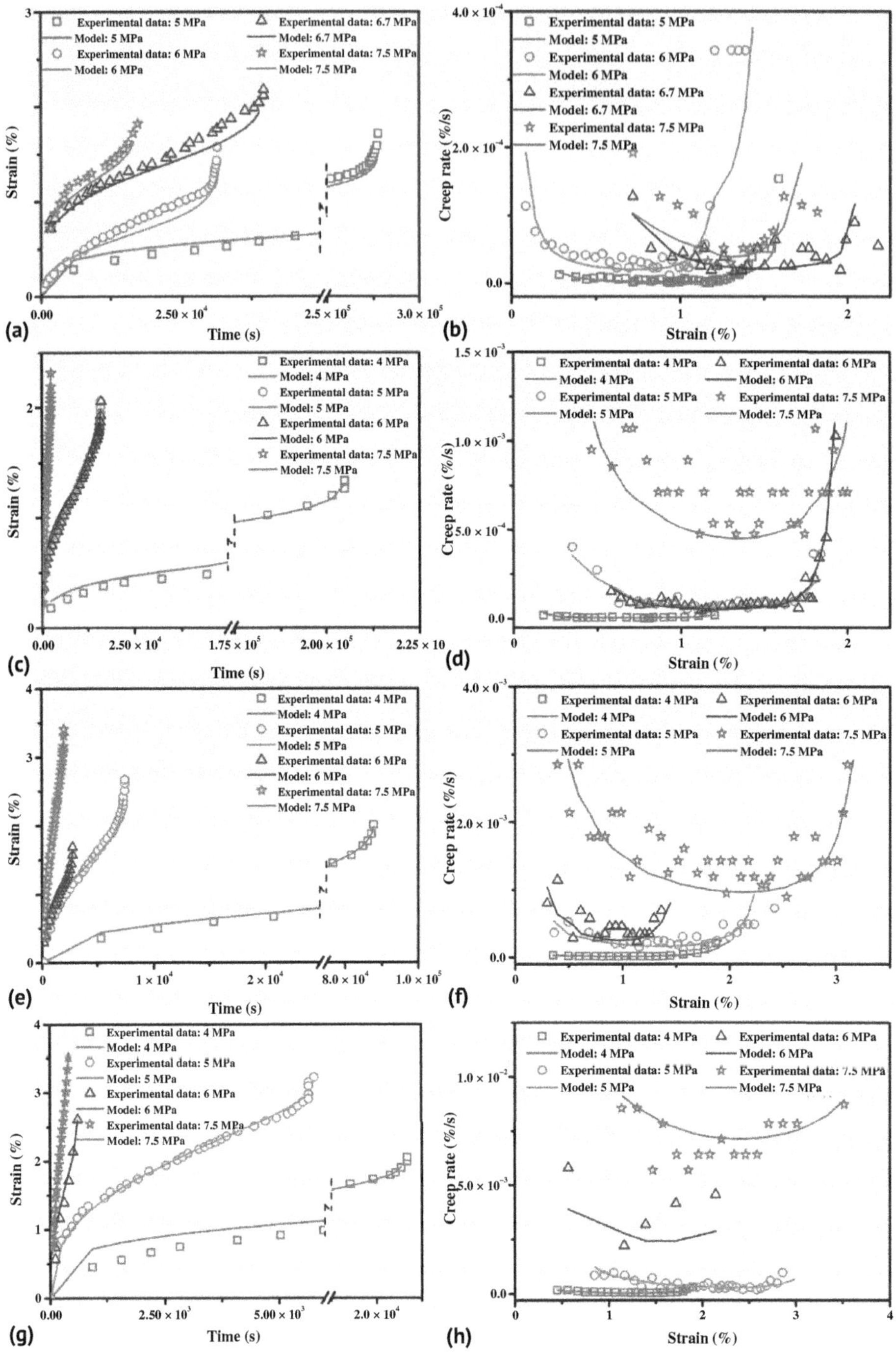

Figure 8.6 Creep behavior and creep rate of sintered nano-silver at various temperatures: (a, c, e, g) creep curves at 100 °C, 125 °C, 150 °C, and 175 °C, respectively; (b, d, f, h) corresponding creep rate curves at 100 °C, 125 °C, 150 °C, and 175 °C, respectively.

Table 8.2 Details of sintered nano-silver creep conditions.

Load	Temperature	100 °C	125 °C	150 °C	175 °C
4 MPa	ϑ		3.8	2.9	4.4
	α		45.0	8.8	0.7
5 MPa	ϑ	4.3	3.3	3.2	2.6
	α	16.3	5.0	2.1	0.2
6 MPa	ϑ	2.9	4.1	4.1	2.6
	α	22.5	2.6	4.0	0.3
6.7 MPa	ϑ	4.3			
	α	1.4			
7.5 MPa	ϑ	5.0	3.4	2.6	1.5
	α	1.7	1.4	0.3	0.2

2. The failure process of the sintered neck was simulated through molecular dynamics, which confirmed that Shockley dislocations dominate the entire failure process of the sintered neck. To some extent, the generation of Stair-rod and Hirth dislocations leads to the hardening of the sintered neck, which helps to hinder the cracks propagation and contributes to the ductile failure of the sintered nano-silver.
3. Based on the damage mechanism of sintered nano-silver, a damage dissipation potential that could predict the entire damage process was proposed. Then, a creep model was established that matched the failure mechanism of sintered nano-silver, and it could well predict the creep and creep rate of sintered nano-silver.

References

1 Li, X., Chen, G., Wang, L. et al. (2013). Creep properties of low-temperature sintered nano-silver lap shear joints. *Mater. Sci. Eng. A* 579: 108–113.

2 Gong, H., Yao, Y., and Yang, Y. (2021). Size effect on the fracture of sintered porous nano-silver joints: experiments and Weibull analysis. *J. Alloy. Compd.* 863: 158611.

3 Siow, K.S. (2012). Mechanical properties of nano-silver joints as die attach materials. *J. Alloy. Compd.* 514: 6–19.

4 Gong, H., Wu, H., and Yao, Y. (2021). Creep of sintered porous micron-silver: nanoindentation experiment and theoretical analysis. *J. Mater. Sci.* 56 (32): 18281–18299.

5 Gong, H., Yang, Y., Wu, H. et al. (2024). Size effect between micron-silver joint and nano-silver joint in die-attached interconnection. *IEEE. T. Comp. Pack. Man.* 14 (11): 1912–1920.

6 Kim, Y.J., Park, B.H., Hyun, S.K., and Nishikawa, H. (2021). The influence of porosity and pore shape on the thermal conductivity of silver sintered joint for die attach. *Mater. Today Commun.* 29: 102772.

7 Tan, Y.S., Li, X., Chen, X. et al. (2020). Feasibility investigation and characterization of low-pressure-assisted sintered-silver bonded large-area DBA plates. *Solder. Surf. Mt. Tech.* 32: 129–136.

8 Chen, C.T., Zhang, Z., Kim, D.J. et al. (2019). Interfacial oxidation protection and thermal-stable sinter Ag joining on bare Cu substrate by single-layer graphene coating. *Appl. Surf. Sci.* 497: 143797.

9 Long, X., Li, Z., Lu, X.Z. et al. (2019). Mechanical behaviour of sintered silver nanoparticles reinforced by SiC microparticles. *Mat. Sci. Eng. A-Struct.* 744: 406–414.

10 Gong, H., Wang, T.Y., Zhu, J.Q. et al. (2022). Compressive experimental analysis and constitutive model of sintered nano-silver. *J. Appl. Mech.-T. ASME.* 90 (3): 031004.

11 Gong, H., Guo, H., Li, S.J. et al. (2023). Compressive failure mechanism of sintered nano-silver. *J. Mater. Res.* 38: 4201–4213.

12 Gadaud, P., Caccuri, V., Bertheau, D. et al. (2016). Ageing sintered silver: relationship between tensile behavior, mechanical properties and the nanoporous structure evolution. *Mat. Sci. Eng. A-Struct.* 669: 379–386.

13 Wakamoto, K., Mochizuki, Y., Otsuka, T. et al. (2020). Temperature dependence on tensile mechanical properties of sintered silver film. *Materials* 13 (18): 4061.

14 Cai, W., Wang, P., and Fan, J. (2020). A variable-order fractional model of tensile and shear behaviors for sintered nano-silver paste used in high power electronics. *Mech. Mater.* 145: 103391.

15 Gong, H., Zheng, Y., Liu, B. et al. (2024). Failure mechanism and theoretical model of high-temperature tensile creep of sintered nano-silver. *J. Mater. Sci: Mater Electron.* 35: 648.

16 Li, M.Y., Xiao, Y., Zhang, Z.H., Yu, J. (2015). Bimodal sintered silver nanoparticle paste with ultrahigh thermal conductivity and shear strength for high temperature thermal interface material applications. *ACS Appl. Mater. Interfaces* 7 (17): 9157–9168.

17 Yao, Y. and Gong, H. (2019). Damage and viscoplastic behavior of sintered nano-silver joints under shear loading. *Eng. Fract. Mech.* 222: 106741.

18 Yu, D.J., Chen, X., Chen, G. et al. (2009). Applying Anand model to low-temperature sintered nanoscale silver paste chip attachment. *Mater. Design* 30 (10): 4574–4579.

19 Chen, G., Zhang, Z.S., Mei, Y.H. et al. (2014). Applying viscoplastic constitutive models to predict ratcheting behavior of sintered nanosilver lap-shear joint. *Mech. Mater.* 72: 61–71.

20 Tan, Y.S., Li, X., Chen, X., and Lu, G.Q. (2018). Elevated-temperature shear creep evolution and life prediction of sintered nano-silver lap-shear joint. *J. Mater. Sci.-Mater. El.* 29 (1): 303–312.

21 Chen, G., Sun, X.H., Nie, P. et al. (2012). High-temperature creep behavior of low-temperature-sintered nano-silver paste films. *J. Electron. Mater.* 41 (4): 782–790.

22 Paknejad, S.A. and Mannan, S.H. (2017). Review of silver nanoparticle based die attach materials for high power/temperature applications. *Microelectron. Reliab.* 70: 1–11.

23 Bonora, N., Ruggiero, A., Esposito, L., and Gentile, D. (2006). CDM modeling of ductile failure in ferritic steels: assessment of the geometry transferability of model parameters. *Int. J. Plast.* 22 (11): 2015–2047.

24 Lemaitre, J. (1996). *A Course on Damage Mechanics*. Springer.

25 Gong, H., Chen, X., Song, Y. et al. (2023). A study of the creep properties and constitutive model of sintered nano-silver: role of loading condition and temperature. *Jom* 75 (9): 3859–3869.

26 Lemaitre, J. and Desmorat, R. (2005). *Engineering Damage Mechanics: Ductile, Creep, Fatigue and Brittle Failures*, 32–33. Springer.

27 Bai, J., Zhao, Y., Shi, J., and He, X. (2022). Damage degradation model of aeolian sand concrete under freeze–thaw cycles based on macro-microscopic perspective. *Constr. Build. Mater.* 327: 126885.

28 Yao, Y., He, X., Keer, L.M., and Fine, M.E. (2015). A continuum damage mechanics-based unified creep and plasticity model for solder materials. *Acta Materialia* 83: 160–168.

29 Bonora, N. (1997). A nonlinear CDM model for ductile failure. *Eng. Fract. Mech.* 58 (1–2): 11–28.

30 Kachanov, L.M. (1958). Time of the rupture process under creep conditions. *Izy Akad, Nank S. S. R. Otd Tech Nauk* 8: 26–31.

31 Sherby, O.D. and Burke, P.M. (1968). Mechanical behavior of crystalline solids at elevated temperature. *Prog. Mater. Sci.* 13: 323–390.

9

Unified Creep-plasticity Model for High-strength Steel and Solder Alloys

9.1 Introduction

In high-temperature environments, metals and alloys frequently exhibit inelastic deformation dominated by creep strain. Material responses vary significantly with temperature; for instance, Sn–3.0Ag–0.5Cu solder (melting point ≈220 °C) demonstrates pronounced time-dependent deformation even at room temperature due to its high homologous temperature. In contrast, most metals and alloys with higher melting points display rate-independent deformation at room temperature, which can be adequately described by classical plasticity theory.

The development of constitutive models for viscoplastic deformation in metals and alloys falls into two main categories: physics-based and phenomenological.

1. Physics-based models incorporate microstructural evolution mechanisms, such as dislocation motion, dynamic recrystallization, and thermal activation, with some also adopting a thermodynamic framework [1–13].
2. Phenomenological models, on the other hand, are widely used due to their empirical nature, enabling deformation prediction with fewer parameters and straightforward calibration [14–20].

To predict the nonlinear deformation of solder joints, Tang and Basaran [21] and Basaran et al. [22] extended the uniaxial creep law by Kashyap and Murty [23] to multiaxial loading conditions; Gomez and Basaran [24] further enhanced the model by incorporating damage effects and nonlinear kinematic hardening, capturing the temperature- and rate-dependent viscoplastic behavior of Sn–Pb solder.

At elevated homologous temperatures, irrecoverable deformation comprises both rate-independent plastic deformation and rate-dependent creep behavior, which collectively contribute to the total inelastic strain. Early non-unified viscoplastic constitutive theories [25, 26] treated creep and plastic strains as separate components, enabling distinct calculations for steel and solder alloys. However, recognizing that both mechanisms stem from dislocation motion, later work introduced a unified creep-plasticity (UCP) constitutive law, integrating these strains into a single inelastic strain measure to better capture the material's rate-dependent response. Further refinements emerged from comparative studies: Lee and Basaran [27] evaluated prevalent creep models against Sn–Pb solder data, synthesizing their strengths into an optimized solder-specific creep law. Meanwhile, Basaran et al. [22] advanced a unified model incorporating the disturbed state concept (DSC), addressing engineering challenges by coupling nonlinear deformation, damage, and fatigue effects into a cohesive framework.

The deformation behavior of materials exhibits distinct rate sensitivities under varying loading conditions, making it essential to accurately characterize the coupled effects of strain rate and temperature. In this chapter, rate sensitivity is treated as an intrinsic material property that varies with temperature. Solid materials typically display three classes of rate-dependent responses: (1) normal rate sensitivity, where flow stress increases with strain rate; (2) zero rate sensitivity (rate-independent behavior); and (3) inverse rate sensitivity, where stress decreases with increasing strain rate at fixed strain [28]. Current modeling approaches quantify these effects either through strain-rate exponents in flow laws [14, 15] or via specialized parameters in kinematic hardening models [16, 29]. Nevertheless, the temperature dependence of rate sensitivity remains an area requiring deeper investigation, particularly for materials operating across wide thermomechanical regimes.

This chapter focuses on developing a unified creep-plasticity constitutive model to characterize how material rate sensitivity evolves with temperature. The proposed approach decomposes the complex rate-sensitivity trend into distinct regimes, each exhibiting monotonic stress–strain rate dependence within specific temperature ranges. By ensuring each segment follows either strictly increasing or decreasing behavior, the model achieves broad applicability in capturing normal rate sensitivity variations across temperatures while enabling separate characterization of different deformation mechanisms. Current formulation intentionally excludes additional complexities such as damage accumulation, kinematic hardening, and fatigue effects to maintain focus on the fundamental rate-temperature relationship. However, following the modular framework established by [19], these mechanisms can be systematically incorporated into the model through subsequent extensions.

9.2 Viscoplastic Constitutive Framework

In certain applications, metals and alloys may undergo substantial deformation, with equivalent strains exceeding 10% and even reaching 60%. To accurately model such large deformations, many researchers employ finite strain theory (Farrokh and Khan, 2009; Molinari and Ravichandran, 2005). The velocity gradient is given by:

$$\boldsymbol{L} = \dot{\boldsymbol{F}}\boldsymbol{F}^{-1} \tag{9.1}$$

where $\boldsymbol{F}$ is the deformation gradient tensor. The rate of deformation $\boldsymbol{D}$ and continuum spin tensor $\boldsymbol{w}$ are derived from the velocity gradient $\boldsymbol{L}$ through its symmetric and antisymmetric parts, respectively:

$$\boldsymbol{D} = \frac{1}{2}\left(\boldsymbol{L} + \boldsymbol{L}^{T}\right) \tag{9.2}$$

$$\boldsymbol{w} = \frac{1}{2}\left(\boldsymbol{L} - \boldsymbol{L}^{T}\right) \tag{9.3}$$

The unified constitutive framework is adopted, wherein the total strain rate tensor $\boldsymbol{D}$ decomposes additively into elastic $\boldsymbol{D}^e$ and inelastic $\boldsymbol{D}^{in}$ components:

$$\boldsymbol{D} = \boldsymbol{D}^e + \boldsymbol{D}^{in} \tag{9.4}$$

To establish the flow rule formulation, the temperature dependence of rate sensitivity must first be characterized. The strain rate sensitivity is quantified through a dimensionless parameter that defines the magnitude of rate dependence. Following McDowell's [15] unified constitutive framework, this is typically represented by an exponent n in the flow function, which serves as a direct measure of rate sensitivity. In order to eliminate the interference of other factors, a basic function that can determine the value of n is proposed:

$$\boldsymbol{D}^{in} = \left(\frac{S_v}{d\,(T)}\right)^n \boldsymbol{N} \tag{9.5}$$

$$S_v = \left\langle \sqrt{\frac{3}{2}}\,\|\boldsymbol{S} - \boldsymbol{\alpha}\| - R \right\rangle \tag{9.6}$$

$$\boldsymbol{N} = \frac{\boldsymbol{S} - \boldsymbol{\alpha}}{\|\boldsymbol{S} - \boldsymbol{\alpha}\|} \tag{9.7}$$

where d is the drag stress, S_v is the overstress, $\boldsymbol{S}$ is the deviatoric stress tensor, $\boldsymbol{\alpha}$ is the back stress tensor, R represents isotropic hardening, $\boldsymbol{N}$ is the direction of deformation. According to the experimental results, the drag stress d in Eq. (9.5) can be defined as a function of temperature to obtain the accurate value of n. In this chapter, only the case where the strain rate is less than 1s^{-1} is analyzed.

Dynamic recovery, a temperature dependent process, plays a significant role in the rate-dependent behavior and hardening response of materials [30]. Consequently, the rate sensitivity of a material is inherently influenced by temperature. During plastic deformation, dislocations accumulate at microstructural obstacles, leading to an increase in dislocation density. As the density rises, the rate of further dislocation accumulation diminishes due to enhanced resistance to motion. This resistance necessitates higher stresses to sustain plastic flow, thereby contributing to strain hardening. Throughout the deformation process, dynamic recovery acts in competition with dislocation accumulation, influencing the evolving microstructure and mechanical response of the material.

At low homologous temperatures, the material microstructure remains stable, and deformation is primarily governed by dislocation motion, exhibiting nearly rate-independent behavior. As temperature increases under fixed strain and strain rate conditions, enhanced dynamic recovery reduces the rate of dislocation accumulation, which in turn lowers the hardening rate. This reduction continues until a balance between hardening and softening is achieved at the peak flow stress. At lower strain rates, this equilibrium is reached more readily, resulting in a lower stress level due to the slower dislocation accumulation. In contrast, higher strain rates lead to faster dislocation buildup, shortening the time required to reach equilibrium and diminishing the influence of dynamic recovery. As a result, the material sustains a higher stress level and displays typical positive strain rate sensitivity.

Dynamic recovery intensifies with increasing temperature and significantly influences dislocation accumulation at elevated temperatures. When the microstructure remains stable, materials typically exhibit greater strain rate sensitivity as temperature rises. Correspondingly, the strain rate sensitivity exponent n tends to decrease with temperature, as illustrated in Figure 9.1. A higher n value indicates lower sensitivity of flow stress to the loading rate, whereas a lower n reflects more pronounced rate dependence.

Following the first stage, the rate sensitivity enters a plateau in the second stage, where its variation with temperature becomes negligible, as indicated in Figure 9.1. This behavior arises because dynamic recovery approaches its saturation level, reflecting the stabilized microstructure at this stage. For certain materials, however, this intermediate stage may be absent. Beyond the second stage, materials exhibit divergent rate-dependent responses. Typically, two distinct trends in the strain rate sensitivity exponent n are observed: a continuous increase, as in the case of Sn–3.0Ag–0.5Cu solder alloy, or a sustained decrease, as seen in high-strength alloy steels (represented by the black line in Figure 9.1). These contrasting behaviors stem from the onset of microstructural instability at elevated temperatures, where differences in microstructural evolution across materials result in varied strain rate sensitivity trends.

At high homologous temperatures, recrystallization becomes prominent in Sn–3.0Ag–0.5Cu solder alloy, leading to the transformation of elongated grains into fine equiaxed structures. This microstructural refinement introduces additional obstacles to dislocation motion, facilitating dislocation accumulation. However, the refined grains also reduce the energy barrier for crack initiation and propagation [31], promoting the formation of microcracks along grain boundaries and interfaces [32]. These micro-cracks disrupt the local stress field and diminish the effectiveness of dynamic recovery, even when coupled with recrystallization, resulting in a reduction of strain rate sensitivity during the third stage.

In contrast, for high-strength steels, the rate sensitivity continues to increase in this temperature regime. This behavior is attributed to ongoing microstructural evolution and the need to reach a new dynamic recovery saturation state. As temperature rises, atomic

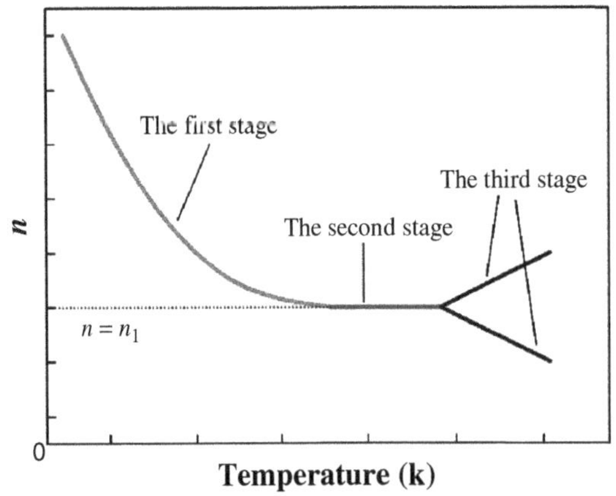

Figure 9.1 Representative evolution trend of n versus temperature.

mobility increases, which enhances dynamic recovery and progressively drives the material toward this new equilibrium, thereby sustaining the upward trend in rate sensitivity.

At a given strain, changes in strain rate for highly rate-sensitive materials can lead to significant stress changes. When the loading rate is altered, the inelastic strain rate responds significantly, leading to notable variations in the calculated stress. The rate sensitivity exponent n in Eq. (9.5), being temperature-independent, is insufficient to capture the evolution of rate sensitivity under varying thermal conditions. To incorporate the influence of temperature on inelastic deformation, a modified flow rule is proposed:

$$\boldsymbol{D}^{in} = \left(\frac{S_v}{d\,(T)}\right)^n r(n,T)\, f(T)\, \boldsymbol{N} \tag{9.8}$$

where $f(T)$ is introduced to account for the influence of temperature on inelastic flow behavior. The temperature-dependent evolution of rate sensitivity is represented by the function $r(n, T)$. Given the complexity of rate sensitivity mechanisms, the form of $r(n, T)$ is not universal and must be tailored to the specific characteristics of different materials.

With a suitable formulation of $r_k(n, T)$, the governing equation for $r(n, T)$ can be decomposed into the following components:

$$r(n,T) = r_1(n,T)\, r_2(n,T) \cdots r_k(n,T) \tag{9.9}$$

Here, the terms on the right-hand side of Eq. (9.9) are mutually independent and can be defined separately based on the observed patterns of rate sensitivity evolution.

From the previous analysis, n decreases as temperature rises in the initial stage, leading to enhanced sensitivity to loading rate. Therefore, $r_1\ (n, T)$ can be expressed as:

$$r_1(n,T) = \tanh^{m_1}\left(t(T)\left(\frac{S_v}{d(T)}\right)^{n+1}\right) \tag{9.10}$$

Here, m_1 is a rate-related exponent, n is assigned the value n_1 (see Figure 9.1), corresponding to the lowest rate sensitivity exponent observed in the initial stage. The function $t(T)$ is defined to adjust the contribution of the second term in the hyperbolic tangent expression. When the temperature-dependent parameter $t(T)$ exceeds a critical threshold, the function $r_1(n, T)$ asymptotically approaches unity, becoming negligible in its influence on strain rate sensitivity (see Figure 9.2). For notational convenience, we subsequently simplify $r_1(n, T)$ to r_1 in the following analysis.

The second-stage inelastic flow maintains a consistent rate dependence, as evidenced by the constant exponent n in Eq. (9.5). Through appropriate parameter selection, where:

1. $t\ (T)$ is sufficiently large to drive $r_1 \to 1$
2. All remaining terms $r_1(n, T)\ (l = 2, 3, \ldots, k)$ are identically unity

So, a mathematical equivalence can be established between Eqs. (9.8) and (9.5). This equivalence permits complete characterization of secondary deformation through direct substitution of Eq. (9.10) into Eq. (9.8), thereby eliminating the need for additional r_1 functional dependencies.

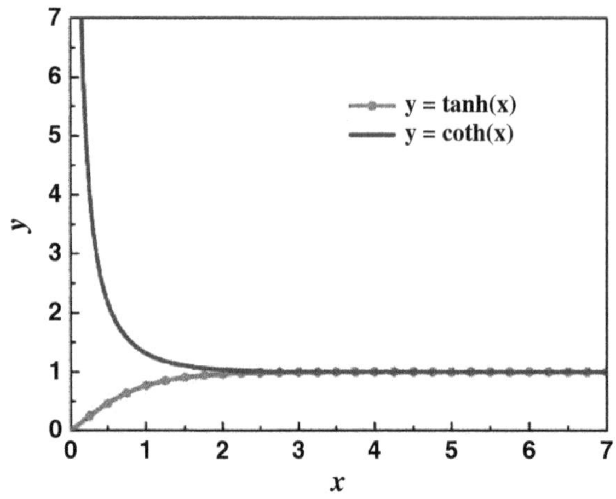

Figure 9.2 The value of the hyperbolic tangent and hyperbolic cotangent functions.

The determination of r_2 follows from the mathematical properties of the hyperbolic tangent (tanh) and hyperbolic cotangent (coth) functions. In the third deformation stage, we consider two distinct thermal regimes:

$$r_2(n, T) = \coth^{m_2}\left(c(T)\left(\frac{S_v}{d(T)}\right)^{n+1}\right) \tag{9.11}$$

or

$$r_2(n, T) = \tanh^{m_2}\left(c(T)\left(\frac{S_v}{d(T)}\right)^{n+1}\right) \tag{9.12}$$

where m_2 governs rate dependence in a manner analogous to m_1. In Eqs. (9.11) and (9.12), the rate exponent n is equivalent to n_1, as illustrated in Figure 9.1. In the third stage, the adjustment function $t(T)$ needs to be set to a very large value to ensure that r_1 is close to 1. When $c(T)$ is large enough, the value of r_2 in Eqs. (9.11) and (9.12) can also be considered equal to 1. Therefore, the $r(n, T)$ term in Eq. (9.8) is eliminated, and the flow rule can describe the viscoplastic deformation in the second stage, and the rate sensitivity remains unchanged (Figure 9.1).

For materials exhibiting enhanced rate sensitivity at elevated temperatures (characterized by a decreasing rate exponent n with increasing temperature), Eq. (9.11) provides an effective framework for modeling the amplified inelastic strain rate response. The formulation incorporates a temperature-dependent scaling function $c(T)$ that captures the intensified stress–strain rate coupling. As shown in Figure 9.2, this results in significantly wider stress amplitude variations under strain rate fluctuations, reflecting the material's strengthened rate-dependent behavior at higher temperatures.

For materials demonstrating positive temperature dependence of rate sensitivity (i.e. increasing n values with temperature) in the third deformation stage, Eq. (9.12) incorporates a hyperbolic tangent function to characterize the inelastic response. Within this thermal regime, the progressive reduction in r_2 values with rising temperature yields two key effects:

1. diminished inelastic strain rates;
2. elevated stress levels. This formulation effectively captures the intrinsic rate sensitivity of materials, as evidenced by its accurate prediction of thermomechanical coupling effects.

The current formulation incorporates a temperature- and rate-dependent drag stress $d(T)$, governed by the following evolution law:

$$d(T) = d_0(T) + (d_\infty(T) - d_0(T))(1 - \exp(-bp^m)) \tag{9.13}$$

where $d_0(T)$ and $d_\infty(T)$ are the initial and saturated drag stress values, respectively, both being temperature-dependent functions, b is a constant governing the transition rate between initial and saturated states, m is the strain-rate sensitivity exponent [33], p denotes the accumulated inelastic strain [29].

The stress rate can be derived by:

$$\dot{\boldsymbol{\sigma}} = \hat{\boldsymbol{\sigma}} + \boldsymbol{w}\boldsymbol{\sigma} - \boldsymbol{\sigma}\boldsymbol{w} \tag{9.14}$$

where $\hat{\boldsymbol{\sigma}}$ is the Jaumann stress rate.

The hardening response of metals and alloys exhibits significant temperature dependence. However, the present study focuses specifically on temperature-dependent rate sensitivity evolution, consequently neglecting rate-independent isotropic and kinematic hardening effects. The isotropic hardening component, when considered, follows the formulation given in Eq. (9.13). Notably, the constitutive framework developed by Yao et al. [19] provides a modular approach for incorporating these rate independent hardening mechanisms when required.

The proposed model establishes a phenomenological unified constitutive framework with three key temperature-dependent parameters:

1. $f(T)$: Controls thermal softening effects
2. $d_0(T)$: Governs initial drag stress behavior
3. $d_\infty(T)$: Determines saturated drag stress response

These functions can be systematically calibrated to capture specific material characteristics across different temperature regimes.

9.3 Application of the Proposed Theory

The proposed constitutive model was validated through uniaxial tensile simulations of two distinct alloy systems: (1) a high-strength steel and (2) Sn–3.0Ag–0.5Cu solder alloy, representing contrasting material behaviors under varying loading conditions. Model predictions were benchmarked against experimental data from Lin et al. [34] and Bai and Chen [16].

The temperature evolution of rate sensitivity was characterized through the flow rule in Eq. (9.5), with the following implementation approach:

1. The rate sensitivity-temperature relationship was segmented into thermodynamically relevant regimes.
2. For each regime, the adjustment function r_1 in Eq. (9.9) was formulated to capture the dominant deformation mechanisms.
3. The piecewise formulation ensures accurate representation of transitions between different rate-sensitivity behaviors.

9.3.1 Applied to High-strength Steel

The proposed constitutive model was validated against uniaxial tensile test data for a representative high-strength steel (chemical composition: 0.450C–0.280Si–0.960Cr–0.630Mn–0.190Mo–0.016P–0.012S–0.014Cu–(bal.)Fe).

Figure 9.3 presents the temperature-dependent evolution of rate sensitivity derived from both experimental measurements and theoretical predictions using Eq. (9.5). The results demonstrate a significant enhancement of strain-rate sensitivity as temperature increases. Notably, the rate exponent n maintains relative stability within the temperature range of 1223–1323 K. For the third deformation stage, where n exhibits a decreasing trend, the flow rule is modified as follows:

$$\mathbf{D}^{in} = \left(\frac{S_v}{d(T)}\right)^n \tanh^{m_1}\left(t(T)\left(\frac{S_v}{d(T)}\right)^{n+1}\right)\coth^{m_2}\left(c(T)\left(\frac{S_v}{d(T)}\right)^{n+1}\right) f(T)\,\mathbf{N} \tag{9.15}$$

It is noted that Eqs. (9.6) and (9.7) are still available for this material. The viscoplastic deformation occurs at an early stage under elevated temperature. Here, the value of yield stress R in Eq. (9.6) is defined as a small constant to ensure the first increment stays in the elastic state. The initial value of drag stress is taken as a constant for simplification, and Eq. (9.13) can be rewritten as: Eqs. (9.6) and (9.7) remain applicable for modeling the material's behavior. At elevated temperatures, viscoplastic deformation initiates in the early loading stage. To maintain numerical stability, the yield stress R in Eq. (9.6) is assigned a sufficiently small constant value, ensuring the initial loading increment remains elastic.

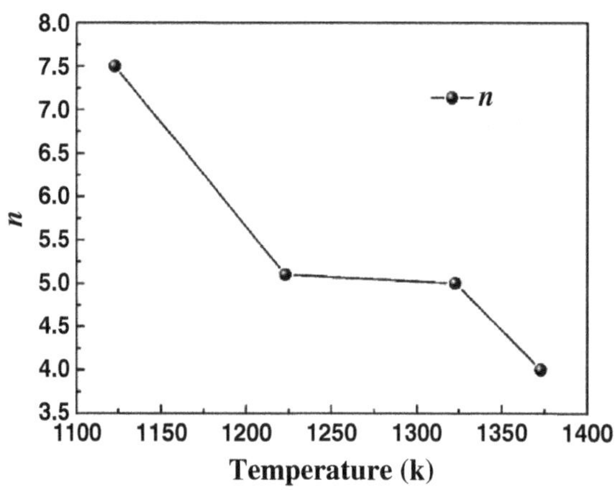

Figure 9.3 Evolution of rate sensitivity with temperature of the high-strength steel.

For simplicity, the initial drag stress d_0 (T) is treated as temperature-independent, leading to a modified form of Eq. (9.13):

$$d(T) = d_0 + (d_\infty (T) - d_0)\,(1 - \exp(-bp^m)) \tag{9.16}$$

The model incorporates a linear thermal softening effect on the drag stress d_∞ (T) saturation limit, expressed as:

$$d_\infty (T) = kT + C \tag{9.17}$$

The formulation of Eq. (9.17) was derived through rigorous analysis of experimental data within the 1123–1373 K temperature range. Extrapolation beyond these thermal boundaries may require appropriate model adjustments to maintain prediction accuracy.

The $f(T)$ in Eq. (9.15) can be expressed as:

$$f(T) = \exp(-T/T_R) \tag{9.18}$$

where T_R is the melting temperature of materials.

The temperature-dependent adjustment functions exhibit opposing trends:

1. t (T) demonstrates a monotonically increasing relationship with temperature, reaching sufficiently large magnitudes in the 1223–1323 K range to drive the hyperbolic tangent term in Eq. (9.15) toward unity.
2. c (T) follows a decreasing trend with rising temperature, maintaining elevated values below 1323 K.

For high-strength alloy steels, these functions adopt the following specific forms:

$$t(T) = \exp(t_0 + t \times \exp(-T/t_1)) \tag{9.19}$$

$$c(T) = \exp(c_0 + c \times \exp(-T/c_1)) \tag{9.20}$$

The complete constitutive framework integrates Eqs. (9.6), (9.7), and (9.15)–(9.20), with model parameters calibrated using the experimental data presented in Table 9.1. Numerical simulations (simulation curves in Figure 9.4) demonstrate the model's predictive capability. For comparative analysis, additional simulations (dotted curves) were performed using the simplified formulation of Eq. (9.5) under varying loading conditions, with each case employing distinct rate sensitivity exponents (n-values).

The developed constitutive model maintains a constant rate-sensitivity exponent n, yet demonstrates satisfactory predictive capability across various loading conditions, as evidenced by the comparative analysis with Eq. (9.5) simulations in Figure 9.4. The model effectively captures the rate-sensitive behavior of high-strength steel, showing excellent agreement with experimental data at 1223 and 1373 K (Figure 9.4b and d). Predictions at strain rates of 0.01 and 0.0001 s^{-1} (Figure 9.4a and c) closely match experimental measurements. However, discrepancies emerge at 0.001 s^{-1} due to the model's fundamental assumption of positive strain-rate sensitivity. The experimental results in Figure 9.4b display characteristic positive rate-sensitivity behavior, where stress amplitude increases progressively with rising strain rates. Conversely, at 1123 and 1323 K (Figure 9.4a and c), the stress variation between 0.001 and 0.01 s^{-1} proves less pronounced than between 0.0001 and 0.001 s^{-1}. This attenuated response contradicts typical positive rate-sensitivity characteristics, explaining the observed prediction inaccuracies in this regime.

Table 9.1 Parameters in the proposed model.

	Symbols	High-strength alloy steel	Sn–3.0Ag–0.5Cu
Rate exponents	n	5	7.1
	m	2/5	1/3
	m_1	1/2	1/3
	m_2	1/2	1/3
Drag stress evolution function	d_0	10	20
	b	4.2	3.8
	k	−0.514	−0.332
	C	789.02	170.34
Adjustment functions	t_0	−7.31703	3.80644
	t	1.204E−4	2.519E−5
	t_1	−101	−28.124
	c_0	5.622	3.0936
	c	1.123E14	3.386E9
	c_1	43.445	19.410

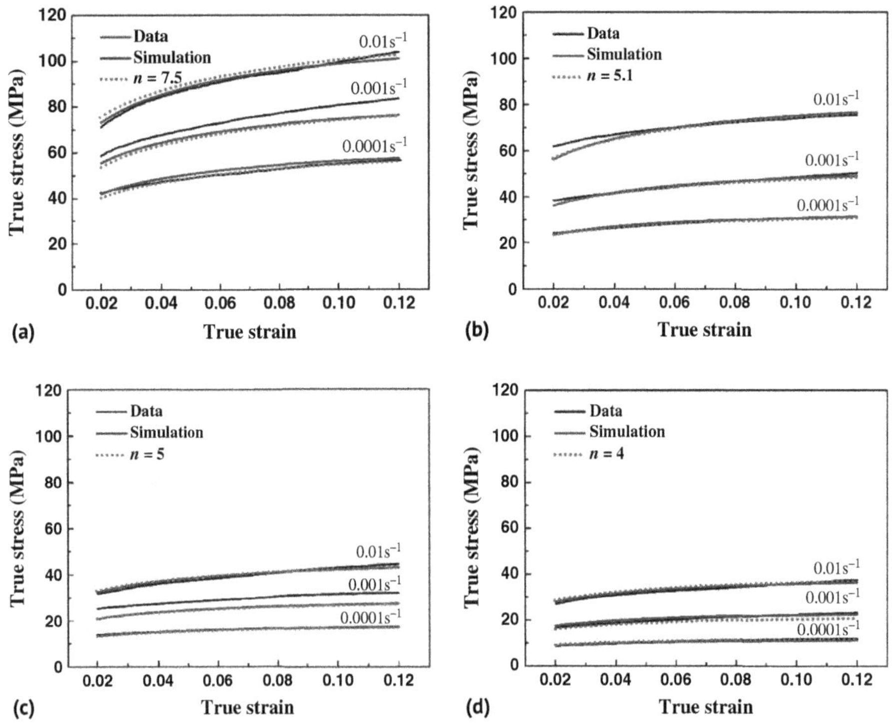

Figure 9.4 Comparisons between experimental data and simulation results of high-strength steel: (a) $T = 1123$ K; (b) $T = 1223$ K; (c) $T = 1323$ K; and (d) $T = 1373$ K.

9.3.2 Applied to Sn–3.0Ag–0.5Cu Solder Alloy

The proposed constitutive framework has been evaluated through numerical simulation of Sn–3.0Ag–0.5Cu solder alloy deformation, a critical material for microelectronic packaging applications. Given the complex thermomechanical loading conditions encountered in service, comprehensive characterization of its temperature- and strain-dependent behavior is particularly important.

Analysis of the rate sensitivity evolution reveals:

1. The temperature dependence of parameter n, derived from Eq. (9.5), displays the characteristic trend illustrated in Figure 9.5.
2. The observed increase in n values during the later second deformation stage necessitates implementation of the flow rule specified in Eq. (9.21).

$$\mathbf{D}^{in} = \left(\frac{S_v}{d(T)}\right)^n \tanh^{m_1}\left(t\,(T)\left(\frac{S_v}{d(T)}\right)^{n+1}\right)\tanh^{m_2}\left(c(T)\left(\frac{S_v}{d(T)}\right)^{n+1}\right) f(T)\,\mathbf{N} \quad (9.21)$$

The constitutive framework developed for high-strength steel has been successfully adapted to model Sn–3.0Ag–0.5Cu solder behavior, with modifications limited to the flow rule formulation. The complete material model for the solder alloy integrates Eqs. (9.6), (9.7), and (9.16)–(9.21), with corresponding parameters provided in Table 9.1.

Model validation demonstrates:

1. Excellent correlation between simulations and experimental data across strain rates (0.0001–0.01 s^{-1}) at various temperatures (Figure 9.6).
2. Accurate prediction of strain-rate jump test results at 25 °C (Figure 9.7).

The proposed formulation effectively captures both the deformation response and rate-sensitivity evolution of Sn–3.0Ag–0.5Cu under diverse loading conditions, demonstrating robust predictive capability.

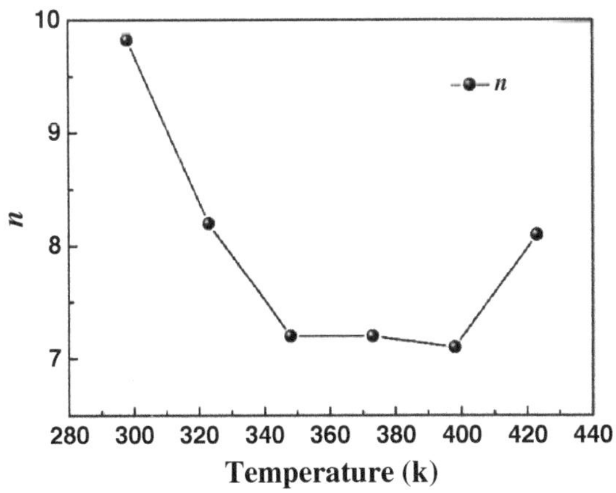

Figure 9.5 Evolution of rate sensitivity of the Sn–3.0Ag–0.5Cu alloy with temperature.

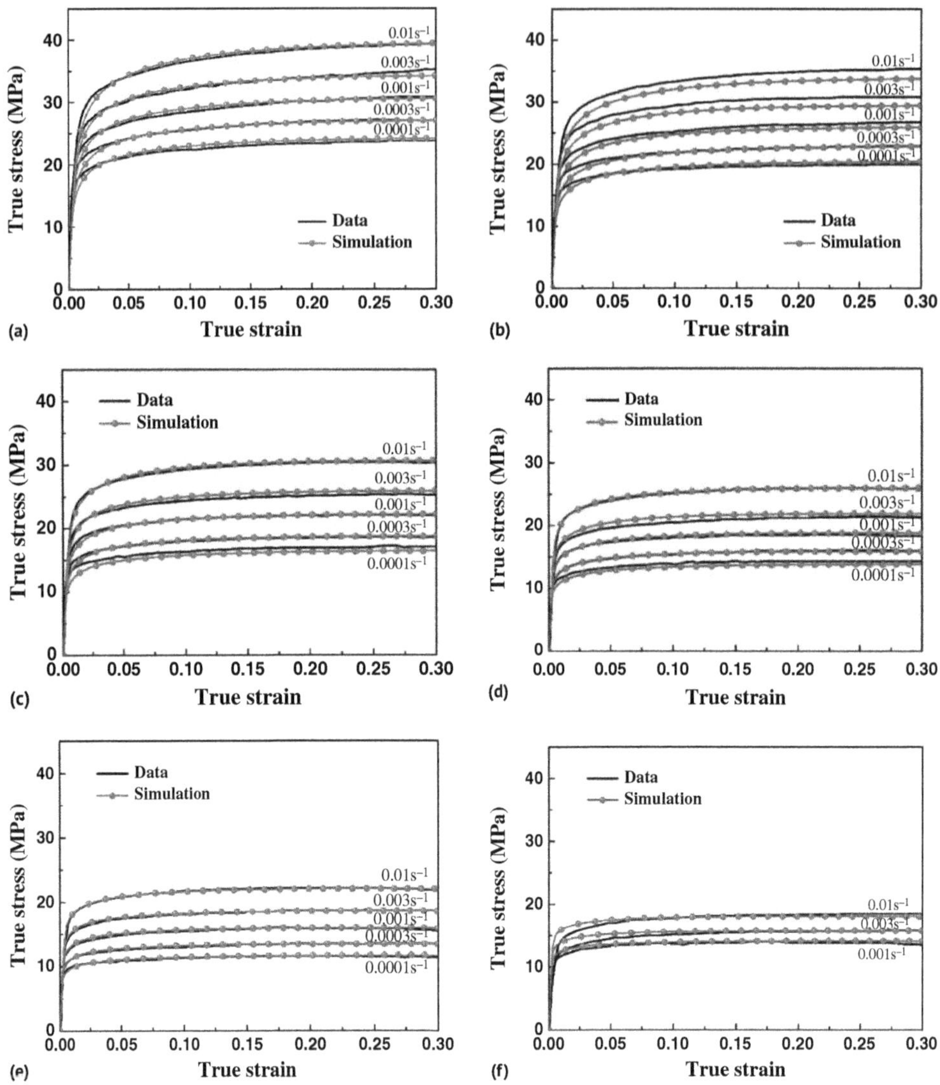

Figure 9.6 Comparisons between experimental data and simulation results of the Sn–3.0Ag–0.5Cu alloy: (a) $T = 25\,°C$; (b) $T = 50\,°C$; (c) $T = 75\,°C$; (d) $T = 100\,°C$; (e) $T = 125\,°C$; and (f) $T = 150\,°C$.

Numerical simulations demonstrate the effectiveness of the proposed drag stress evolution law (Eq. 9.16) in characterizing the material's hardening behavior. The analysis reveals a direct proportionality between the saturation drag stress $d_\infty\,(T)$ and the peak stress observed in experiments. As evidenced by the experimental results, both the maximum stress and consequently $d_\infty\,(T)$ exhibit a decreasing trend with rising temperature. This thermal softening behavior is quantitatively captured by the negative slope parameter k in Eq. (9.17).

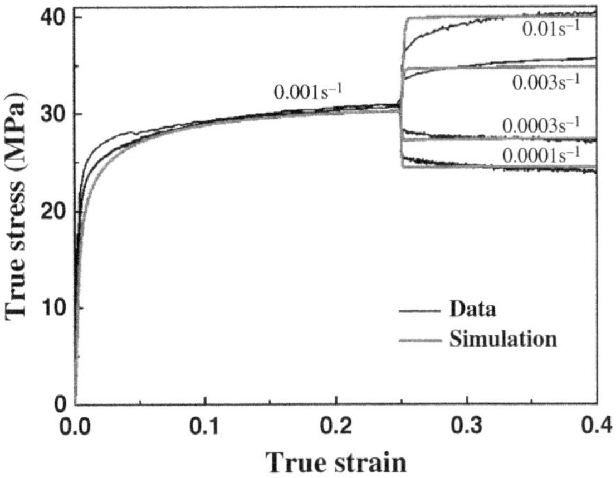

Figure 9.7 The strain rate jump tests data and simulation results of the Sn–3.0Ag–0.5Cu alloy at 25 °C.

9.4 Conclusion

This study systematically investigates the temperature-dependent strain rate sensitivity of various metals and alloys, which exhibit complex material-specific behaviors. A novel analytical approach is developed to characterize the evolution of rate sensitivity with temperature, employing hyperbolic tangent and cotangent functions within a unified constitutive framework. The model incorporates:

1. **Temperature-dependent functions** to capture thermal effects.
2. **Drag stress evolution laws** for hardening behavior.
3. **Adjustment equations** calibrated from experimental data.

Validation through uniaxial tensile simulations of high-strength steel and Sn–3.0Ag–0.5Cu solder demonstrates the model's accuracy in predicting rate sensitivity across diverse loading conditions and temperature regimes. Comparative analysis with experimental measurements confirms the framework's robustness in capturing thermomechanical coupling effects.

References

1 Abed, F.H., Ranganathan, S.I., and Serry, M.A. (2014). Constitutive modeling of nitrogen-alloyed austenitic stainless steel at low and high strain rates and temperatures. *Mechanics of Materials* 77: 142–157.

2 Bernstein, B. and Fong, J.T. (1993). A nonequilibrium thermodynamic theory of viscoplastic materials. *Journal of Applied Physics* 74: 2220–2228.

3 Chaboche, J.-L. (1997). Thermodynamic formulation of constitutive equations and application to the viscoplasticity and viscoelasticity of metals and polymers. *International Journal of Solids and Structures* 34: 2239–2254.

4 Dong, Y., Kang, G., and Yu, C. (2014). A dislocation-based cyclic polycrystalline visco-plastic constitutive model for ratchetting of metals with face-centered cubic crystal structure. *Computational Materials Science* 91: 75–82.

5 Hall, R.B. (2008). Combined thermodynamics approach for anisotropic, finite deformation overstress models of viscoplasticity. *International Journal of Engineering Science* 46: 119–130.

6 Helm, D. (2006). Stress computation in finite thermoviscoplasticity. *International Journal of Plasticity* 22: 1699–1727.

7 Klepaczko, J.R. (1991). Physical-state variables — the key to constitutive modeling in dynamic plasticity. *Nuclear Engineering and Design* 127: 103–115.

8 Lee, W.-S. and Lin, C.-F. (2001). Impact properties and microstructure evolution of 304L stainless steel. *Materials Science and Engineering: A* 308: 124–135.

9 Luo, J., Li, M., Li, X., and Shi, Y. (2010). Constitutive model for high temperature deformation of titanium alloys using internal state variables. *Mechanics of Materials* 42: 157–165.

10 Scheidler, M. and Wright, T.W. (2001). A continuum framework for finite viscoplasticity. *International Journal of Plasticity* 17: 1033–1085.

11 Shutov, A.V. and Ihlemann, J. (2012). A viscoplasticity model with an enhanced control of the yield surface distortion. *International Journal of Plasticity* 39: 152–167.

12 Voyiadjis, G.Z. and Abed, F.H. (2005). Microstructural based models for bcc and fcc metals with temperature and strain rate dependency. *Mechanics of Materials* 37: 355–378.

13 Zhang, Y., Volinsky, A.A., Xu, Q.-Q. et al. (2015). Deformation behavior and microstructure evolution of the Cu-2Ni-0.5Si-0.15Ag alloy during hot compression. *Metallurgical and Materials Transactions A* 46: 5871–5876.

14 Anand, L. (1985). Constitutive equations for hot-working of metals. *International Journal of Plasticity* 1: 213–231.

15 McDowell, D.L., Miller, M.P., and Brooks, D.C. (1994). A unified creep-plasticity theory for solder alloys. *Fatigue of Electronic Materials* ASTM STP:42–59.

16 Bai, N. and Chen, X. (2009). A new unified constitutive model with short- and long-range back stress for lead-free solders of Sn–3Ag–0.5Cu and Sn–0.7Cu. *International Journal of Plasticity* 25: 2181–2203.

17 Johnson, G.R. and Cook, W.H. (1983). A constitutive model and data for metals subjected to large strains, high strain rates and high temperatures. *Engineering Fracture Mechanics* 21: 541–548.

18 Rusinek, A., Zaera, R., and Klepaczko, J.R. (2007). Constitutive relations in 3-D for a wide range of strain rates and temperatures – application to mild steels. *International Journal of Solids and Structures* 44: 5611–5634.

19 Yao, Y., He, X., Keer, L.M., and Fine, M.E. (2015). A continuum damage mechanics-based unified creep and plasticity model for solder materials. *Acta Materialia* 83: 160–168.

20 Yao, Y. and Keer, L.M. (2013). Cohesive fracture mechanics based numerical analysis to BGA packaging and lead free solders under drop impact. *Microelectronics Reliability* 53: 629–637.

21 Tang, H. and Basaran, C. (2003). A damage mechanics-based fatigue life prediction model for solder joints. *Journal of Electronic Packaging* 125: 120–125.

22 Basaran, C., Desai, C.S., and Kundu, T. (1998). Thermomechanical finite element analysis of problems in electronic packaging using the disturbed state concept: Part 1—Theory and formulation. *Journal of Electronic Packaging* 120: 41–47.

23 Kashyap, B.P. and Murty, G.S. (1981). Experimental constitutive relations for the high temperature deformation of a Pb-Sn eutectic alloy. *Materials Science and Engineering* 50: 205–213.

24 Gomez, J. and Basaran, C. (2006). Damage mechanics constitutive model for Pb/Sn solder joints incorporating nonlinear kinematic hardening and rate dependent effects using a return mapping integration algorithm. *Mechanics of Materials* 38: 585–598.

25 Contesti, E. and Cailletaud, G. (1989). Description of creep-plasticity interaction with non-unified constitutive equations: application to an austenitic stainless steel. *Nuclear Engineering and Design* 116: 265–280.

26 Darveaux, R. (2002). Effect of simulation methodology on solder joint crack growth correlation and fatigue life prediction. *Journal of Electronic Packaging* 124: 147–154.

27 Lee, Y. and Basaran, C. (2011). A creep model for solder alloys. *Journal of Electronic Packaging* 133: 1696–1699.

28 Cailletaud, G. and Saï, K. (1995). Study of plastic/viscoplastic models with various inelastic mechanisms. *International Journal of Plasticity* 11: 991–1005.

29 Ho, K. (2009). A unified constitutive law for cyclic viscoplasticity. *International Journal of Solids and Structures* 46: 1007–1018.

30 Argon, A. (2007). *Strengthening Mechanisms in Crystal Plasticity*. Oxford University Press.

31 Hokka, J., Mattila, T.T., Xu, H., and Paulasto-Kröckel, M. (2013). Thermal cycling reliability of Sn-Ag-Cu solder interconnections—Part 2: Failure mechanisms. *Journal of Electronic Materials* 42: 963–972.

32 Shengmin, W. (2005). Damage based fatigue criterion for solders in electronic packaging. *IEEE Transactions on Components and Packaging Technologies* 28: 435–440.

33 Chaboche, J.L. (2008). A review of some plasticity and viscoplasticity constitutive theories. *International Journal of Plasticity* 24: 1642–1693.

34 Lin, Y.C., Chen, X.-M., and Liu, G. (2010). A modified Johnson–Cook model for tensile behaviors of typical high-strength alloy steel. *Materials Science and Engineering: A* 527: 6980–6986.

10

A Multiscale Framework for the Constitutive Modeling of Rock

10.1 Introduction

When designing and safety-assessing underground structures such as tunnels, mines, reservoirs, and nuclear waste repositories, it is crucial to clearly understand the mechanical characteristics of the formation rock. However, rocks are inherently heterogeneous materials with complex internal structures spanning multiple spatial and temporal scales—from mineral grains and microcracks to large-scale discontinuities and faults. This complexity poses significant challenges for constitutive modeling.

Most traditional constitutive models rely on empirical data fitting, with limited consideration of the underlying physical mechanisms [1–4]. While these models are computationally efficient and many of them have been applied in practical engineering, they tend to oversimplify the material behavior by assuming homogeneity and isotropy at the macroscopic scale. This simplification becomes problematic in scenarios where localized damage, anisotropy, or time-dependent processes are dominant. Furthermore, these models generally lack the ability to account for evolving microstructural changes.

In recent years, micromechanical modeling has emerged as a powerful alternative to phenomenological approaches, owing to its capacity to explicitly incorporate microstructural morphology and the intrinsic deformation mechanisms of individual constituent phases. Guéry et al. and Jiang et al. [5, 6] extended Hill's incremental homogenization scheme—originally formulated to estimate the effective properties of metallic composites—to geomaterials by embedding essential features such as pressure sensitivity and plastic dilatancy. By assigning physically representative constitutive models to each phase, they established a robust elastoplastic damage framework. This formulation was subsequently enhanced by Shen et al. [7], who introduced a two-scale homogenization strategy. At the microscale, the clay matrix is homogenized while explicitly considering pores as an evolving internal phase. The resulting effective behavior is then upscaled through Hill's method to obtain the macroscopic response. Huang et al. [8] further generalized this framework by incorporating time-dependent deformation mechanisms across multiple scales. By distinguishing the kinetics of different processes, their model offers a unified description of both short- and long-term mechanical responses in argillite.

All of the aforementioned micromechanical models are fundamentally grounded in two classical theoretical frameworks: Eshelby's equivalent inclusion theory and Hill's incremental theory. Eshelby's solution lays a solid foundation for evaluating the stress and strain

Constitutive Models of Solid Materials: From Mechanical Principles to Engineering Applications, First Edition.

fields within inclusions embedded in an infinite medium, which is essential for capturing the interaction between different constituent phases. Building upon this, Hill's incremental method extends the linear to nonlinear, path-dependent behavior of materials (such as plasticity and damage) by solving the local constitutive equations for each phase and averaging the resulting stress response over the RVE at each load increment. Combining these two theories allows for a rigorous and flexible framework to incorporate the complex microstructural features and phase-specific deformation mechanisms.

The primary objective of this chapter is to develop a multiscale framework for the constitutive modeling of rock materials. The discussion begins with fundamental aspects of rock behavior, emphasizing the inherent heterogeneity, anisotropy, and the critical influence of microstructural features across multiple spatial scales. The multiscale modeling framework is then introduced, including key concepts such as scale separation, homogenization, and Hill's incremental method. A pressure-dependent plasticity model is developed through thermodynamic formulation and elastoplastic constitutive relations. Attention is then directed toward the application of the multiscale constitutive model to COx argillite, aiming to capture its complex mechanical behavior across different scales. The numerical aspects, including the local integration algorithms, consistent tangent moduli derivation, FEM implementation, are systematically studied. Finally, the model's validity is examined through numerical validations.

10.2 Fundamentals of Rock Behavior

10.2.1 Heterogeneity and Anisotropy of Rock

Rocks are inherently heterogeneous and anisotropic due to their geological formation processes, which involve sedimentation, metamorphism, tectonic activity, and mineralogical transformations over geological time. At the microscale, they are composed of varying mineral grains with distinct crystallographic orientations, chemical compositions, and mechanical properties. These grains are bonded together in complex patterns that introduce local heterogeneities and directional dependencies in strength and stiffness.

Anisotropy arises from both the intrinsic properties of the minerals and their spatial arrangement, such as preferred grain orientations or aligned microcracks. Additionally, stratification, foliation, or joint sets at larger scales contribute to directional mechanical behavior, where properties like Young's modulus, shear strength, and fracture toughness differ significantly with loading direction.

10.2.2 Mechanical Behavior Across Scales

The mechanical behavior of rock is governed by mechanisms that span across multiple scales:

- **Microscale (grain scale):** At this level, deformation involves mechanisms such as intergranular sliding and intragranular plasticity.
- **Mesoscale (microstructure scale):** This scale includes features such as pores, microcracks, heterogeneity of minerals, and mineral interfaces.

- **Macroscale (rock sample scale or laboratory scale):** At this level, the mechanical behavior is influenced by larger-scale structures. For a constitutive model based on laboratory data, only the effect of bedding planes is taken into account.

10.2.3 Role of Microstructure in Governing Macroscopic Behavior

The overall mechanical response of rock is closely linked to its internal structure, with grain shape and pore distribution being especially influential. Grain morphology—including grain size, shape, and spatial orientation—directly influences how stress is transmitted and concentrated within the material. For instance, elongated or aligned grains can introduce directional stiffness and strength, leading to anisotropic deformation responses. Grain boundaries, especially those with weak bonding or irregular contact surfaces, serve as potential sites for debonding and crack initiation under mechanical loading.

In parallel, the distribution and size of pores significantly affect the compressibility, strength, and damage evolution of the rock. Rocks with high porosity or well-connected pore networks are more prone to volumetric deformation and early dilatancy, particularly under deviatoric stress. These microstructural characteristics contribute to a reduction in effective load transfer paths and enhance the likelihood of microcrack formation and interaction, ultimately expediting stiffness deterioration and failure.

Taken together, the interplay between grain morphology and pore distribution governs the localization of deformation and the transition from microcracking to macroscopic failure. Therefore, explicitly incorporating these features into constitutive models is not merely an enhancement, but a necessity for accurately capturing the stress–strain behavior, anisotropy, and long-term deformation characteristics of geomaterials under complex loading conditions.

10.3 Multiscale Modeling Framework

10.3.1 Concept of Scale Separation and Homogenization

In multiscale modeling, the concept of scale separation is fundamental. It is based on the idea that different physical phenomena operate at different length scales, and the behavior at a larger scale (macroscale) can be described by the averaged or homogenized behavior of the smaller-scale components (microscale). This separation allows for the modeling of complex materials in a more tractable manner by focusing on different scales individually and then linking them through appropriate transition rules.

As presented in Chapter 5, homogenization entails averaging the microscale features of a heterogeneous material to obtain equivalent macroscopic properties that govern its effective mechanical response. Within rock mechanics frameworks, homogenization allows for the representation of a heterogeneous rock mass (composed of grains, pores, microcracks, etc.) by a set of effective parameters, such as effective elasticity, plasticity, or damage laws, which can be applied at the larger scale of the engineering structure.

10.3.2 Eshelby's Equivalent Inclusion Theory and Effective Properties

Consider a heterogeneous material Ω consisting of a matrix phase Ω_0 and ellipsoidal inclusions $\Omega_1, \Omega_2, \ldots, \Omega_N$, as shown in Figure 10.1. Let the volume fraction of inclusion Ω_r be denoted as $f_r = \Omega_r/\Omega$. It is evident that

$$\sum_{r=0}^{N} f_r = 1 \tag{10.1}$$

Assume the stiffness tensor of phase Ω_r is represented by $\boldsymbol{L}_r (r = 0, 1, \ldots, N)$. The objective is to evaluate the effective elastic modulus of the composite by incorporating the stiffness characteristics and volume fractions of both the matrix and inclusion phases. Although the multiphase configuration shown in Figure 10.1 offers a generalized description, practical constitutive modeling for rock-like materials often simplifies the composite into two or three distinct components—typically including a continuous matrix, one or more reinforcing elements, and defects such as voids or cracks that act as weakening phases.

Consider the case where the composite is subjected to a specified displacement field along its boundaries

$$\boldsymbol{u}|_S = \boldsymbol{E}\boldsymbol{x} \tag{10.2}$$

and the traction boundary condition, namely

$$\boldsymbol{\sigma} \cdot \boldsymbol{n}|_S = \sum \boldsymbol{n} \tag{10.3}$$

where $\boldsymbol{E}$ and $\sum$ are constant second-order tensors and are related by

$$\sum = \boldsymbol{L}^{\text{hom}} : \boldsymbol{E} \tag{10.4}$$

where $\boldsymbol{L}^{\text{hom}}$ is the fourth-order effective stiffness taking into account the mechanical behavior of inhomogeneities and their morphology.

To derive $\boldsymbol{L}^{\text{hom}}$, Eshelby's equivalent inclusion theory is adopted, which could determine the local strain (stress) under the action of a displacement boundary condition or a

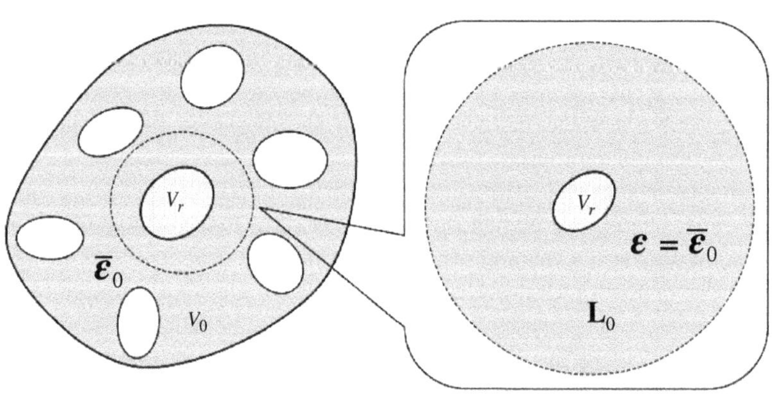

Figure 10.1 Composite.

traction boundary condition. Taking the displacement boundary condition as an example, the equivalent inclusion equation is

$$\boldsymbol{L}_r : (\boldsymbol{\varepsilon}^0 + \boldsymbol{\varepsilon}_r^d) = \boldsymbol{L}_0 : (\boldsymbol{\varepsilon}^0 + \boldsymbol{\varepsilon}_r^d - \boldsymbol{\varepsilon}_r^*) \tag{10.5}$$

where $\boldsymbol{\varepsilon}_r^d$ is the disturbed strain due to the presence of the inhomogeneity r and $\boldsymbol{\varepsilon}_r^*$ is the corresponding eigenstrain. According to Eshelby's solution,

$$\boldsymbol{\varepsilon}_r^d = \boldsymbol{S} : \boldsymbol{\varepsilon}_r^* \tag{10.6}$$

where $\boldsymbol{S}$ denotes the Eshelby tensor [9], which is determined by the shape of the embedded inhomogeneities and the elastic properties of the surrounding matrix.

Substituting Eq. (10.6) into Eq. (10.5) yields the eigenstrain

$$\boldsymbol{\varepsilon}_r^* = [(\boldsymbol{L}_r - \boldsymbol{L}_0) : \boldsymbol{S} + \boldsymbol{L}_0]^{-1} : (\boldsymbol{L}_r - \boldsymbol{L}_0) : \boldsymbol{\varepsilon}^0 \tag{10.7}$$

Accordingly, the total strain experienced by the inhomogeneity rth may be formulated as:

$$\boldsymbol{\varepsilon}_r = \boldsymbol{\varepsilon}^0 + \boldsymbol{\varepsilon}^d = \left[\boldsymbol{I} + \boldsymbol{S} : \boldsymbol{L}_0^{-1} : (\boldsymbol{L}_r - \boldsymbol{L}_0)\right]^{-1} : \boldsymbol{\varepsilon}^0 = \boldsymbol{A}_r : \boldsymbol{\varepsilon}^0 \tag{10.8}$$

where $\boldsymbol{A}_r$ is the so-called concentration tensor. The corresponding stress in the $\boldsymbol{r}$th inhomogeneity is

$$\boldsymbol{\sigma}_r = \boldsymbol{L}_r : \boldsymbol{\varepsilon}_r = \boldsymbol{L}_r : \boldsymbol{A}_r : \boldsymbol{E} \tag{10.9}$$

The average stress and strain over a heterogeneous material Ω is defined as:

$$\sum = \frac{1}{\Omega} \int_D \sigma dV \quad \boldsymbol{E} = \frac{1}{\Omega} \int_D \varepsilon dV \tag{10.10}$$

Assuming that the stress and strain is homogeneous in inhomogeneities, Eq. (10.10) can be rewritten in discrete form as

$$\sum = \sum_{r=0}^{N} f_r \boldsymbol{\sigma}_r \quad \boldsymbol{E} = \sum_{r-0}^{N} f_r \boldsymbol{\varepsilon}_r \tag{10.11}$$

Substituting Eq. (10.9) into Eq. (10.11) yields

$$\sum = \sum_{r=0}^{N} f_r \boldsymbol{L}_r : \boldsymbol{A}_r : \boldsymbol{E} = \boldsymbol{L}^{\text{hom}} : \boldsymbol{E} \tag{10.12}$$

In summary, micromechanics fundamentally aims to determine local concentration tensors. Once these tensors are established, the effective properties can be directly derived using Eq. (10.12).

10.3.3 Hill's Incremental Method

The nonlinear homogenization approach originally proposed by Hill is briefly introduced in this section. Within the framework of incremental formulation, the average stress and strain rate in the rth inhomogeneity are connected through the following relationship:

$$\dot{\boldsymbol{\sigma}}_r = \boldsymbol{C}_r^{\text{tg}} : \dot{\boldsymbol{\varepsilon}}_r \tag{10.13}$$

where $\boldsymbol{C}_r^{\mathrm{tg}}$ denotes the continuum tangent modulus. In order to extend homogenization schemes originally formulated for linear systems to the nonlinear regime, it is typically postulated that the mechanical fields (e.g. stress, strain) are uniform within each individual constituent phase. Although this uniformity assumption deviates from the true heterogeneous nature of inelastic deformation, it remains a tractable and effective approximation. When coupled with auxiliary numerical procedures, such as isotropic projection or regularization strategies, the micromechanical framework can yield effective macroscopic responses that closely align with experimental observations and full-field simulations.

Analogously, the overall mechanical behavior of composite materials is governed by the homogenized parameters obtained from:

$$\dot{\boldsymbol{\Sigma}} = \boldsymbol{C}^{\mathrm{hom}} : \dot{\boldsymbol{E}} \tag{10.14}$$

where $\dot{\boldsymbol{\Sigma}}$ and $\dot{\boldsymbol{E}}$ denote the macroscopic strain and strain rate, respectively. $\boldsymbol{C}^{\mathrm{hom}}$ is an effective tangent operator, which governs the constitutive response at the macroscopic level, is yet to be determined. Hill further proposed that the local average strain $\dot{\varepsilon}$ within a given phase is related to the macroscopic strain $\dot{\boldsymbol{E}}$ via a strain localization tensor $\boldsymbol{A}_\gamma$:

$$\dot{\varepsilon}_r = \boldsymbol{A}_r : \dot{\boldsymbol{E}} \tag{10.15}$$

The expression of $\boldsymbol{A}_r$ can be determined using Eq. (10.8) with the stiffness tensor $\boldsymbol{L}$ replaced by continuum tangent operator. In nonlinear homogenization, given that the stiffness tensor $\boldsymbol{C}_0^{\mathrm{tg}}$ is typically anisotropic, the computation of the corresponding Eshelby tensor $\boldsymbol{S}$ cannot be performed analytically in general and thus necessitates numerical implementation [10].

Substituting Eq. (10.15) into Eq. (10.13) and considering the micro-to-macro stress rate relation $\langle \dot{\sigma}_r \rangle = \dot{\sum}$, the resulting expression yields the effective tangent operator:

$$\boldsymbol{C}^{\mathrm{hom}} = \langle \boldsymbol{C}_r^{\mathrm{tg}} : \boldsymbol{A}_r \rangle = \sum_{r=0}^{r} f_r \boldsymbol{C}_r^{\mathrm{tg}} : \boldsymbol{A}_r \tag{10.16}$$

where $\langle \cdot \rangle$ represents the volume-averaged value of the specified tensorial quantity within the considered domain.

In practical computational implementations, constitutive relations are discretized over prescribed time steps, during which the incremental changes in stress and strain are connected via a numerically consistent tangent stiffness operator, denoted as:

$$d\sigma_r = \boldsymbol{C}_r^{\mathrm{alg}} : d\varepsilon_r \tag{10.17}$$

where $\boldsymbol{C}_r^{\mathrm{alg}}$ is a consistent tangent operator. According to Doghri [11], the consistent tangent operator plays a crucial role in nonlinear homogenization schemes, as its use typically leads to a softer macroscopic response compared to other approximations. Moreover, making use of the consistent tangent operator improves the stability of the local constitutive law integration and the homogenization procedure. Therefore, the consistent tangent operator is used in the following calculation. All the quantities related to $\boldsymbol{C}^{\mathrm{tg}}$ are then changed to $\boldsymbol{C}^{\mathrm{alg}}$.

10.4 Pressure Dependent Plasticity Model

To accurately describe the mechanical behavior of rock-like materials, which fundamentally differ from metals, a pressure-dependent plasticity model is indispensable. While metals typically yield under deviatoric (shear) stresses and can be modeled using pressure-independent plasticity, rocks exhibit significant sensitivity to hydrostatic pressure. At low confining pressures, rocks tend to fail in a brittle manner through cracking or fracturing; however, under high confining pressures, they may undergo more ductile deformation, displaying increased strength and plasticity. This pressure-dependent yielding implies that the strength and deformation of rocks are strongly influenced by the stress state, requiring plasticity models like the Mohr–Coulomb or Drucker–Prager criteria that explicitly account for confining pressure effects.

10.4.1 Thermodynamic Formulation

The total free energy governing elastoplastic behavior is the sum of the recoverable elastic strain energy and the stored (locked) energy associated with plastic hardening:

$$W(\varepsilon, \varepsilon^p, \gamma^p) = \frac{1}{2}(\varepsilon - \varepsilon^p) : \boldsymbol{C} : (\varepsilon - \varepsilon^p) + W^\gamma(\gamma^p) \tag{10.18}$$

where $\boldsymbol{\varepsilon}^p$ is the plastic strain tensor, and γ^p is the scalar internal variable characterizing isotropic hardening. The isotropic stiffness tensor $\boldsymbol{C} = 3k\boldsymbol{J} + 2\mu\boldsymbol{K}$

The state laws are derived by taking the standard derivatives of the free energy with respect to the corresponding internal variables, namely:

$$\sigma(\varepsilon, \varepsilon^p, \gamma^p) = \frac{\partial W}{\partial \varepsilon} = \boldsymbol{C} : (\boldsymbol{\varepsilon} - \boldsymbol{\varepsilon}^p) \tag{10.19}$$

$$\boldsymbol{F}^p(\varepsilon, \varepsilon^p, \gamma^p) = -\frac{\partial W}{\partial \varepsilon^p} = \sigma \tag{10.20}$$

$$\boldsymbol{F}^\gamma(\varepsilon, \varepsilon^p, \gamma^p) = -\frac{\partial W}{\partial \gamma^p} = -\frac{\partial W^p}{\partial \gamma^p} \tag{10.21}$$

10.4.2 Elastoplastic Constitutive Relations

Nonlinear constitutive formulations commonly employ rate forms of the governing equations, which facilitate incremental numerical integration for path-dependent processes such as plastic deformation and damage progression. Within this framework, the additive decomposition of the strain rate tensor is typically assumed, which separates the total strain rate into its elastic and plastic components, namely:

$$\dot{\boldsymbol{\varepsilon}} = \dot{\boldsymbol{\varepsilon}}^e + \dot{\boldsymbol{\varepsilon}}^p \tag{10.22}$$

In pressure-dependent plasticity, the yield function is typically formulated in terms of the first and second invariants of the stress tensor, and can be expressed as

$$f(p, q, H^\alpha) = 0 \tag{10.23}$$

Here, p and q are the hydrostatic stress and equivalent stress, respectively:

$$p = -\frac{1}{3}\boldsymbol{\sigma} : \boldsymbol{\delta} \quad \boldsymbol{s} = \boldsymbol{\sigma} + p\boldsymbol{\delta} \quad q = \sqrt{\frac{3}{2}\boldsymbol{s} : \boldsymbol{s}} \tag{10.24}$$

where H^α ($\alpha = 0, 1, \ldots, n$) represents a set of scalar internal state variables, $\boldsymbol{\delta}$ denotes the second-order identity tensor, and $\boldsymbol{s}$ is the deviatoric part of the stress tensor. The function f is defined in such a way that the material behaves elastically as long as $f < 0$.

A nonassociated plasticity is assumed, which is realized by introducing a new plastic potential, $g(p, q, H^\alpha)$, different from that of the yield function. In this case, the flow rule is written:

$$\dot{\boldsymbol{\varepsilon}}^p = \lambda \frac{\partial g}{\partial \boldsymbol{\sigma}} \tag{10.25}$$

where λ is a positive scalar. Using Eq. (10.24), the flow rule becomes

$$\dot{\boldsymbol{\varepsilon}}^p = \lambda\left(-\frac{1}{3}\frac{\partial g}{\partial p}\boldsymbol{\delta} + \frac{\partial g}{\partial q}\boldsymbol{n}\right) = \frac{1}{3}\dot{\varepsilon}_p\boldsymbol{\delta} + \dot{\varepsilon}_q\boldsymbol{n} \tag{10.26}$$

where

$$\dot{\varepsilon}_p = -\lambda\frac{\partial g}{\partial p} \quad \dot{\varepsilon}_q = \lambda\frac{\partial g}{\partial q} \quad \boldsymbol{n} = \frac{3}{2q}\boldsymbol{s}$$

The plasticity framework is fully defined upon specifying the evolution equations governing the internal state variables, denoted by

$$\dot{H}^\alpha = h^\alpha\left(\boldsymbol{\sigma}, \dot{\boldsymbol{\varepsilon}}^p, H^\beta\right) \tag{10.27}$$

10.5 Application of the Multiscale Constitutive Modeling to the COx Argillite

10.5.1 Mineralogy and the Mechanical Response of the COx Argillite

The subject of investigation in this chapter is the Callovo–Oxfordian (COx) argillite, provided by the French National Radioactive Waste Management Agency (ANDRA). The composition and microstructure of COx argillite were analyzed by X-ray diffraction (XRD) and scanning electron microscopy (SEM), and the results are shown in Figure 10.2.

As Figure 10.2 shows, three main mineral phases are identified at the mesoscopic scale: quartz (23%), calcite (27%), and clay matrix (45%) together with a small portion of feldspars, pyrite, and iron oxides (5%). Mercury intrusion porosimetry (MIP) further reveals that the clay matrix comprises both interparticle pores and a mixture of clay minerals (primarily 65% illite-smectite interstratified phases, 30–35% illite, and up to 5% kaolinite and chlorite), which together form the dominant porosity of the COx argillite [12]. For clarity, the porous clay matrix, calcite, and quartz are denoted as phases 1, 2, and 3, respectively. A schematic representation of the COx argillite composition is provided in Figure 10.3.

A range of experimental studies has been conducted on COx argillite, including investigations into its short-term and long-term mechanical behavior [13], hydro-mechanical coupling [14], oedometric compression, and thermo-mechanical responses [15], among

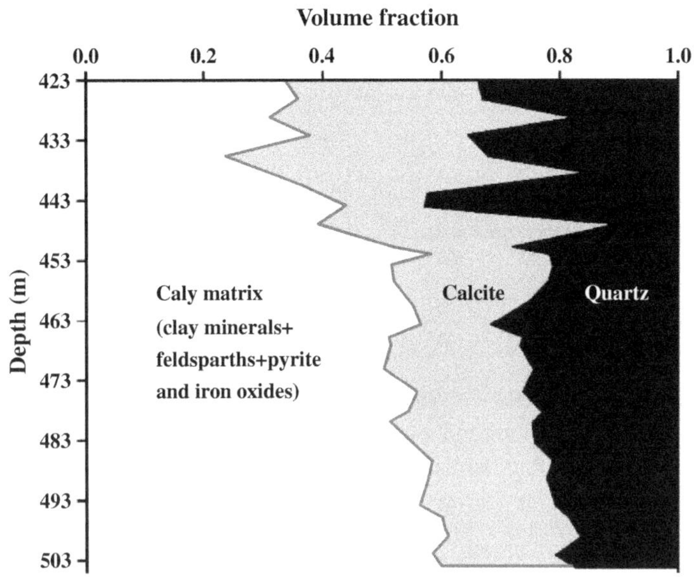

Figure 10.2 Relationship between volume fraction of different phases and depth in COx argillite. *Source*: Adapted from [12].

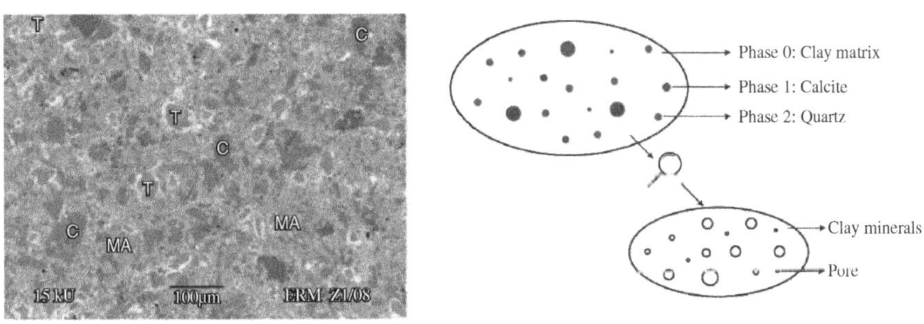

(a) (b)

Figure 10.3 (a) COx argillite sample: C-calcite grains, T-quartz grains, MA-clay matrix [12]; and (b) the composition of COx argillite.

others. These experimental results provide a robust foundation for the subsequent numerical modeling. Key features of its short-term mechanical response, directly relevant to our simulation efforts, are documented in [1]. The study highlights that the COx argillite exhibits limited anisotropy, with irreversible deformation mainly governed by mechanisms such as clay particle sliding, void nucleation, and growth. Moreover, mineralogical composition significantly affects its plastic behavior: higher calcite content leads to increased brittleness, whereas a greater proportion of porous clay promotes ductile characteristics [1].

10.5.2 Modelling of the Porous Clay Matrix by the extended GTN Model

To incorporate the influence of confining pressure on yielding behavior, the solid phase of the clay matrix is modeled as plastically compressible. Accordingly, the von Mises yield criterion traditionally employed in the derivation of the GTN model is substituted with the Drucker–Prager yield criterion:

$$\bar{f}(\bar{p}, \bar{q}, \bar{\varepsilon}^p) = \bar{p} - \alpha\bar{q} - \sigma_y(\bar{\varepsilon}^p) \tag{10.28}$$

where

$$\bar{p} = -\frac{1}{3}\bar{\sigma} : \delta \quad \bar{\mathbf{s}} = \bar{\sigma} + \bar{p}\delta \quad q = \sqrt{\frac{3}{2}\bar{\mathbf{s}} : \bar{\mathbf{s}}}$$

and a nonassociated plastic potential:

$$\bar{g}(\bar{p}, \bar{q}, \bar{\varepsilon}^p) = \bar{p} - \beta\bar{q} - \sigma_y(\bar{\varepsilon}^p) \tag{10.29}$$

$\bar{\sigma}$ denotes the microscopic stress associated with the solid phase of the porous clay matrix. α is the friction coefficient, while β represents the dilatancy coefficient governing the volumetric component of the plastic strain, which equals:

$$\mathrm{tr}(d\bar{\varepsilon}^p) = d\bar{\lambda}\left(\frac{3}{2}\frac{\bar{\mathbf{s}}}{\bar{q}} + \frac{1}{3}\beta\delta\right) : \delta = \beta d\bar{\lambda} \tag{10.30}$$

where $d\bar{\lambda}$ is a strictly positive scalar determined in accordance with the Kuhn–Tucker complementarity conditions [16]. $d\bar{\varepsilon}^p$ denotes the plastic strain rate of the solid phase within the porous clay matrix. For frictional materials, the microscopic equivalent plastic strain is commonly defined as:

$$d\bar{\varepsilon}^p = \sqrt{\frac{2}{3}\mathrm{dev}\,(d\bar{\varepsilon}^p) : \mathrm{dev}\,(d\bar{\varepsilon}^p)} = d\bar{\lambda} \tag{10.31}$$

where $\mathrm{dev}(d\bar{\varepsilon}^p)$ denotes the deviatoric component of $d\bar{\varepsilon}^p$. By combining Eqs. (10.30) and (10.31), the volumetric plastic strain rate is related to the microscopic equivalent plastic strain through the following expression:

$$\mathrm{tr}(d\bar{\varepsilon}^p) = \beta d\bar{\varepsilon}^p \tag{10.32}$$

Considering that the solid phase of the porous clay matrix satisfies Eq. (10.28), Lazzeri and Bucknall [17] extended the yield criterion of the GTN model as follows:

$$f(p, q, \bar{\varepsilon}^p, \phi) = \left(\frac{q}{\sigma_y}\right)^2 - \frac{\alpha p}{\sigma_y}\left(2 + \frac{\alpha p}{\sigma_y}\right) + 2q_1\phi\cosh\left(-\frac{3}{2}\frac{q_2 p}{\sigma_y}\right) - (1 + q_3\phi^2) = 0 \tag{10.33}$$

where

$$p = -\frac{1}{3}\sigma : \delta \quad \mathbf{s} = \sigma + p\delta \quad q = \sqrt{\frac{3}{2}\mathbf{s} : \mathbf{s}}$$

In Eq. (10.33), p and q are hydrostatic stress and equivalent stress, respectively. ϕ is the interparticle porosity of the clay matrix. q_1, q_2, and q_3 are parameters, which always equal one. It is noteworthy that Eq. (10.33) simplifies to the GTN model as ϕ approaches zero, and reduces to the Drucker–Prager model when α tends to zero. $\sigma_y(\bar{\varepsilon}^p)$ is the yield stress of the fully dense clay matrix (i.e. $\phi = 0$), expressed as a function of the microscopic equivalent plastic strain $\bar{\varepsilon}^p$. In the subsequent numerical simulations, an exponential function is employed to characterize the hardening behavior of the COx argillite:

$$\sigma_y = \sigma_{ym} - \left(\sigma_{ym} - \sigma_{y0}\right)e^{-b\bar{\varepsilon}^p} \tag{10.34}$$

where σ_{y0} represents the initial yield stress, typically very low, while σ_{ym} denotes the asymptotic yield stress at saturation. The parameter b controls the rate at which the material approaches this saturation level.

It is widely recognized that the plastic flow of geomaterials such as rocks and soils deviates from the normality (associated) flow rule, exhibiting a transition from compressibility to plastic dilation. Therefore, a nonassociated flow rule is commonly employed to capture the inelastic volumetric deformation. Since this study focuses on the feasibility of the micro–macro constitutive model, a simplified modification of the yield function in Eq. (10.33) is proposed for the plastic potential function:

$$g(p, q, \bar{\varepsilon}^p, \phi) = \left(\frac{q}{\sigma_y}\right)^2 - \frac{\beta p}{\sigma_y}\left(2 + \frac{\beta p}{\sigma_y}\right) + 2q_1\phi\cosh\left(-\frac{3}{2}\frac{q_2 p}{\sigma_y}\right) - (1 + q_3\phi^2) \tag{10.35}$$

Then, the plastic strain rate can be defined as:

$$d\varepsilon^p = d\lambda\frac{\partial g}{\partial \boldsymbol{\sigma}} = \frac{1}{3}d\varepsilon_p\boldsymbol{\delta} + \frac{3}{2q}d\varepsilon_q\boldsymbol{s} \tag{10.36}$$

where

$$d\varepsilon_p = -d\lambda\frac{\partial g}{\partial p} \quad d\varepsilon_q = d\lambda\frac{\partial g}{\partial q}$$

To fully define the local constitutive model, it is essential to specify the evolution laws for the state variables $\bar{\varepsilon}^p$ and ϕ. For $\bar{\varepsilon}^p$, its variation is assumed to depend on the accumulated equivalent plastic work, expressed as:

$$(1 - \phi)\sigma_y d\bar{\varepsilon}^p = \boldsymbol{\sigma} : d\varepsilon^p \tag{10.37}$$

For most geomaterials, changes in porosity typically arise from two primary mechanisms: (1) the growth of pre-existing voids under various loading paths, and (2) the nucleation of new voids. However, due to the limited availability of microscopic analyses and experimental data, the underlying mechanisms governing these processes, as well as the associated parameters, remain difficult to characterize with accuracy.

Therefore, the second question is not investigated in this chapter. The change due to porosity is related to the volume change of the solid part of the matrix by applying kinematic compatibility:

$$d\phi = (1 - \phi)\left(\mathrm{tr}(d\varepsilon^p) - \mathrm{tr}\left(d\bar{\varepsilon}^p\right)\right) = (1 - \phi)\left(d\varepsilon_p - \beta d\bar{\varepsilon}^p\right) \tag{10.38}$$

In summary, the constitutive behavior of the material is described by the following system of equations:

$$\boldsymbol{\sigma} = \boldsymbol{C} : (\varepsilon - \varepsilon^p) \tag{10.39}$$

$$d\boldsymbol{\varepsilon}^p = d\lambda \frac{\partial g}{\partial \boldsymbol{\sigma}} = \frac{1}{3} d\varepsilon_p \boldsymbol{\delta} + \frac{3}{2q} d\varepsilon_q \mathbf{s} \tag{10.40}$$

$$f(p, q, \bar{\varepsilon}^p, \phi) = 0 \tag{10.41}$$

$$d\bar{\varepsilon}^p = \frac{\boldsymbol{\sigma} : d\boldsymbol{\varepsilon}^p}{(1-\phi)\sigma_y} \tag{10.42}$$

$$d\phi = (1-\phi)\left(d\varepsilon_p - \beta d\bar{\varepsilon}^p\right) \tag{10.43}$$

The system comprises nine independent equations (Eqs. 10.39–10.43) and nine unknowns (ε_{ij}^p, λ, $\bar{\varepsilon}^p$ and ϕ), forming a closed set of equations. It is solved numerically using the backward Euler method.

10.6 Computational Aspects

This section provides a detailed description of the computational procedures, including both the local integration algorithm and the homogenization scheme. The local integration is performed using an elastic predictor/return-mapping algorithm, while the homogenized response, particularly the effective tangent operator in Eq. (10.16), is computed using a fixed-point iteration method.

10.6.1 Local Integration Algorithm and the Consistent Tangent Moduli

Based on the strain increment $\Delta\varepsilon$ and the known state variables at time step t, the local integration process aims to update all internal variables at the next time step $t + \Delta t$. Simultaneously, the consistent tangent moduli are calculated, which play a critical role in maintaining the convergence efficiency and numerical stability of the homogenization algorithm [11]. This process is carried out using the backward Euler method. Accordingly, Eqs. (10.39)–(10.43) are reformulated, with all variables (unless otherwise noted) referring to their values at time step $t + \Delta t$:

$$\boldsymbol{\sigma} = \mathbf{C} : \left(\boldsymbol{\varepsilon}_t + \boldsymbol{\varepsilon} - \boldsymbol{\varepsilon}_t^p\right) - \mathbf{C} : \Delta\boldsymbol{\varepsilon}^p = \boldsymbol{\sigma}^{\text{trial}} - \mathbf{C} : \Delta\boldsymbol{\varepsilon}^p \tag{10.44}$$

$$\Delta\boldsymbol{\varepsilon}^p = \frac{1}{3}\Delta\varepsilon_p \boldsymbol{\delta} + \frac{3}{2q}\Delta\varepsilon_q \mathbf{s} \tag{10.45}$$

$$f(p, q, \bar{\varepsilon}^p, \phi) = 0 \tag{10.46}$$

$$\Delta\bar{\varepsilon}^p = \frac{q\Delta\varepsilon_q - p\Delta\varepsilon_p}{(1-\phi)\sigma_y} \tag{10.47}$$

$$\Delta\phi = (1-\phi)\left(\Delta\varepsilon_p - \beta\Delta\bar{\varepsilon}^p\right) \tag{10.48}$$

where $\boldsymbol{\sigma}^{\text{trial}} = \boldsymbol{C} : \boldsymbol{\varepsilon}^{\text{trial}} = \boldsymbol{C} : (\boldsymbol{\varepsilon}_t + \Delta\boldsymbol{\varepsilon} - \boldsymbol{\varepsilon}_t^p)$ is the elastic predictor. Considering that the elastic stiffness tensor $\boldsymbol{C}$ is isotropic, and using Eqs. (10.45) and (10.44) is recast into:

$$\boldsymbol{\sigma} = \boldsymbol{\sigma}^{\text{trial}} - K\Delta\varepsilon_p\boldsymbol{\delta} - 3G\Delta\varepsilon\frac{\boldsymbol{s}}{q} \tag{10.49}$$

Within the deviatoric stress space, Eq. (10.49) can be rewritten as:

$$\boldsymbol{s} = \boldsymbol{s}^{\text{trial}} - 3G\Delta\varepsilon\frac{\boldsymbol{s}}{q} \tag{10.50}$$

A self-inner product operation on the transformed form of Eq. (10.50) leads to the following result:

$$q + 3G\Delta\varepsilon_q = q^{\text{trial}} \tag{10.51}$$

where $q^{\text{trial}} = \sqrt{3\boldsymbol{s}^{\text{trial}} : \boldsymbol{s}^{\text{trial}}/2}$. Then, combining Eqs. (10.50) and (10.51), the following relationship can be known:

$$\frac{s}{q} = \frac{s^{\text{trial}}}{q^{\text{trial}}} \tag{10.52}$$

This feature significantly simplifies the computation. Eq. (10.45) now involves only two independent variables instead of eight. Likewise, Eq. (10.49) indicates that the stress tensor at time step $t + \Delta t$ can be fully determined by the two quantities $\Delta\varepsilon_p$ and $\Delta\varepsilon_q$. By projecting Eq. (10.49) onto $\boldsymbol{\delta}$ and $3s/2q$, more explicit relationships between the stress invariants p and q with $\Delta\varepsilon_p$ and $\Delta\varepsilon_q$ are as follows:

$$p = p^{\text{trial}} + K\Delta\varepsilon_p \quad q = q^{\text{trial}} - 3G\Delta\varepsilon_q \tag{10.53}$$

where $p^{\text{trial}} = -\sigma^{\text{trial}} : \delta/3$. In practical implementation, an alternative formulation of Eq. (10.45) is often adopted for computational convenience, namely:

$$\Delta\varepsilon_p\frac{\partial g}{\partial q} + \Delta\varepsilon_q\frac{\partial g}{\partial p} = 0 \tag{10.54}$$

Equation (10.54) can be obtained from the definitions of $\Delta\varepsilon_p$ and $\Delta\varepsilon_q$ by eliminating $\Delta\lambda$. In summary, the local integration algorithm is reduced to solving a system of four nonlinear equations with $\Delta\varepsilon_p$ $\Delta\varepsilon_q$ $\Delta\bar{\varepsilon}^p$ and $\Delta\phi$:

$$\Delta\varepsilon_p\frac{\partial g}{\partial q} + \Delta\varepsilon_q\frac{\partial g}{\partial p} = \text{res}(\varepsilon^p) = 0 \tag{10.55}$$

$$f(p, q, \bar{\varepsilon}^p, \phi) = \text{res}(f) = 0 \tag{10.56}$$

$$\Delta\bar{\varepsilon}^p - \frac{q\Delta\varepsilon_q - p\Delta\varepsilon_p}{(1-\phi)\sigma_y} = \text{res}(\Delta\bar{\varepsilon}^p) = 0 \tag{10.57}$$

$$\Delta\phi - (1-\phi)(\Delta\varepsilon_p - \beta\Delta\bar{\varepsilon}^p) = \text{res}(\Delta\phi) = 0 \tag{10.58}$$

where "res" denotes the residual associated with each governing equation. The resulting nonlinear system is solved using the Newton–Raphson iterative scheme. At each iteration step k, the increments of the four unknown variables are computed by solving the following linearized system:

$$\begin{bmatrix} A_{11} & A_{12} & A_{13} & A_{14} \\ A_{21} & A_{22} & A_{23} & A_{24} \\ A_{31} & A_{32} & A_{33} & A_{34} \\ A_{41} & A_{42} & A_{43} & A_{44} \end{bmatrix} \begin{bmatrix} \delta\Delta\varepsilon_p \\ \delta\Delta\varepsilon_q \\ \delta\Delta\bar{\varepsilon}^p \\ \delta\Delta\phi \end{bmatrix} = - \begin{bmatrix} \mathrm{res}(\varepsilon^p) \\ \mathrm{res}(f) \\ \mathrm{res}(\Delta\bar{\varepsilon}^p) \\ \mathrm{res}(\Delta\phi) \end{bmatrix} \tag{10.59}$$

After completing the local integration using the backward Euler method, the consistent tangent modulus can be obtained by differentiating the solution variables of the nonlinear algebraic system obtained from Eq. (10.55) to (10.58):

$$\begin{bmatrix} A_{11} & A_{12} & A_{13} & A_{14} \\ A_{21} & A_{22} & A_{23} & A_{24} \\ A_{31} & A_{32} & A_{33} & A_{34} \\ A_{41} & A_{42} & A_{43} & A_{44} \end{bmatrix} \begin{bmatrix} d\Delta\varepsilon_p \\ d\Delta\varepsilon_q \\ d\Delta\bar{\varepsilon}^p \\ d\Delta\phi \end{bmatrix} = \begin{bmatrix} -\Delta\varepsilon_q \dfrac{\partial^2 g}{\partial p^2} dp^{\mathrm{trial}} - \Delta\varepsilon_p \dfrac{\partial^2 g}{\partial q^2} dq^{\mathrm{trial}} \\ -\dfrac{\partial f}{\partial p} dp^{\mathrm{trial}} - \dfrac{\partial f}{\partial q} dq^{\mathrm{trial}} \\ \dfrac{-\Delta\varepsilon_p}{(1-f)\sigma_y} dp^{\mathrm{trial}} + \dfrac{\Delta\varepsilon_q}{(1-f)\sigma_y} dq^{\mathrm{trial}} \\ 0 \end{bmatrix} \tag{10.60}$$

Through the resolution of Eq. (10.60) and subsequent term collection, a linear dependence of $d\Delta\varepsilon_p$ and $d\Delta\varepsilon_q$ on dp^{trial} and dq^{trial} is established as:

$$d\Delta\varepsilon_p = b_{11} dp^{\mathrm{trial}} + b_{12} dq^{\mathrm{trial}} \quad d\Delta\varepsilon_q = b_{21} dp^{\mathrm{trial}} + b_{22} dq^{\mathrm{trial}} \tag{10.61}$$

where b_{11}, b_{12}, b_{21}, b_{22} denote coefficient parameters. The consistent tangent moduli are subsequently obtained by differentiating Eq. (10.49), in conjunction with Eq. (10.61), and are expressed as:

$$\begin{aligned} \boldsymbol{C}^{\mathrm{alg}} &= \boldsymbol{C} + b_{11} K^2 \boldsymbol{\delta} \otimes \boldsymbol{\delta} - 2 b_{12} K G \boldsymbol{n}^{\mathrm{trial}} \otimes \boldsymbol{\delta} \\ &\quad + 2 b_{21} K G \boldsymbol{\delta} \otimes \boldsymbol{n}^{\mathrm{trial}} - 4 G^2 b_{22} \boldsymbol{n}^{\mathrm{trial}} \otimes \boldsymbol{n}^{\mathrm{trial}} \\ &\quad - \frac{4 G^2 \Delta\varepsilon_q}{q^{\mathrm{trial}}} \left(\frac{3}{2} \boldsymbol{K} - \boldsymbol{n}^{\mathrm{trial}} \otimes \boldsymbol{n}^{\mathrm{trial}} \right) \end{aligned} \tag{10.62}$$

where $n^{\mathrm{trial}} = 3 s^{\mathrm{trial}} / 2 q^{\mathrm{trial}}$. Even with the associated flow laws, the consistent tangent modulus for porous materials is inherently asymmetric. This asymmetry arises from the coupling between the deviatoric and hydrostatic responses. It is noteworthy that in the derivation of Eq. (10.62), no additional matrix inversion is required, in contrast to the approach of Aravas [18]. This difference stems from the fact that the consistent tangent modulus is derived based on the variation of the test stress rather than the stress itself. Therefore, the derivation is applicable to materials that behave in a rigid-plastic manner.

10.6.2 Numerical Implementation of the Homogenization Procedure

Given the macroscopic stress Σ_t and all solution-dependent state variables at time step t, the objective of the homogenization procedure is to compute the effective tangent operator and the updated macroscopic variables $\Sigma_{t+\Delta t}$ corresponding to the applied macroscopic strain increment $\Delta \boldsymbol{E}$.

Initialize the average strain in the calcite and the quartz:

$$\Delta\varepsilon_1^{(0)} = \Delta \boldsymbol{E}, \Delta\varepsilon_2^{(0)} = \Delta \boldsymbol{E}.$$

1. For every iteration (i), the average strain in the clay matrix is determined as:

$$\Delta\varepsilon_0^{(i)} = \frac{\Delta E - f_1\Delta\varepsilon_1^{(i)} - f_2\Delta\varepsilon_2^{(i)}}{1 - f_1 - f_2}$$

2. Determine the consistent tangent moduli of the calcite and the quartz. In this paper, they simply equal their elastic moduli, namely

$$C_1^{\text{alg},(i)} = C_1, C_2^{\text{alg},(i)} = C_2.$$

3. With the latest updated $\Delta\varepsilon_0^{(i)}$ and all solution dependent variables at time step t, determine the consistent tangent moduli $C_0^{\text{alg},(i)}$ of the clay matrix according to the local integration procedure described in Section 10.5.1.
4. Computer the Eshelby tensor $\boldsymbol{S}^{(i)}$ with the consistent tangent moduli $C_0^{\text{alg},(i)}$, or isotropic approximation of the consistent tangent moduli as described in Section 10.6.1 if necessary.
5. Computer strain localization tensors of the calcite $\boldsymbol{A}_r^{(i)}$

$$B_r^{(i)} = \left[\boldsymbol{I} + \boldsymbol{S}^{(i)} : \left(\boldsymbol{C}_r^{\text{alg},(i)} : \left(\boldsymbol{C}_0^{\text{alg},(i)}\right)^{-1} - \boldsymbol{I}\right)\right]^{-1}$$

$$A_2^{(i)} = \boldsymbol{B}_2^{(i)} : \left(\sum_{r=0}^{2} f_r \boldsymbol{B}_r^{(i)}\right)^{-1}$$

$$A_1^{(i)} = \boldsymbol{B}_1^{(i)} : \left(\sum_{r=0}^{2} f_r \boldsymbol{B}_r^{(i)}\right)^{-1}$$

6. Check the convergence of the strain in the calcite and the quartz by computing the residues:

$$\text{res}(\Delta\varepsilon_1) = \boldsymbol{A}_1^{(i)} : \Delta\boldsymbol{E} - \Delta\varepsilon_1^{(i)}$$
$$\text{res}(\Delta\varepsilon_2) = \boldsymbol{A}_2^{(i)} : \Delta\boldsymbol{E} - \Delta\varepsilon_2^{(i)}$$

7. If $\|\text{res}\,(\Delta\varepsilon_1)\| < \text{TOL} \times \|\Delta\boldsymbol{E}\|$ and $\|\text{res}\,(\Delta\varepsilon_2)\| < \text{TOL} \times \|\Delta\boldsymbol{E}\|$, then exit the loop. $\|\cdot\|$ is the Euclidean norm and TOL is a prescribed tolerance, which equals 10^{-3} in our program.
8. If the convergence check fails, new iteration is needed by using the updated strain:

$$\Delta\varepsilon_1^{(i+1)} = \Delta\varepsilon_1^{(i)} + \text{res}(\Delta\varepsilon_1)$$
$$\Delta\varepsilon_2^{(i+1)} = \Delta\varepsilon_2^{(i)} + \text{res}(\Delta\varepsilon_2)$$

The iteration should repeat until the convergence is obtained.

Once convergence is achieved, calculate the effective (homogenized) stiffness tensor and update the macroscopic stress accordingly:

$$\boldsymbol{C}^{\text{hom}} = \sum_{r=0}^{N} f_r \boldsymbol{C}_r^{\text{alg}} : \boldsymbol{A}_r$$
$$\Sigma = \Sigma_t + \Delta\Sigma = \Sigma_t + \boldsymbol{C}^{\text{hom}} : \Delta E$$

10.7 Numerical Validations

To evaluate the performance of the proposed model, two types of validation are performed in this section. First, the model predictions are compared with reference results from a representative two-phase RVE simulation. Second, the model predictions are compared with existing experimental data.

10.7.1 Comparison Against FE Analysis on a Two-phase Unit Cell

Figure 10.4 illustrates the periodic composite architecture comprising hexagonal unit cells, each containing a spherical reinforcement. For such periodically structured composites, the predominant numerical approach involves FE analysis of a representative unit cell subjected to periodic boundary constraints, enabling efficient homogenization of macroscopic properties.

To optimize computational efficiency, the hexagonal unit cell is commonly simplified to an axisymmetric cylindrical representation. Leveraging geometric symmetry, the analysis domain is further reduced to one quadrant of the symmetry plane in the FE implementation.

The numerical simulations are carried out using ABAQUS 6.14. The geometry of the model consists of a rectangular domain with height L and width R. For a given inclusion volume fraction $f(f = 0.1)$, the radius of the inclusion is $r = \sqrt[3]{1.5fLR^2}$. The matrix phase is modeled using the micro-macro constitutive framework described in Section 10.4, while the inclusion phase is assumed to exhibit linear isotropic elastic behavior. The left and bottom edges of the model are fixed. Vertical displacement is applied on the top edge while the right edge is constrained to have the same radial displacement to avoid a gap between

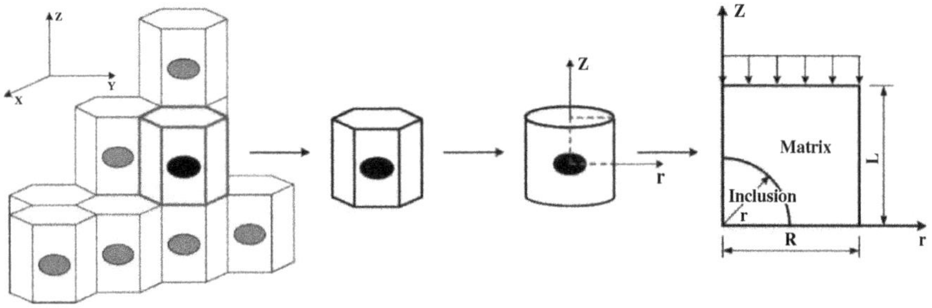

Figure 10.4 Two-phase unit cell: schematic diagram of a composite material with a regularly arranged microstructure and its simplified numerical model.

neighboring unit cells (periodic boundary conditions). The horizontal displacement of the right edge is coupled to the top-right node with the help of the "*Equation" command in ABAQUS. The computational domain is discretized using 3376 CAX8R elements and 1081 nodes, with a globally defined mesh size of $0.03R$.

The parameters of the matrix are as follows:

$$E = 3000\,\text{MPa},\ \mu = 0.3,\ \sigma_{y0} = 1\,\text{MPa},\ b = 250,\ \alpha = 0.4,\ \beta = 0.2,\ \sigma_{ym} = 25\,\text{MPa},\ \phi = 0.25$$

The parameters of the inclusion are as follows: $E = 98000\,\text{MPa}$, $\mu = 0.15$. In the finite element simulations, the automatic time incrementation technique is employed, with an initial increment of 0.1% and a maximum allowable increment of 1%. A total macroscopic strain of 2% is applied, and the macroscopic stress and strain are evaluated by

$$\boldsymbol{\Sigma} = \frac{1}{V} \sum_{g} \upsilon^g \boldsymbol{\sigma}^g$$

$$\boldsymbol{E} = \frac{1}{V} \sum_{g} \upsilon^g \boldsymbol{\varepsilon}^g$$

where ν^g denotes the volume associated with a specific Gauss integration point (available as the output variable "IVOL" in ABAQUS). The total model volume V (or area in the 2D case) is obtained by summing over all ν^g. The superscript g refers to quantities evaluated at the Gauss integration points.

The entire computational workflow can be automated through a Python script. Figure 10.5 presents the numerical results obtained from both finite element simulations and the proposed homogenization approach. The labels "Hom-iso" and "Hom-ani" refer respectively to the homogenization method employing an isotropic approximation of the anisotropic consistent tangent moduli and the one considering the full anisotropic form. The isotropic component of a fourth-order tensor is extracted using the following expression:

$$\boldsymbol{C}^{\text{iso}} = \left(\boldsymbol{J} :: \boldsymbol{C}^{\text{alg}}\right) \boldsymbol{J} + \frac{1}{5} \left(\boldsymbol{K} :: \boldsymbol{C}^{\text{alg}}\right) \boldsymbol{K}$$

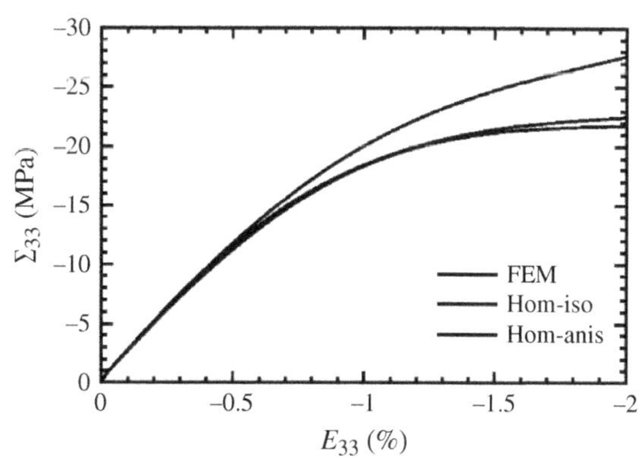

Figure 10.5 Macroscopic axial stress of the unit cell predicted by the proposed micromechanical model and FE computations.

Here, the superscript "iso" denotes the isotropic part of the tensor. Based on the isotropic approximation, the corresponding Eshelby tensor can be evaluated analytically. As illustrated in Figure 10.5, the use of isotropic consistent moduli yields predictions that align more closely with the reference results. Consequently, this approximation will be adopted in the subsequent validation against experimental data.

It should be emphasized that the isotropic approximation lacks a rigorous physical or mathematical foundation. Its adoption is primarily motivated by the improved agreement between the predictions of the homogenization scheme and those obtained from finite element simulations. A more detailed discussion of this issue can be found in [19].

10.7.2 Experimental Validation

As the second step of model validation, the predictions obtained from the proposed micromechanical framework are compared with experimental results on COx argillite samples exhibiting various mineralogical compositions. A major difficulty in implementing micromechanical models lies in the identification of constitutive parameters, which remains an unresolved issue due to the limited availability of microscale testing techniques. Nevertheless, certain elastic properties can still be estimated with reasonable confidence. For instance, the elastic moduli of quartz and calcite are well-documented in material property handbooks and monographs [20].

Given that the macroscopic elastic behavior of the samples has already been characterized through hydrostatic and triaxial compression experiments, the elastic constants of the porous clay matrix can be effectively back-calculated using analytical formulations based on the M-T homogenization scheme [21]. Additionally, the initial intergranular porosity is quantified by MIP, which provides reliable input for the micromechanical analysis. As for plasticity-related parameters, only the initial yield strength has been briefly mentioned in the literature, and it is generally acknowledged to be very low. Other plastic parameters, which are difficult to measure directly, are calibrated using the indirect parameter identification method proposed by Jiang et al. [19]. In this approach, a representative uniaxial or triaxial loading test is selected as a reference case, and a trial-and-error fitting procedure is employed to adjust model outputs until a satisfactory agreement with experimental stress–strain responses is achieved. The calibrated set of material parameters is then regarded as the reference dataset.

These reference parameters are subsequently used to simulate the mechanical response of other samples with varying mineral compositions under different loading conditions. It should be emphasized that in all further simulations, the calibrated parameters are kept constant, and only the volume fractions of constituent phases are adjusted to reflect changes in material composition. A complete list of the reference parameters used in the simulations is provided in Table 10.1.

Figures 10.6–10.12 illustrate the comparison between the numerical predictions generated by the proposed micromechanical model and the corresponding experimental measurements. The loading is applied along the three directions. E_{33} and E_{11} represent the axial

Table 10.1 Reference parameters used in experimental validation.

	Phase 0 Clay	Phase 1 Calcite	Phase 2 Quartz
Elastic parameters	$E = 2$ GPa	$E = 95$ GPa	$E = 100$ GPa
	$\mu = 0.2$	$\mu = 0.27$	$\mu = 0.06$
Plastic parameters	$\alpha = 0.7$		
	$\beta = 0.65$		
	$\sigma_{y0} = 1$ MPa		
	$\sigma_{ym} = 29$ MPa		
	$b = 175$		
Porosity	$\phi = 0.25$		

and radial strains, respectively. For conventional triaxial compression tests on cylindrical specimens, the volumetric strain E_v is equivalent to $E_{33} + 2E_{11}$ due to $E_{11} = E_{22}$. Overall, a satisfactory agreement is observed between the model predictions and experimental data. The micromechanical model demonstrates a strong capability to capture the influence of varying mineralogical compositions, representing a notable improvement over traditional phenomenological approaches that often neglect such microstructural details. Figure 10.13 and Figure 10.14 provide simulated results for the evolution of porosity at different depths within the specimen. As experimental data for porosity variation are unavailable for direct comparison, the provided results serve primarily a qualitative purpose. The simulations suggest that porosity decreases under confining pressure, reflecting a pore collapse or plastic compaction behavior in the early loading stage. Nevertheless, the overall change in porosity throughout the loading process remains relatively limited, which is attributed to the relatively low strain levels (up to 0.02).

It is acknowledged that the current model still requires further refinement, particularly in predicting volumetric strains, where noticeable discrepancies remain. One plausible reason is that inaccuracies in the predicted radial strain, which contributes twice to the volumetric strain calculation, may significantly amplify the overall deviation between the numerical and experimental results. Drawing on insights from published studies and our own numerical investigations, potential improvements to the model can be achieved by introducing strain-dependent constitutive features. Specifically, the friction coefficient α and the dilatancy coefficient β may be formulated as functions of the microscopic equivalent plastic strain, thereby capturing the evolving nature of plastic deformation. Alternatively, damage mechanisms can be incorporated into the framework, such as the progressive deterioration of elastic moduli and yield strength due to microstructural degradation. However, incorporating such improvements necessitates the inclusion of additional material parameters, which inevitably increases the model's complexity and imposes greater demands on experimental calibration.

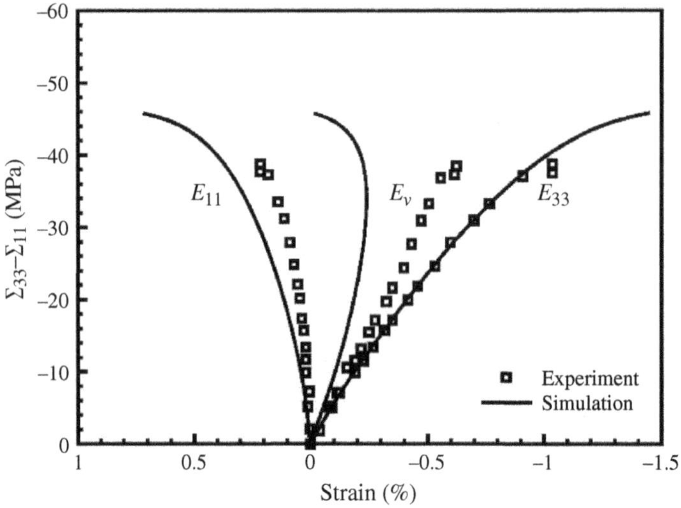

Figure 10.6 Comparison of triaxial compression test data with model predictions (confining pressure 5 MPa, depth = 451.5 m, sample composition: $f_0 = 0.49$, $f_1 = 0.19$, $f_2 = 0.32$).

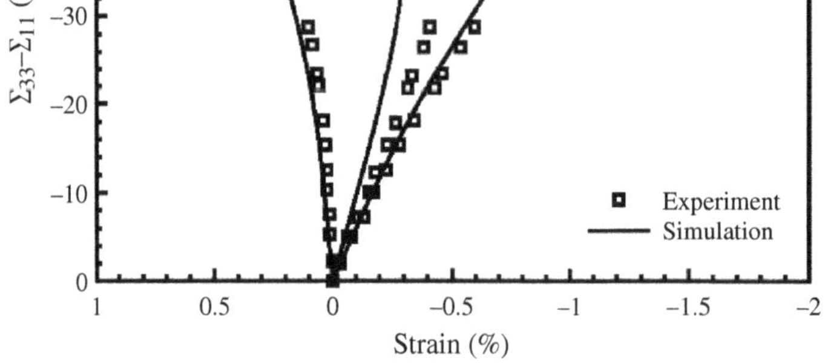

Figure 10.7 Comparison of triaxial compression test data with model predictions (confining pressure 10 MPa, depth = 451.4 m, sample composition: $f_0 = 0.47$, $f_1 = 0.31$, $f_2 = 0.22$).

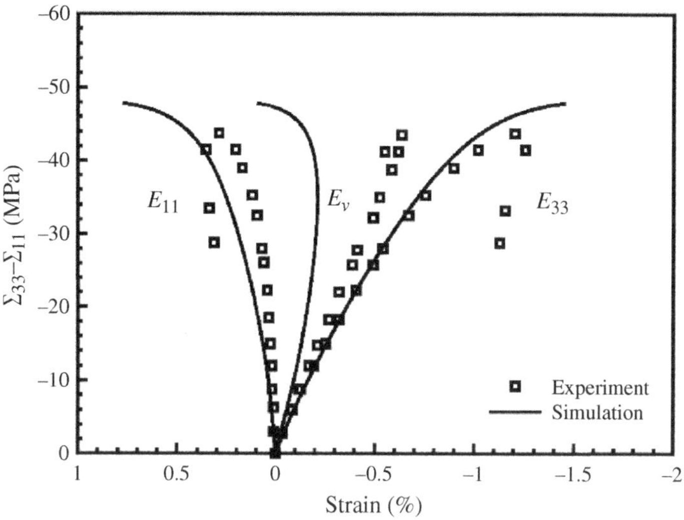

Figure 10.8 Comparison of triaxial compression test data with model predictions (confining pressure 5 MPa, depth = 469.0 m, sample composition: $f_0 = 0.44$, $f_1 = 0.33$, $f_2 = 0.23$).

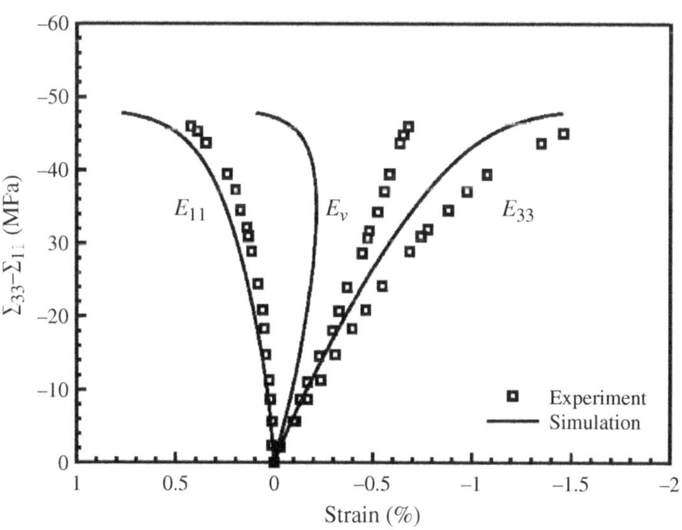

Figure 10.9 Comparison of triaxial compression test data with model predictions (confining pressure 10 MPa, depth = 469.1 m, sample composition: $f_0 = 0.55$, $f_1 = 0.23$, $f_2 = 0.22$).

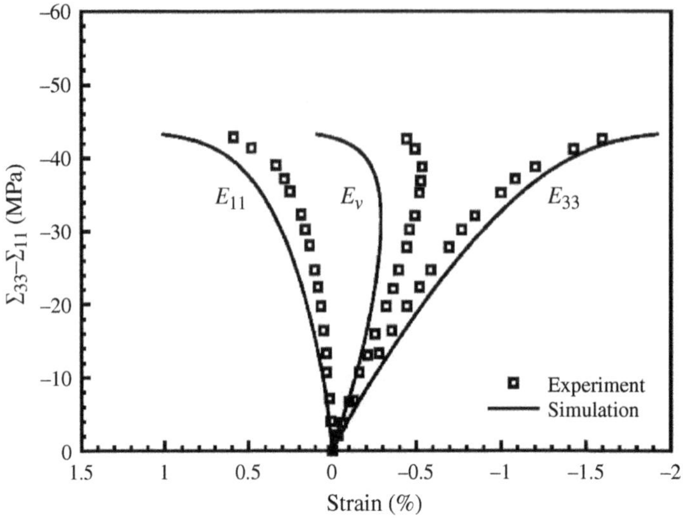

Figure 10.10 Comparison of triaxial compression test data with model predictions (confining pressure 5 MPa, depth = 482.0 m, sample composition: $f_0 = 0.60$, $f_1 = 0.26$, $f_2 = 0.14$).

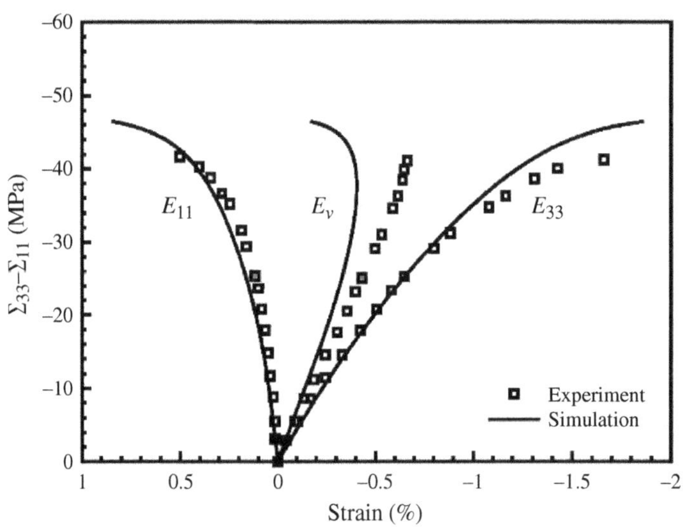

Figure 10.11 Comparison of triaxial compression test data with model predictions (confining pressure 10 MPa, depth = 482.0 m, sample composition: $f_0 = 0.60$, $f_1 = 0.26$, $f_2 = 0.14$).

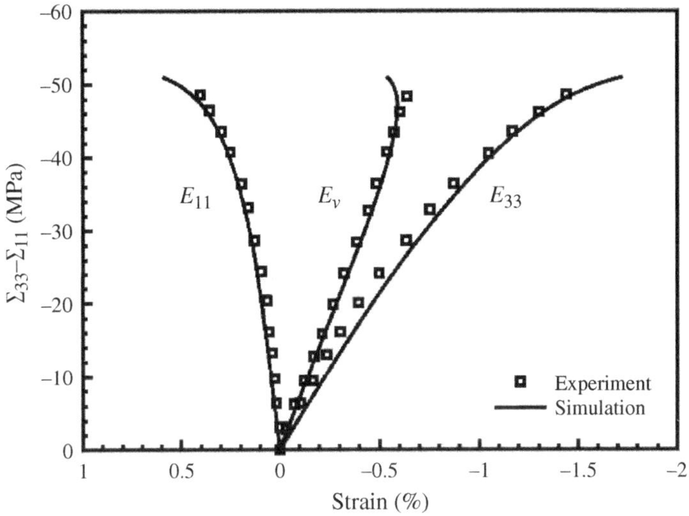

Figure 10.12 Comparison of triaxial compression test data with model predictions (confining pressure 20 MPa, depth = 482.0 m, sample composition: $f_0 = 0.60$, $f_1 = 0.26$, $f_2 = 0.14$).

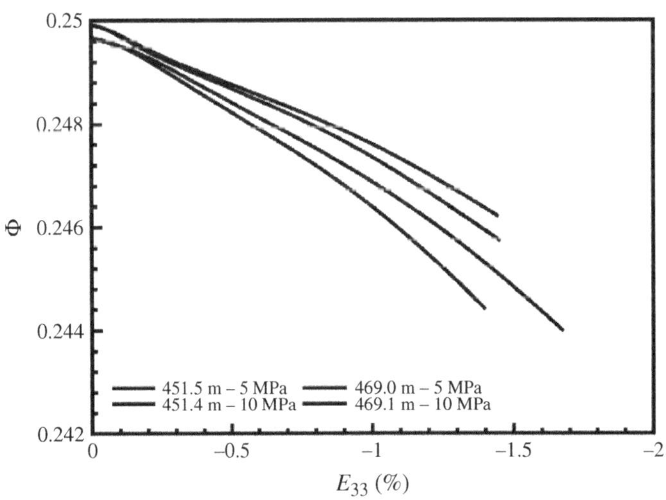

Figure 10.13 The evolution of porosity at the depths of 451 m and 469 m.

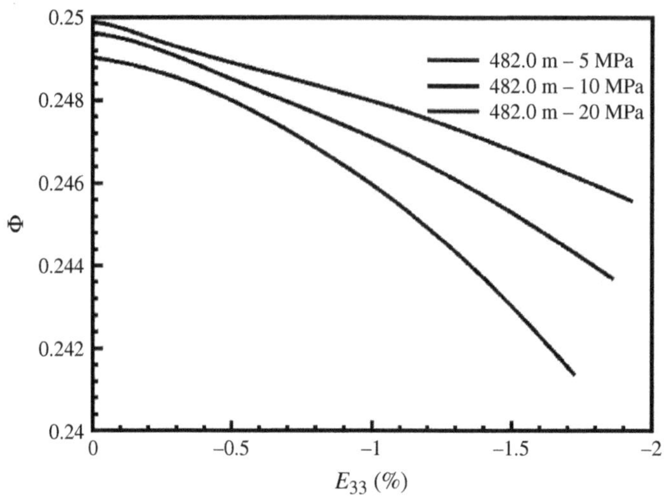

Figure 10.14 The evolution of porosity at a depth of 482 m.

10.8 Conclusion

According to the micromechanical model proposed by González et al. [22], the accuracy of the prediction depends on the linear homogenization model and the hardening capacity of the matrix material. For materials with low hardening capacity, deformation will occur rapidly and preferentially at the interface between the matrix and the inclusion, eventually forming a shear surface (with a high hardening rate, which also indicates the high value of b). The progressive development of local heterogeneities within the matrix can markedly impact the macroscopic behavior. Hence, further validation is required to assess the effectiveness of the chosen reference parameters.

The geometric model introduced in Section 10.7.1 is employed again, with a simplification applied to the COx argillite as a two-phase system. Specifically, quartz and calcite are combined into a single phase, whose elastic properties are taken as the average of their respective values, denoted by $E = 97.5$ MPa and $\nu = 0.165$. Figure 10.15 presents the macroscopic stress–strain responses for varying values of b. The proposed micromechanical model tends to overestimate the macroscopic response as b increases. Nevertheless, it effectively captures the overall response at relatively low hardening rates, lending further support to the suitability of the adopted reference parameters. However, predictions diverge significantly once the hardening rate surpasses a critical threshold, reflecting a decrease in the hardening capacity of the matrix material.

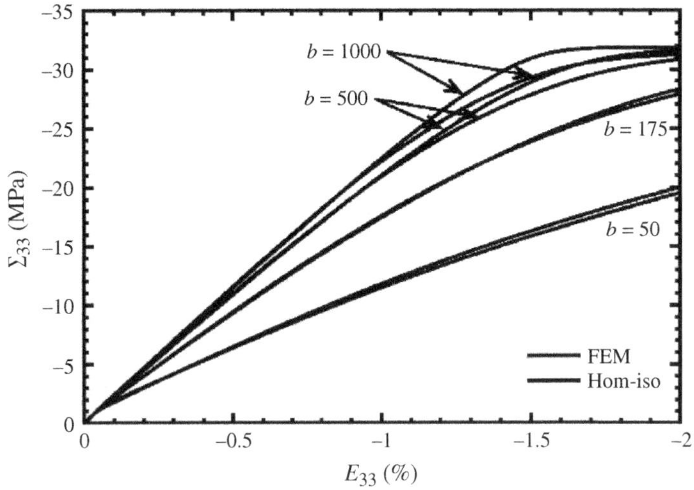

Figure 10.15 Macroscopic axial stress predicted by the proposed micromechanical model and FE computations on the unit cell with different values of *b*.

References

1 Chiarelli, A.S., Shao, J.F., and Hoteit, N. (2003). Modeling of elastoplastic damage behavior of a claystone. *International Journal of Plasticity* 19 (1): 23–45.

2 Shao, J.F., Jia, Y., Kondo, D., and Chiarelli, A.S. (2006). A coupled elastoplastic damage model for semi-brittle materials and extension to unsaturated conditions. *Mechanics of Materials* 38 (3): 218–32.

3 Hoxha, D., Giraud, A., Homand, F., and Auvray, C. (2007). Saturated and unsaturated behaviour modelling of Meuse Haute Marne argillite. *International Journal of Plasticity* 23 (5): 733–766.

4 Jia, Y., Bian, H.B., Su, K.et al. (2010). Elastoplastic damage modelling of desaturation and resaturation in argillites. *International Journal for Numerical and Analytical Methods in Geomechanics* 34 (2): 187–220.

5 Guéry, A.A., Cormery, F., Shao, J.F., and Kondo, D. (2008). A micromechanical model of elastoplastic and damage behavior of a cohesive geomaterial. *International Journal of Solids and Structures* 45 (5): 1406–1429.

6 Jiang, T., Guéry, A.A., Kondo, D., and Shao, J.F. (2009). Multi-scale modeling for inelastic behavior of a cohesive geomaterial. *Mechanics Research Communications* 36 (6): 673–681.

7 Shen, W.Q., Shao, J.F., Kondo, D., and Gatmiri, B. (2012). A micro-macro model for clayey rocks with a plastic compressible porous matrix. *International Journal of Plasticity* 36: 64–85.

8 Huang, Y., Shen, W.Q., Shao, J.F. et al. (2014). Multi-scale modeling of time-dependent behavior of claystones with a viscoplastic compressible porous matrix. *Mechanics of Materials* 79: 25–34.

9 Eshelby, J.D. (1957). The determination of the elastic field of an ellipsoidal inclusion and related problems. *Proceedings of the Royal Society of London* 241 (1226): 376–396.

10 Gavazzi, A. and Lagoudas, D. (1990). On the numerical evaluation of Eshelby tensor and its ap plication to elastoplastic fibrous composites. *Computational Mechanics* 7 (1): 13–19.
11 Doghri, I. and Ouaar, A. (2003). Homogenization of two-phase elasto-plastic composite materials and structures: study of tangent operators, cyclic plasticity and numerical algorithms. *International Journal of Solids and Structures* 40 (7): 1681–1712.
12 Robinet, J. (2008). Minéralogie, porosité et diffusion des solutés dans l'argilite du Callovo Oxfordien de Bure (Meuse, Haute-Marne, France) de l'échelle centimétrique à micrométrique PhD thesis Université de Poitiers.
13 Chiarelli, A.S. (2000). Experimental investigation and constitutive modeling of coupled elastoplastic damage in hard claystones PhD thesis University of Lille.
14 Armand, G., Conil, N., Talandier, J., and Seyedi, D.M. (2017). Fundamental aspects of the hydro mechanical behaviour of Callovo-Oxfordian claystone: from experimental studies to model calibration and validation. *Computers and Geotechnics* 85: 277–286.
15 Menaceur, H., Delage, P., Tang, A., and Conil, N. (2015). The thermo-mechanical behaviour of the Callovo-Oxfordian claystone. *International Journal of Rock Mechanics and Mining Sciences* 78 (78): 290–303.
16 Simó, J.C. and Hughes, T.J.R. (1998). *Computational Inelasticity*. New York: Springer Verlag.
17 Lazzeri, A. and Bucknall, C.B. (1993). Dilatational bands in rubber-toughened polymers. *Journal of Materials Science* 28: 6799–6808.
18 Aravas, N. (1987). On the numerical integration of a class of pressure-dependent plasticity models. *International Journal for Numerical Methods in Engineering* 24 (7): 1395–1416.
19 Jiang, T. and Shao, J.F. (2009). On the incremental approach for nonlinear homogenization of composite and influence of isotropization. *Computational Materials Science* 46 (2): 447–451.
20 Lide, D.R. (2004). *Handbook of Chemistry and Physics*. CRC Press.
21 Mura, T. (1987). *Micromechanics of Defects in Solids*. Springer.
22 González, C., Segurado, J., and Lorca, L. (2004). Numerical simulation of elasto-plastic deformation of composites: evolution of stress micro fields and implications for homogenization models. *Journal of the Mechanics and Physics of Solids* 52: 1573–1593.

11

Development of a High Temperature Constitutive Model for Concrete Based on Elastoplastic Theory

11.1 Introduction

Concrete is a fundamental material in civil engineering. Precise constitutive modeling plays a critical role in finite element analysis (FEA) and structural capacity assessments, enabling accurate representation of its nonlinear mechanical response under various loading scenarios. Concrete inherently contains microstructural defects such as pores and microcracks, which contribute to strain hardening, softening, and pronounced asymmetry between tensile and compressive behaviors. Most constitutive models combine plasticity and continuum damage mechanics to describe this inelastic behavior. At elevated temperatures, thermal damage further intensifies the nonlinearity. Nechnech [1] proposed a thermo-mechanical model incorporating mechanical and thermal damage with transient thermal creep, but it is restricted to plane stress and fails to capture tension–compression asymmetry or general transient creep accurately. To overcome these issues, Gernay et al. [2–4] developed a multiaxial model for fire-exposed concrete, applicable to various stress states. However, limitations remain in the explicit treatment of thermal and creep strains, particularly in modeling transient thermal creep.

In this chapter, we develop an enhanced transient thermal creep model that considers the effects of moisture evaporation, dehydration of chemically bound water, and decomposition reactions occurring in concrete at high temperatures. The model is integrated within an elastoplastic damage framework to more precisely simulate the thermomechanical behavior of concrete under fire exposure. The enhanced constitutive model is implemented via a user-defined material subroutine (UMAT) within the ABAQUS software, facilitating advanced numerical simulations and subsequent validation.

11.2 Constitutive Theory of Concrete Based on Thermodynamic Framework

11.2.1 Constitutive Theory of Concrete

Most concrete materials exhibit significantly different compressive and tensile behaviors under uniaxial loading conditions, as shown in Figure 11.1. Therefore, different yield conditions, plastic flow, hardening, and damage models should be developed to more accurately

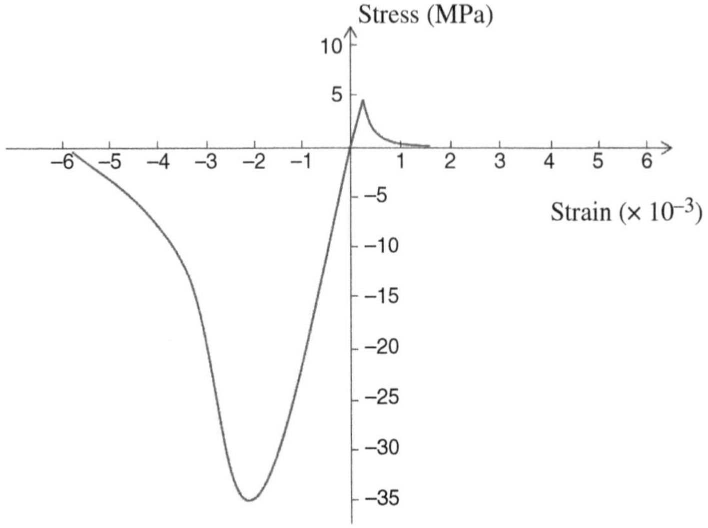

Figure 11.1 Uniaxially compressed concrete and the anisotropy of concrete in tension and compression.

characterize concrete behavior under different stress conditions, reflecting the differences in the compressive and tensile properties of concrete.

Under compression, concrete exhibits highly nonlinear and obvious plastic behavior. As shown in Figure 11.2, the left figure shows the relationship between the axial strain ε_1 and transverse strain ε_2 with stress under compression, and the right figure shows the relationship between the volume strain $\varepsilon_{\mathrm{vol}}$ (equal to the sum of the strain components in the three main directions) and stress. When the stress exceeds 30% of the compressive strength, the concrete stress–strain curve exhibits increasing nonlinearity, microcracks begin to germinate in the cement matrix, and the concrete enters the plastic hardening stage. Finally, the material reaches the peak stress, that is, the maximum compressive strength f_c. The axial strain corresponding to the maximum compressive strength is called the peak stress strain ε_{c1}. As shown in the right figure of Figure 11.2, when the concrete attains the peak stress, the volume strain $\varepsilon_{\mathrm{vol}}$ becomes positive, which means that the material exhibits volume expansion. This volume expansion under compressive load is called shear dilatancy.

A notable feature of concrete is that its tensile strength is lower than its compressive strength. Typically, the ratio of the uniaxial tensile strength to the compressive strength of concrete is between 0.05 and 0.10 [5]. By adding toughening materials such as steel fibers, polyethylene fibers, and polyvinyl alcohol fibers to cement, the tensile strength and ultimate tensile strain of concrete can be greatly improved. The tensile strength of ultra-high performance concrete (UHPC) and engineered cement-based materials (ECC) exhibits substantially higher tensile properties compared to conventional concrete. ECC is frequently employed for structural reinforcement and repair due to its superior tensile capacity and effective crack-healing performance.

The mechanical response of concrete under uniaxial tension is shown in Figure 11.3 [6]. Before reaching the peak stress f_t, the strain and stress have a linear elastic relationship.

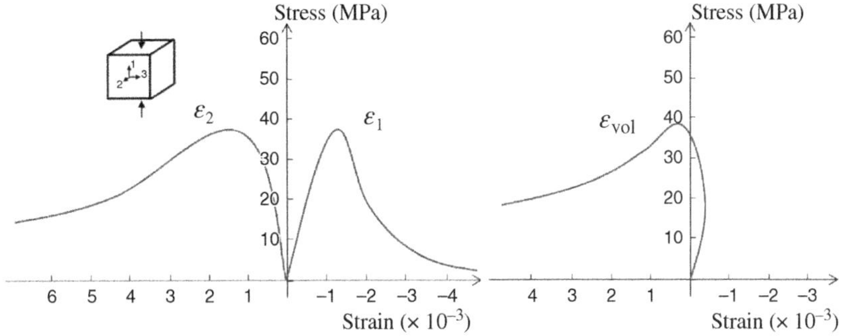

Figure 11.2 Stress–strain curve of concrete under uniaxial compression.

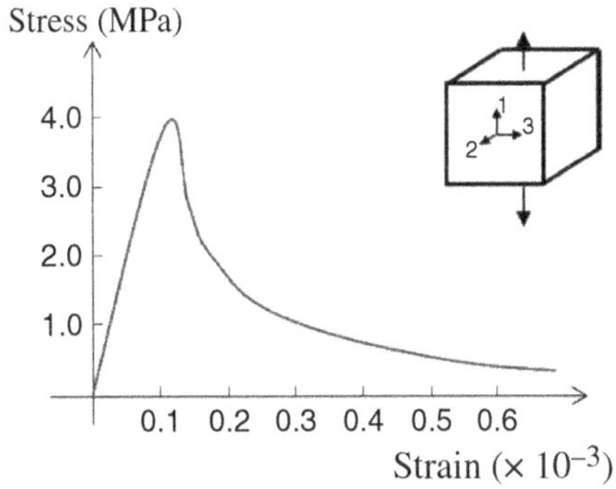

Figure 11.3 Stress–strain curve of concrete under uniaxial tension.

After the peak stress, the stress decreases rapidly, and there is basically no stress hardening stage (except for ECC). Following a sudden and significant stress drop, the rate of stress reduction gradually diminishes. Finally, compared to compressive loading, the strain ε_t at the time of concrete failure in uniaxial tension is relatively small.

Due to the differences in the compressive and tensile mechanical behaviors of concrete, constitutive models typically consist of two components: separate yield criteria, plastic flow laws, and hardening equations for the different tensile and compressive properties of concrete to describe its plastic and hardening behaviors. Based on plasticity theory and damage mechanics methods, Nechnech et al. [1] and Gernay et al. [7] combined isotropic damage with plasticity models to propose the basic framework of concrete constitutive model and extended it to high temperature. Based on this, Yao et al. [8] proposed a multi-axial high-temperature constitutive model for concrete within a thermodynamic theory framework. At high temperatures, the total strain tensor of concrete can be decomposed into two parts:

$$\varepsilon_{\text{tot}} = \varepsilon_{\text{th}} + \varepsilon_m \tag{11.1}$$

where ε_{tot} is the total strain, ε_{th} is the free thermal strain caused by temperature, and ε_m is the external load strain. They are all second-order tensors.

In Eq. (11.1), the temperature-induced free thermal strain ε_{th} is usually considered to be a nonlinear function of temperature. This nonlinearity results from the varying thermal responses between different aggregates and the cement matrix in concrete and the physical and chemical changes at high temperatures [9]. Experimental analyses indicate that the free thermal strain is a proportional function of the thermal expansion coefficient α and can be calculated by the following formula:

$$\dot{\varepsilon}_{\text{th}} = \alpha(T)\,\dot{T}\boldsymbol{I} \tag{11.2}$$

where α represents the coefficient of thermal expansion; T represents temperature; $\boldsymbol{I}$ is a unit second-order tensor; $\dot{T}$ represents the rate of change of temperature with time. The thermal expansion coefficient α is a function of temperature and changes with the temperature T. Youssef and Moftah [10] defined the calculation formula for the thermal expansion coefficient of different aggregates through experiments.

In Equation (11.1), the strain ε_m caused by the external load is usually composed of multiple parts. According to the model given by Schneider [11] and Li [12], the strain caused by the external load can usually be divided into elastic strain, plastic strain, high temperature transient creep and basic creep, as shown in Eq. (11.3):

$$\varepsilon_m = \varepsilon_e + \varepsilon_p + \varepsilon_{\text{td}} + \varepsilon_{\text{cr}} \tag{11.3}$$

where ε_e is the elastic strain; ε_p is the plastic strain; ε_{td} is the transient creep strain; ε_{cr} is the basic creep strain. They are also second-order tensors.

Transient creep ε_{td} is the difference in strain between different load paths at high temperatures. As shown in Figure 11.4, the presence of transient creep in concrete results in a significant difference between the strain produced when concrete is first subjected to high-temperature loading followed by compressive loading, and the strain generated by concrete first subjected to compressive load and then heated.

Transient creep is equal to the difference between the total strain of concrete under external load and the thermal free expansion strain. It is a kind of deformation unique to concrete materials at high temperatures, and its generation mechanism is very complex. Unlike basic creep strain, the development of transient creep strain is largely independent of time; it occurs rapidly during temperature changes and quickly reaches a stable and irreversible state. It is related to the load rate and has a significant impact on the stress–strain relationship of concrete at high temperatures. Transient creep should be considered in the fire

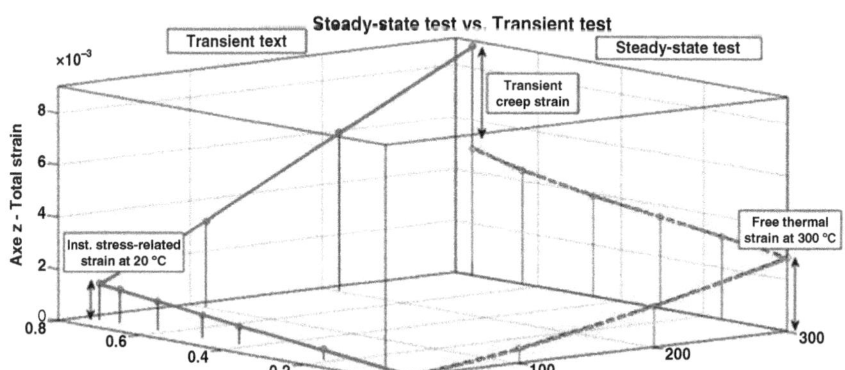

Figure 11.4 Transient creep strains generated by different load paths.

analysis of any concrete compression structure [12, 13]. Otherwise, it may lead to unsafe results for the column structure exposed to fire on three sides. In the model of Anderberg and Thelandsson [14], it is assumed that the high-temperature transient creep strain of concrete is proportional to the stress and free thermal strain, and it is believed that the size of the transient creep is related to the type of aggregate (siliceous or carbonate). De Borst and Peeters [15] believe that transient creep is generated in cement paste, so transient creep strain may not be physically related to free thermal expansion. Torelli et al. [16] found that lightweight concrete with zero free thermal strain will produce the same transient creep strain as normal-weight concrete, thus proving that there is little relationship between high-temperature transient creep strain and free thermal strain. Nielsen et al. [17] believed that transient creep strain is linearly proportional to temperature rather than to free thermal strain. Gernay et al. [2, 7] established an empirical model of external load, temperature, and transient creep strain. The list interpolation method enables a simple and rapid calculation of high-temperature transient creep. However, it is impossible to further accurately characterize the strains generated by drying, dehydration of bound water, and decomposition of hydrates during transient creep at high temperature.

Another calculation method is to implicitly include transient creep strain in the constitutive model, such as the European standard EC2 model. Transient creep strain lacks explicit calculation terms but is implicitly incorporated within the total strain. This model has demonstrated satisfactory results in modeling and analyzing concrete structural elements in fire, but it should be noted that the temperature rise conditions used in these tests are standard fires, and the specimens are simple structural members. Khoury et al. [13] proposed the concept of load-induced thermal strain (LITS) of concrete at high temperatures, including transitional thermal creep, drying creep, basic creep, and elastic strain, and pointed out that transient creep is the sum of transitional thermal creep and drying creep, and is the main component of LITS. Terro [18] combined experimental results to establish an empirical formula for load-induced thermal strain (LITS) by data fitting. LITS was considered as a function of temperature T, load level σ, and aggregate volume ratio v_a. The regression curve of the test data with a load ratio of 0.3 was used as the "benchmark" to calculate LITS:

$$\text{LITS}\,(T, \sigma = 0.3f_c, 20) = (A_0 + A_1T + A_2T_2 + A_3T_3 + A_4T_4) \times 10^{-6} \tag{11.4}$$

$$\text{LITS}\,(T, \sigma = 0.3f_c, 20) = 1.48 \times 10^{-6}\,(B_0 + B_1T + B_2T_2 + B_3T_3 + B_4T_4 + B_5T_5) \tag{11.5}$$

where A_i is a parameter applicable to carbonate aggregate concrete; B_i is a parameter applicable to silicate aggregate concrete.

When the aggregates of concrete are carbonate aggregates and lightweight aggregates, Eq. (11.4) applies. When the aggregates of concrete are silicate aggregates, Eq. (11.5) applies. The values of A_0 to A_4 are −43.87, 2.73, 6.35 × 10^{-2}, −2. 19 × 10^{-4}, 2.77 × 10^{-7}. The values of A_0 to B_5 are −1098.50, 39.21, −0.43, 2.44 × 10^{-3}, −6.27 × 10^{-6}, 5.95 × 10^{-9}.

When the load ratio is not equal to 0.3, *LITS* is calculated based on Eq. (11.4) or (11.5) using the following formula:

$$\text{LITS}\,(T,\sigma) = \text{LITS}\,(T,\sigma = 0.3f_{c,20}) \times \left(0.032 + 3.226\frac{\sigma}{f_{c,20}}\right) \tag{11.6}$$

Equation (11.6) is applicable to concrete with an aggregate content of 65%. When υa is not equal to 65%, it can be calculated according to the following formula:

$$\text{LITS}\,(T,\sigma)\,|\upsilon_a = \text{LITS}\,(T,\sigma)|\;65\% \times \left(\frac{\upsilon_a}{65\%}\right) \tag{11.7}$$

In Terro's empirical formula, *LITS* is assumed to be a linear function of external load and a nonlinear function of temperature. The model also accounts for the influence of aggregate volume fraction on transient creep strain; however, a physically explicit model that accurately reflects the mechanical and physical processes underlying transient creep strain remains lacking. Section 11.3 presents a high-temperature transient creep model to accurately describe the mechanical behavior of concrete under thermal loading.

Under high temperature, complex physical and chemical changes occur inside concrete, resulting in irreversible microcracks in the microstructure and various damages. Therefore, material damage effects should be incorporated when describing the mechanical behavior of concrete at elevated temperatures. The mechanical behavior of concrete is usually described and studied using the method of continuum damage mechanics (CDM). It is generally accepted that the damage to concrete can be characterized by the degradation of its elastic modulus. Within the framework of CDM, many scholars have conducted extensive research on the constitutive model of concrete. The damage is considered to be isotropic [19–21] or anisotropic [22–24]. In order to better describe the irreversible deformation and dilatancy of concrete during compression, Gernay et al. [2, 7] combined the CDM method with plasticity theory and proposed a plastic damage constitutive equation suitable for concrete.

To incorporate damage parameters into the constitutive model, it is first necessary to introduce the concepts of effective stress and effective stress space. According to the Lemaitre strain equivalence principle, the strain generated when the material is damaged is equivalent to the strain in the undamaged area, that is, $\varepsilon_{ij} = \bar{\varepsilon}_{ij}$. Effective stress refers to the stress acting on the undamaged portion of the material. The deformation state of any damaged material can be described by the constitutive relation of the original undamaged material. By replacing the Cauchy stress with the effective stress in the constitutive relation, the resulting stress space defined by the effective stress is referred to as the effective stress space. In the effective stress space, the stress–strain relationship in the elastic stage can be expressed as [1]:

$$\sigma = E : \varepsilon_e \tag{11.8}$$

$$\bar{\sigma} = \bar{E} : \varepsilon_e \tag{11.9}$$

where $\bar{\sigma}$ is the stress in the effective stress space/MPa; E, $\bar{E}$ are the nominal fourth-order stiffness tensor and the effective fourth-order stiffness tensor respectively. ε_e is the elastic strain of concrete, and the elastic strain at high temperature $\varepsilon_e = \varepsilon_{\text{tot}} - \varepsilon_{\text{th}} - \varepsilon_p - \varepsilon_{\text{td}}$.

The nominal stress corresponds to the effective stress, that is, the stress calculated on the effective cross section of the specimen without considering geometric discontinuities (such as holes, grooves, strips, corrugations, etc.). The nominal stress and effective stress can be converted by the following formula:

$$\sigma = (I - D) : \overline{\sigma} \tag{11.10}$$

$$E = (I - D) : \bar{E} \tag{11.11}$$

where σ is the stress in the nominal stress space, also called Cauchy stress; D represents the fourth-order damage stiffness tensor; and I represents the unit fourth-order tensor.

Damage in concrete materials at elevated temperatures can be primarily categorized into two types: thermal damage Λ caused by temperature and mechanical damage D induced by external loading. Therefore, the nominal stress expression of concrete at high temperatures, Eq. (11.10), can be converted to [1]:

$$\sigma = (I - \Lambda)\,(I - D) : \overline{\sigma} \tag{11.12}$$

where Λ is the temperature damage tensor; D is the external load damage tensor.

Since concrete exhibits significant anisotropy between tension and compression, distinct damage behaviors under tensile and compressive loads have been observed experimentally. Therefore, the damage caused by external loads needs to be divided into two parts for calculation, namely, tensile damage d_t and compressive damage d_c. In order to simplify the calculation, this chapter adopts the isotropic damage theory and assumes that the tensile and compressive behaviors of concrete are decoupled from each other. The calculation formula of the stress tensor is as follows:

$$\boldsymbol{\sigma} = (\boldsymbol{I} - \boldsymbol{D})\,\boldsymbol{\sigma} = (\boldsymbol{I} - d_t\boldsymbol{P}^+ - d_c\boldsymbol{P}^-) \tag{11.13}$$

$$\boldsymbol{P}^+ = \Sigma_i H\,(\,\bar{\sigma}_i)(p_{ii} \otimes p_{ii}),\, \boldsymbol{P}^- = \boldsymbol{I} - \boldsymbol{P}^+ \tag{11.14}$$

$$p_{ij} = p_{ji} = \frac{1}{2}\left(n_i \otimes n_j + n_j \otimes n_i\right) \tag{11.15}$$

where d_c represents the compressive damage scalar; d_t represents the tensile damage scalar; $\boldsymbol{P}^-, \boldsymbol{P}^+$ are fourth-order projection tensors; $H\,(\,\bar{\sigma}_i)$ is the Heaviside function with $\bar{\sigma}_i$ as the independent variable, 0 when $\bar{\sigma}_i$ is less than 0, and 1 when $\bar{\sigma}_i$ is greater than 0; n_i represents the ith normalized eigenvector.

Based on the above formula, the damage caused by tensile and compressive loads on concrete at high temperatures can be calculated, and the corresponding nominal stress and effective stress conversion relationship can be obtained. However, once the effective stress reaches a critical threshold, the concrete enters the plastic stage and begins to yield. At this time, irreversible plastic strain develops within the concrete, and the material strength initially increases within a certain range, then deteriorates, ultimately leading to failure. Therefore, it is also necessary to define the yield equation, flow law, and hardening law after the concrete enters the plastic stage to accurately describe the plastic behavior of concrete.

11.2.2 Thermodynamic Equations

Many scholars have studied the mechanical properties of concrete through uniaxial, biaxial, triaxial, and confining pressure tests, and proposed a variety of yield equations and hardening criteria based on the test results. Among the more classical models are the Drucker–Prager and Mohr–Coulomb yield criteria, both of which have been widely applied in the structural analysis of concrete and rock materials. The mechanical behavior of concrete at high temperature involves two irreversible thermodynamic processes, namely plastic deformation and damage. Plastic deformation arises from dislocation movements between cement-based material crystal grains and slip along microcrack surfaces, both of which are linked to the current plastic internal variables. Damage results from the progressive growth and evolution of microcracks within concrete under external loading and elevated temperature, correlating with the current damage variables, damage state, and temperature. Considering the differences in the tensile and compressive damage processes of concrete, the Helmholtz free energy of concrete at high temperature is expressed as [25–27]:

$$\Psi = \Psi\left(\boldsymbol{\varepsilon_e}, \theta, k_p, d_{\mathrm{th}}, d_m\right) \geq 0 \tag{11.16}$$

where Ψ represents Helmholtz free energy; θ represents thermodynamic temperature; k_p is an internal variable related to plastic deformation; d_{th} is a thermal damage variable, which represents the damage caused by temperature under no-load conditions; and d_m is the total damage variable caused by external loads.

Since the physical mechanisms underlying the plastic and elastic stages are fundamentally distinct and independent, the Helmholtz free energy of concrete can be decomposed into elastic and plastic parts. Temperature and damage have an impact on both elastic and plastic processes, so it is necessary to couple the temperature and damage variables with the free energy of the two processes. Therefore, Eq. (11.16) can be rewritten as follows:

$$\Psi = \Psi_e\left(\boldsymbol{\varepsilon_e}, \theta, k_p, d_{\mathrm{th}}, d_m\right) + \Psi_p\left(\boldsymbol{\varepsilon_e}, \theta, k_p, d_{\mathrm{th}}, d_m\right) \tag{11.17}$$

According to the research of Nechnech et al. [1], the elastic Helmholtz free energy at different temperatures is related to elastic strain and thermodynamic changes. Therefore, the formula for the elastic Helmholtz free energy of concrete at a certain thermodynamic temperature θ is as follows:

$$\Psi_e\left(\boldsymbol{\varepsilon_e}, \theta, d_t\right) = \frac{1}{2}\boldsymbol{\varepsilon_e} : \boldsymbol{E} : \boldsymbol{\varepsilon_e} - 3\left(\theta - \theta_0\right)\alpha \boldsymbol{I} : \boldsymbol{K} : \boldsymbol{\varepsilon_e} - \frac{1}{2}C\frac{\left(\theta - \theta_0\right)^2}{\theta_0} \tag{11.18}$$

where θ_0 is the thermodynamic temperature of the system at the initial state; K represents the bulk modulus, a fourth-order tensor; I is the unit second-order tensor; α is the thermal expansion coefficient; C represents the specific heat capacity; $\boldsymbol{E}$ is the fourth-order stiffness tensor of the material, which is related to the initial material stiffness $\boldsymbol{E}_0$, $\boldsymbol{E} = (1 - d_m)(1 - d_{\mathrm{th}})\,\boldsymbol{E}_0$.

Partially differentiating Eq. (11.17) yields:

$$\dot{\Psi} = \frac{\partial \Psi e}{\partial \boldsymbol{\varepsilon_e}}\dot{\boldsymbol{\varepsilon}}_{\boldsymbol{e}} + \frac{\partial \Psi_p}{\partial k_p}\dot{k}_p + \frac{\partial \Psi}{\partial d_{\mathrm{th}}}\dot{d}_{\mathrm{th}} + \frac{\partial \Psi}{\partial d_m}\dot{d}_m + \frac{\partial \Psi}{\partial \theta}\dot{\theta} \tag{11.19}$$

On this basis, the Clausius–Duhem inequality is introduced as a constitutive constraint. The Clausius–Duhem inequality, also known as the dissipation inequality, is an expression of the second law of thermodynamics in continuum mechanics. It is employed to characterize irreversible thermodynamic processes and commonly serves to verify whether a material's constitutive relations comply with thermodynamic principles. After the Helmholtz free energy of the material is introduced, its form is:

$$\boldsymbol{\sigma} : \dot{\varepsilon} - \dot{\Psi} \geq 0 \tag{11.20}$$

Substituting the expression of $\dot{\Psi}$ into the Clausius–Duhem inequality:

$$\boldsymbol{\sigma} : \dot{\varepsilon} - \left(\frac{\partial \Psi e}{\partial \boldsymbol{\varepsilon_e}} \dot{\boldsymbol{\varepsilon}}_{\boldsymbol{e}} + \frac{\partial \Psi_p}{\partial k_p} \dot{k}_p + \frac{\partial \Psi}{\partial d_{\text{th}}} \dot{d}_{\text{th}} + \frac{\partial \Psi}{\partial d_m} \dot{d}_m + \frac{\partial \Psi}{\partial \theta} \dot{\theta} \right) \geq 0 \tag{11.21}$$

After rearranging the above formula:

$$\begin{aligned} &\boldsymbol{\sigma} : \dot{\boldsymbol{\varepsilon}}_{\boldsymbol{e}} + \boldsymbol{\sigma} : \dot{\boldsymbol{\varepsilon}}_{\boldsymbol{p}} + \boldsymbol{\sigma} : \dot{\varepsilon}_{\text{td}} + \boldsymbol{\sigma} : \dot{\varepsilon}_{\text{th}} \\ &- \frac{\partial \Psi e}{\partial \boldsymbol{\varepsilon_e}} \dot{\boldsymbol{\varepsilon}}_{\boldsymbol{e}} - \frac{\partial \Psi_p}{\partial k_p} \dot{k}_p - \frac{\partial \Psi}{\partial d_{\text{th}}} \dot{d}_{\text{th}} - \frac{\partial \Psi}{\partial d_m} \dot{d}_m - \frac{\partial \Psi}{\partial \theta} \dot{\theta} \geq 0 \end{aligned} \tag{11.22}$$

After further sorting, the result is obtained as:

$$\begin{aligned} &\left(\boldsymbol{\sigma} - \frac{\partial \Psi e}{\partial \boldsymbol{\varepsilon_e}} \right) \dot{\boldsymbol{\varepsilon}}_{\boldsymbol{e}} + \boldsymbol{\sigma} : \dot{\boldsymbol{\varepsilon}}_{\boldsymbol{p}} - \frac{\partial \Psi_p}{\partial k_p} \dot{k}_p + \boldsymbol{\sigma} : \dot{\boldsymbol{\varepsilon}}_{\mathbf{td}} \\ &- \frac{\partial \Psi}{\partial d_{\text{th}}} \dot{d}_{\text{th}} + \boldsymbol{\sigma} : \dot{\boldsymbol{\varepsilon}}_{\mathbf{th}} - \frac{\partial \Psi}{\partial \theta} \dot{\theta} - \frac{\partial \Psi}{\partial d_m} \dot{d}_m \geq 0 \end{aligned} \tag{11.23}$$

Since the elastic strain rate $\dot{\boldsymbol{\varepsilon}}_{\boldsymbol{e}}$ is arbitrary, in order to satisfy the basic law of thermodynamic dissipation, Eq. (11.23) should satisfy:

$$\boldsymbol{\sigma} = \frac{\partial \Psi e}{\partial \boldsymbol{\varepsilon_e}} \tag{11.24}$$

$$\Pi_p = \boldsymbol{\sigma} : \dot{\boldsymbol{\varepsilon}}_{\boldsymbol{p}} - \frac{\partial \Psi_p}{\partial k_p} \dot{k}_p \geq 0 \tag{11.25}$$

$$\Pi_{\text{th}} = \boldsymbol{\sigma} : \dot{\boldsymbol{\varepsilon}}_{\mathbf{td}} + \boldsymbol{\sigma} : \dot{\boldsymbol{\varepsilon}}_{\mathbf{th}} - \frac{\partial \Psi}{\partial \theta} \dot{\theta} - \frac{\partial \Psi}{\partial d_{\text{th}}} \dot{d}_{\text{th}} \geq 0 \tag{11.26}$$

$$\Pi_d = -\frac{\partial \Psi}{\partial d_m} \dot{d}_m \geq 0 \tag{11.27}$$

In this way, some constitutive formulas related to concrete at high temperatures are obtained. Among them, Eq. (11.24) is the generalized constitutive equation of concrete, and the stress is related to the elastic strain and elastic Helmholtz free energy; Eq. (11.25) is the plastic dissipation potential of concrete. Since plastic deformation is an irreversible process, Eq. (11.25) is nonnegative; Eqs. (11.26) and (11.27) are the heat dissipation potential and damage dissipation potential, respectively, which are also nonnegative values.

Then, the total Helmholtz energy dissipation equation generated by concrete at high temperature is:

$$\Pi = \Pi_p + \Pi_d + \Pi_{\text{th}} \tag{11.28}$$

Four energy dissipation potential functions are introduced: plastic dissipation potential F_p, damage dissipation potential F_{td}, and thermal dissipation potential F_{th}. Construct the Lagrange objective function:

$$\Omega = \Pi - \dot{\lambda}_p F_p - \dot{\lambda}_{\text{td}} F_{\text{td}} - \dot{\lambda}_{\text{th}} F_{\text{th}} \tag{11.29}$$

Substituting Eqs. (11.25) to (11.27) into the above equation, when Eq. (11.29) satisfies the principle of minimum potential energy (also known as the principle of maximum dissipation), it can be deduced that:

$$\frac{\partial \Omega}{\partial \sigma} = 0 \tag{11.30}$$

After sorting, the outcome is:

$$\dot{\boldsymbol{\varepsilon}}_{\boldsymbol{p}} + \dot{\boldsymbol{\varepsilon}}_{\mathbf{th}} + \dot{\boldsymbol{\varepsilon}}_{\mathbf{td}} - \dot{\lambda}_p \frac{\partial F_p}{\partial \sigma} - \dot{\lambda}_{\text{th}} \frac{\partial F_{\text{th}}}{\partial \sigma} - \dot{\lambda}_{\text{td}} \frac{\partial F_{\text{td}}}{\partial \sigma} \tag{11.31}$$

Since the strain is arbitrary, the following relationship can be derived from Eq. (11.31):

$$\dot{\boldsymbol{\varepsilon}}_{\boldsymbol{p}} = \dot{\lambda}_p \frac{\partial F_p}{\partial \sigma} \tag{11.32}$$

$$\dot{\boldsymbol{\varepsilon}}_{\mathbf{th}} = \dot{\lambda}_{\text{th}} \frac{\partial F_{\text{th}}}{\partial \sigma} \tag{11.33}$$

$$\dot{\boldsymbol{\varepsilon}}_{\mathbf{td}} = \dot{\lambda}_{\text{td}} \frac{\partial F_{\text{td}}}{\partial \sigma} \tag{11.34}$$

Among them, Eq. (11.32) is the plastic flow law, and λ_p is the plastic multiplier. Eqs. (11.33) and (11.34) are generalized calculation formulas for thermal free deformation and transient creep of concrete at high temperatures. Similarly, it can be assumed that the plastic energy release rate and damage energy release rate are Y_p and Y_d, and let:

$$Y_p = -\frac{\partial \Psi_p}{\partial k_p} \tag{11.35}$$

$$Y_d = -\frac{\partial \Psi}{\partial d_m} \tag{11.36}$$

Substitute Eqs. (11.35) and (11.36) into Eq. (11.29) and, according to the principle of minimum potential energy, find the partial derivatives of Y_p and Y_d respectively, and satisfy the boundary conditions $\partial \Omega / \partial Y_p = 0$ and $\partial \Omega / \partial Y_d = 0$:

$$\dot{k}_p = \dot{\lambda}_p \frac{\partial F_p}{\partial Y_p} \tag{11.37}$$

$$\dot{d}_m = \dot{\lambda}_d \frac{\partial F_d}{\partial Y_d} \tag{11.38}$$

Among them, the hardening/softening law of concrete at high temperature can be obtained Eq. (11.37), λ_p is the plasticity multiplier, k_p is the plasticity internal variable, and is also the hardening variable. Eq. (11.38) is the damage evolution law of concrete, and λ_d is the damage consistency multiplier. Therefore, based on the theoretical framework of irreversible thermodynamics, the plastic flow law, hardening/softening law, and damage evolution equation for concrete at high temperature are derived. Subsequently, this chapter will derive the relevant equations of the constitutive model of concrete at high temperature based on the specific yield law.

11.2.3 Yield Criterion and Hardening Law

The plastic damage constitutive model of concrete should include the yield law, hardening and softening law, damage evolution equation, etc. Based on the above derivation, it is necessary to first determine the yield criterion of concrete at high temperature, and derive the plastic flow law, hardening/softening law, and damage evolution law according to Eqs. (11.32), (11.37), (11.38), and other formulas. Owing to the tension-compression anisotropy of concrete, this chapter adopts the Drucker–Prager yield criterion for the compressive constitutive model, while tensile yield behavior is described using the Rankine criterion [3, 8, 25, 26]. The plastic part removes the influence of damage and is established in the effective stress space. The function is as follows:

$$\begin{cases} F_t\left(\overline{\sigma}, k_p^t, T\right) = \bar{\sigma}_I - \bar{\tau}_t\left(k_p^t, T\right) \leq 0 \\ F_c\left(\overline{\sigma}, k_p^c, T\right) = \sqrt{3\bar{J}_2} + \alpha \bar{I}_1 - (1-\alpha)\,\bar{\tau}_c\left(k_p^c, T\right) \leq 0 \end{cases} \tag{11.39}$$

where F_t and F_c represent the tensile and compressive yield functions of concrete, respectively; $\bar{\sigma}_I$ is the maximum principal stress of effective stress; $\bar{\tau}_i$ is the hardening function, $i = t$ or c, representing the tensile hardening function and the compressive hardening function, respectively; k_p^i is the internal plastic variable and also the hardening variable, $i = t$ or c, representing the tensile hardening variable and the compressive hardening variable, respectively; $\bar{I}_1$ is the first invariant of effective stress; $\bar{J}_2$ is the invariant of the second deviator of effective stress; α is the internal friction coefficient, which is related to the ratio of the biaxial compressive strength to the uniaxial compressive strength of concrete and can be determined by the following formula:

$$\alpha = \frac{f_b - f_c}{2f_b - f_c} \tag{11.40}$$

where f_b is the biaxial compressive strength of concrete; f_c is the uniaxial compressive strength of concrete.

$\bar{I}_1$ and $\bar{J}_2$ in the effective stress space represent the first invariant of the stress tensor and the second invariant of the deviatoric stress tensor, respectively. When the stress space coordinates are transformed, the values of $\bar{I}_1$ and $\bar{J}_2$ remain unchanged and can be calculated by normal stress or principal stress:

$$\bar{I}_1 = \bar{\sigma}_{11} + \bar{\sigma}_{22} + \bar{\sigma}_{33} \tag{11.41}$$

$$\bar{J}_2 = \frac{1}{6}\left[(\bar{\sigma}_{11}-\bar{\sigma}_{22})^2+(\bar{\sigma}_{22}-\bar{\sigma}_{33})^2+(\bar{\sigma}_{33}-\bar{\sigma}_{11})^2+6\left(\bar{\sigma}_{12}^2+\bar{\sigma}_{13}^2+\bar{\sigma}_{23}^2\right)\right] \\ = \frac{1}{6}\left[(\bar{\sigma}_1-\bar{\sigma}_2)^2+(\bar{\sigma}_2-\bar{\sigma}_3)^2+(\bar{\sigma}_3-\bar{\sigma}_1)^2\right] \tag{11.42}$$

where $\bar{\sigma}_{11}, \bar{\sigma}_{22}, \bar{\sigma}_{33}$ are normal stresses in the effective stress space; $\bar{\sigma}_1, \bar{\sigma}_2, \bar{\sigma}_3$ are three principal stresses in the effective stress space, $\bar{\sigma}_1$ is the largest, and $\bar{\sigma}_3$ is the smallest.

In the effective stress space, the yield surface of the Drucker–Prager yield function is an elliptical cone open surface, as shown in Figure 11.5. Similar to the Mises yield law, the cross-section of the deviatoric plane perpendicular to the hydrostatic pressure axis ($\sigma_1 = \sigma_2 = \sigma_3$) is circular. After combining with the yield surface composed of the Rankine law, the closed concrete tension and compression yield surface is obtained.

Take the yield surface composed of Rankine and Drucker–Prager yield functions in two-dimensional stress space for analysis. Assuming that the third principal stress $\sigma_3 = 0$, the yield surface composed of the two yield functions is ellipsoidal, as shown in Figure 11.6. Within the yield surface range, the material is in the elastic stage. As the load increases, the stress gradually increases until F_t or F_c is greater than 0, and the material enters the plastic stage. A distinct difference exists between the tensile and compressive plastic behaviors of concrete. The compressive yield surface range of concrete is larger, and the performance is significantly better than the tensile performance. During hardening, the stress at the time of yielding increases, and the yield surface range increases; conversely, during softening, the yield stress and yield surface both decrease.

Plastic flow law is the equation that controls the evolution of plastic flow. When the stress state of the material reaches yield, plastic deformation will occur. At this time, it is necessary to calculate the generated plastic strain tensor according to the plastic flow law to determine the plastic deformation component and size. From Eq. (11.32), it can be seen that the core of determining the plastic flow law is to clarify the plastic potential function G. There are generally two choices of plastic potential. When the direction of the plastic deformation of the material's plastic deformation aligns with the gradient of the yield function in stress space (that is, the partial derivative of the yield function to stress), the plastic potential is the same as the yield function, and the flow law is the associated flow law. When the direction of plastic deformation does not align with the gradient of the yield function in stress

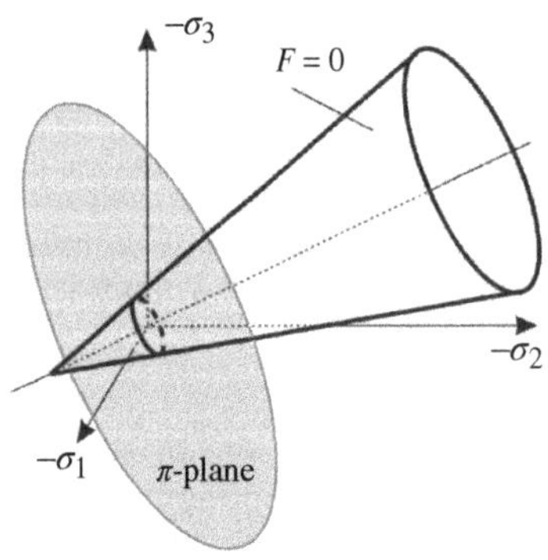

Figure 11.5 Drucker–Prager yield surface in effective stress space.

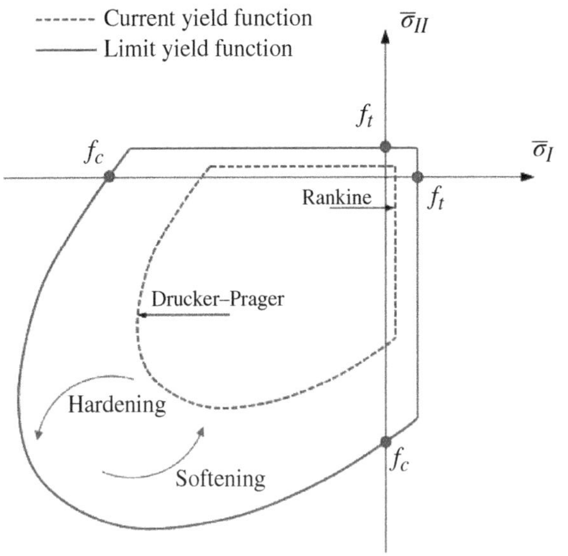

Figure 11.6 Rankine and Drucker–Prager yield surfaces in plane stress space.

space, the plastic potential differs from the yield function, resulting in a nonassociated flow rule. For concrete, when it is under compression, its yield surface is affected by the hydrostatic pressure, and there is an obvious shear dilatancy phenomenon. Plastic deformation is accompanied by volume change, and its direction typically does not coincide with the gradient of the yield function. Therefore, it is necessary to use the nonassociated flow law to define the plastic flow direction. When it is under tension, the associated flow law can be used. Therefore, the concrete plastic flow law adopted in this chapter is as follows [25–27]:

$$\begin{cases} \dot{\varepsilon}_p^t = \dot{\lambda}_p^t = \dot{\lambda}_p^t \dfrac{\partial F_t}{\partial \bar{\sigma}} \\ \dot{\varepsilon}_p^c = \dot{\lambda}_p^c \dfrac{\partial G_c}{\partial \bar{\sigma}} \end{cases} \tag{11.43}$$

where λ_p^t, λ_p^c are the plastic multipliers under tension and compression respectively, and their rate forms are used in the formula; G_t represents the plastic potential of concrete under tension, which is equal to the tensile yield function F_t when the associated flow law is used; G_c represents the plastic potential of concrete under compression, which is related to the shear expansion coefficient α_g of concrete, and its formula can be expressed as follows [25–27]:

$$G_c = \sqrt{3\bar{J}_2} + \alpha_g \bar{I}_1 \tag{11.44}$$

And α_g is usually taken as 0.25.

The total plastic strain tensor of concrete consists of two parts: tension and compression. The strain rate formula is as follows:

$$\dot{\boldsymbol{\varepsilon}}_{\boldsymbol{p}} = \dot{\varepsilon}_p^t + \dot{\varepsilon}_p^c = \dot{\lambda}_p^t \frac{\partial F_t}{\partial \bar{\sigma}} + \dot{\lambda}_p^c \frac{\partial G_c}{\partial \bar{\sigma}} \tag{11.45}$$

By introducing the Kuhn–Tucker condition and determining the loading and unloading criteria, the values of the plastic multipliers $\dot{\lambda}_p^t$ and $\dot{\lambda}_p^c$ can be obtained:

$$\begin{cases} \dot{\lambda}_p^j \geq 0, F_j\left(\bar{\sigma}, k_p^j\right) \leq 0, \dot{\lambda}_p^j F_j\left(\bar{\sigma}, k_p^j\right) = 0, j = t \\ c\dot{\lambda}_p^j \dot{F}_j\left(\bar{\sigma}, k_p^j\right) = 0, j = t, c \end{cases} \tag{11.46}$$

where k_p^t and k_p^c are plastic internal variables, which characterize the degree of material yield. At the same time, the plastic hardening/softening function is an equation related to the plastic internal variables, and k_p^t and k_p^c are also hardening variables. In this chapter, only isotropic softening and hardening are analyzed, and it is believed that the hardening variables k_p^t and k_p^c control the evolution of the Rankine and Drucker–Prager yield surfaces respectively.

There are usually two commonly used plastic internal variables, namely, accumulated plastic strain and plastic work. When concrete enters the hardening stage, k_p^t and k_p^c are generated and greater than zero. As the hardening variable increases, the hardening functions $\bar{\tau}_t$ and $\bar{\tau}_c$ also increase. The stress needs to exceed the initial yield point to reach a new stress level before it can re-enter the yield. The new yield point can be called the subsequent yield point or strengthening point. When the material is in the strengthening stage, the strength corresponding to the subsequent yield point is usually greater than the initial yield point. After reaching the maximum strength, concrete enters the softening stage, and the strength corresponding to the subsequent yield point begins to decrease. The accumulated plastic strain can be taken as the hardening parameter, and $\bar{\tau}$ is its related function, with the unit of MPa. This chapter uses the accumulated plastic strain as the plastic internal variable of concrete, which is defined as:

$$\dot{k}_p = \dot{p} = \sqrt{\frac{2}{3}\dot{\boldsymbol{\varepsilon}}_{\boldsymbol{p}} : \dot{\boldsymbol{\varepsilon}}_{\boldsymbol{p}}} \tag{11.47}$$

where $\dot{p}$ is the cumulative plastic strain rate; $\dot{\boldsymbol{\varepsilon}}_{\boldsymbol{p}}$ is the plastic strain rate.

Next, the expressions of the hardening functions in the effective stress space are clarified. Through uniaxial tension and compression tests, the hardening strengths τ_t and τ_c in the nominal stress space can be obtained. The relationship between the hardening function $\bar{\tau}_j$ in the effective stress space and the hardening function τ_j in the nominal stress space is as follows

$$\tau_j\left(k_p^j, T\right) = (1 - d_j)\,\bar{\tau}_j\left(k_p^j, T\right) \tag{11.48}$$

where $j = t, c$. In the uniaxial tensile test of concrete, as shown in Figure 11.7 the concrete first enters the elastic stage. After reaching the peak stress f_t, the material directly enters the softening stage until failure. The decrease of concrete's tensile strength with increasing strain follows a curve that asymptotically approaches the 0-stress level. Since the decrease in the tensile strength of concrete is considered to be caused by damage, in the effective stress space, after the concrete stress reaches the peak value, it can be considered that the value of $\bar{\tau}_t$ is always equal to f_t, that is:

$$\bar{\tau}_t\left(k_p^t, T\right) = f_t\,(T) \tag{11.49}$$

where $f_t\,(T)$ represents the tensile strength of concrete at a temperature T.

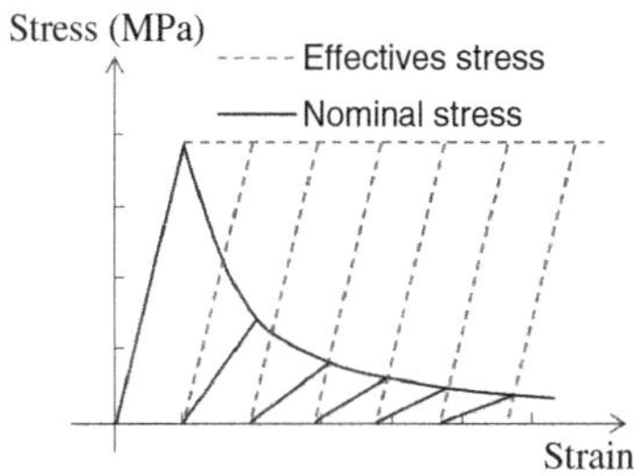

Figure 11.7 Uniaxial tensile stress–strain curve of concrete.

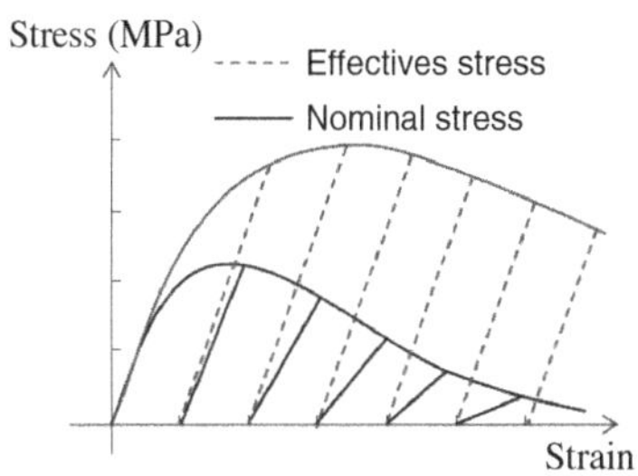

Figure 11.8 Stress–strain curve of concrete under uniaxial compression.

Unlike uniaxial tension, when concrete is subjected to uniaxial compression, it first enters the elastic stage. When the stress reaches the elastic compression limit strength f_{c0}, the concrete enters the hardening stage, and plastic deformation occurs. When the stress reaches the compressive strength f_c, the concrete begins to enter the softening stage, and the subsequent yield strength begins to decrease until it is destroyed, as shown in Figure 11.8. The curve $\bar{\tau}_c$ in the effective stress space needs to be obtained by inverse deduction of the damage variable. Therefore, the hardening function of concrete under compression conditions is divided into two stages, and its formula is as follows:

$$\tau_c\left(k_p^c, T\right) = \begin{cases} f_{c0}(T) + \dfrac{2\left(f_c(T) \quad f_{c0}(T)\right) k_p^c}{k_p^{c1}\left(1 + \left(k_p^c / k_p^{c1}\right)^2\right)} & if \quad k_p^c > k_p^{c1} \\ f_c(T)\left(1 + b_c\left(k_p^c - k_p^{c1}\right)\right) \exp\left(-b_c\left(k_p^c - k_p^{c1}\right)\right) & if \quad k_p^c > k_p^{c1} \end{cases} \tag{11.50}$$

where $f_c(T)$ represents the compressive strength of concrete at temperature T, that is, the peak compressive strength; $f_{c0}(T)$ represents the compressive elastic limit strength of concrete at temperature T; k_p^{c1} is the accumulated plastic strain corresponding to the peak compressive strength of concrete $f_c(T)$; b_c is the hardening parameter based on the energy method, which can be determined by experiment.

11.3 Damage Model of Concrete at High Temperature

The damage evolution equation is a function established based on damage parameters to characterize the entire damage process. Evaluation of concrete damage at high temperatures involves identifying the governing factors, specifying the associated damage types, and formulating both the damage criterion and the evolution equation of the damage variable. Damage to concrete at elevated temperatures primarily consists of two types: thermal damage caused by temperature and mechanical damage induced by external loads. High temperature damage is usually characterized by changes in elastic modulus, while the commonly used parameters for damage caused by external loads include strain, stress, energy, and energy release rate.

11.3.1 External Load Damage Variables of UHPC

External loads can cause slippage between crystal particles of concrete matrix materials, generate microcracks and micropores at the microscopic scale, reduce the effective area of the material, and thus reduce the strength and stiffness of the material. Gawin and Schrefler et al. [28–32] established a thermo-elastic-mechanical damage constitutive model, which has been applied in concrete structures in some practical projects. Ulm et al. [33, 34] proposed a thermo-chemical-plastic constitutive model based on the three-parameter Willam–Warnke elastic-plastic yield criterion. Wu et al. [21] used tensile damage variables and shear damage variables to describe the plastic damage constitutive relationship of the degradation of macroscopic mechanical properties of concrete, derived the expression of damage energy release rate, and established a damage criterion based on damage energy release rate that conforms to the principles of thermodynamics, but this method has many parameters. Based on the work of Wu et al. [21], Yao et al. [25] used energy loss rate as the generalized force, considered the influence of different toughening fiber types and contents on damage evolution, and proposed a damage evolution model for fiber concrete under high temperature. Oliver et al. [35–38] proposed an empirical formula for concrete uniaxial tension and compression damage using strain as the damage variable, and it has been widely used. Its form is as follows:

$$d_t = 1 - \frac{\varepsilon_0}{\widetilde{\varepsilon}}(1 - A^+) - \frac{A^+}{\mathrm{EXP}\,[B^+(\widetilde{\varepsilon} - \varepsilon_0)]} \tag{11.51}$$

$$d_t = 1 - \frac{\varepsilon_0}{\widetilde{\varepsilon}}(1 - A^-) - \frac{A^-}{\mathrm{EXP}\,[B^-(\widetilde{\varepsilon} - \varepsilon_0)]} \tag{11.52}$$

where A^+ and B^+ are tensile test parameters, which control the hardening and softening sections of the tensile stress–strain curve, respectively. Similarly, A^- and B^- are compressive test parameters, which are related to the rising section and softening stage of the compressive stress–strain curve, respectively. The above parameters can be measured through experiments. Since both concrete plasticity and damage processes are r associated with the initiation and evolution of microcracks and microvoids, they may share a common thermodynamic driving mechanism. Specifically, the generation of plasticity simultaneously induces damage, with both processes being irreversible. Nechnech et al. [1] regarded the damage parameters of concrete as a function of the hardening parameters and gave a unified expression of damage for concrete tensile and compressive:

$$1 - D_x = \mathrm{EXP}\,(-c_x k_x) \tag{11.53}$$

where the subscript x is used to distinguish between tension and compression. When tension is applied, $x = t$, and when compression is applied, $x = c$. As mentioned above, the hardening parameter k_x of concrete is equal to the equivalent plastic strain. Therefore, in the theory of Nechnech et al. [1], it is believed that the plastic stage is accompanied by the generation of damage. This chapter uses the hardening parameter k_x as the internal variable and adopts the empirical formula based on experimental results proposed by Gernay et al. [7]. The formulas for the damage variables d_t and d_c are as follows:

$$\begin{cases} d_t = 1 - \left[\frac{1}{2}\mathrm{EXP}\,(-a_t k_t) + \frac{1}{2}\mathrm{EXP}\,(-6a_t k_t)\right] \\ d_c = 1 - \mathrm{EXP}\,(-a_c k_c) \end{cases} \tag{11.54}$$

where a_t and a_c are both dimensionless parameters and can be measured and determined through experiments.

For UHPC, the concrete damage variable evolution model should also consider the influence of fibers. Due to the crack-resistance effect of steel fibers, the elastic modulus of UHPC reinforced with steel fibers is still higher than that of ordinary concrete at the same temperature. Yao et al. [39] proposed a fiber influence coefficient to represent the weakening effect of steel fibers on matrix damage. The calculation method is as follows:

$$\Lambda_f\,(T) = \eta \mathrm{EXP}\left(1 - \frac{E_f\,(T)}{E_c\,(T)}\right) \tag{11.55}$$

where $E_f\,(T)$ represents the elastic modulus of fiber-reinforced UHPC at temperature T; $E_c\,(T)$ represents the elastic modulus of plain UHPC at temperature T; and η represents the influence coefficient of fiber content. The fiber content influence coefficient η is related to the fiber content V_f. Yao et al. [25] gave the value formula:

$$\eta = 1 - \frac{4}{\pi^2} V_f \tag{11.56}$$

11.3.2 Heat Damage

In addition to the damage to concrete caused by external loads, physical and chemical reactions such as free water evaporation, bound water dehydration, and hydrate decomposition inside concrete at high temperatures may cause microcracks and micro-voids in the matrix microstructure, resulting in a decrease in the stiffness and strength of the material. Since the mechanism of temperature damage in concrete is different from that of metals, the processes of free water evaporation, bound water dehydration, and hydrate decomposition are irreversible. Therefore, the recovery of concrete's performance after the stiffness and compressive strength decrease at high temperatures is very small. It is generally believed that high-temperature damage to concrete is irreversible.

High-temperature damage to concrete is usually expressed as the ratio of the elastic modulus at high temperature to the elastic modulus at room temperature. Ulm et al. [33, 34] gave the corresponding formula:

$$d_{\text{th}} = 1 - \frac{E\,(T)}{E} \tag{11.57}$$

Aslani et al. [40, 41] proposed a formula for the elastic modulus of ordinary concrete at high temperatures:

$$E\,(T) = E \begin{cases} 1.0 & 20\,°\text{C} \le T \le 100\,°\text{C} \\ 1.01 - 0.0015T & 100\,°\text{C} \le T \le 400\,°\text{C} \\ 0.78 - 0.00096T & 400\,°\text{C} \le T \le 800\,°\text{C} \end{cases} \tag{11.58}$$

The elastic modulus of UHPC at high temperature is given by Zheng et al. [39]:

$$E\,(T) = E\ [-0.012 + 1.089\text{EXP}\,(-0.00388T)] \tag{11.59}$$

As shown in Figure 11.9a, when the temperature exceeds 350 °C, the compressive strength of concrete begins to decrease rapidly, the elastic stage shortens and the plastic stage lengthens, and the peak strain corresponding to the peak stress increases significantly. As shown in Figure 11.9b, at around 200 °C, the strength of concrete may increase, and the strength increase of UHPC is more obvious. This is because the water-cement ratio of UHPC is relatively low, the hydration reaction of cement clinker during the curing stage is usually not sufficient, and there are a large number of unhydrated particles in the cement matrix.

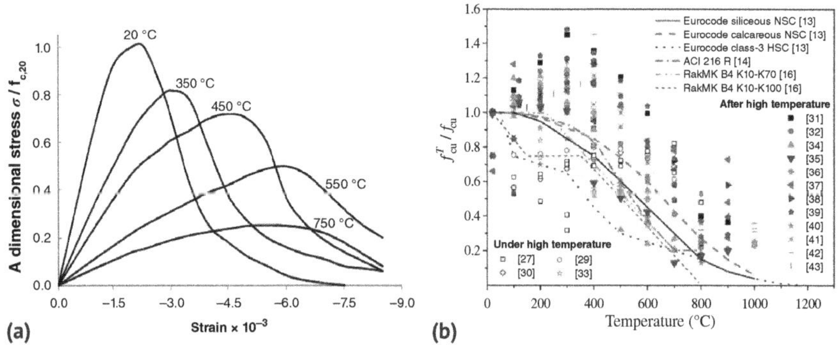

Figure 11.9 Uniaxial compressive behavior of concrete at high temperature. *Source*: [42]/with permission of Elsevier.

At high temperatures, these unhydrated particles (especially Ca $(OH)_2$ crystals) will combine with silica in the active powder (silica fume, fly ash, slag, etc.) to react with volcanic ash to form dense hydration products, which improve the strength of concrete.

According to European Standard BS EN 1992-1-2, the formula for high temperature compressive strength of ordinary concrete is:

$$f_c(T) = \begin{cases} f_c & T \leq 100\,°\text{C} \\ f_c(1.067 - 0.00067T) & 100\,°\text{C} \leq T \leq 400\,°\text{C} \\ f_c(1.44 - 0.0016T) & T \geq 400\,°\text{C} \end{cases} \tag{11.60}$$

Different toughening fibers are often added to UHPC, and its compressive strength at high temperature needs to be calculated based on the fiber type. When the added toughening fiber is the steel fibers, its compressive strength can be calculated using the following formula [42]:

$$\frac{f_c(T)}{f_c} = \begin{cases} 1.06 - 3.04\left(\frac{T}{1000}\right) & 20\,°\text{C} \leq T \leq 100\,°\text{C} \\ 0.796 - 0.644\left(\frac{T}{1000}\right) + 3.08\left(\frac{T}{1000}\right)^2 - 3.62\left(\frac{T}{1000}\right)^3 & 100\,°\text{C} \leq T \leq 800\,°\text{C} \end{cases} \tag{11.61}$$

When using polypropylene (PP) fiber, its compressive strength at high temperature can be calculated using the following formula [42]:

$$\frac{f_c(T)}{f_c} = \begin{cases} 1.109 - 5.43\left(\frac{T}{1000}\right) & 20\,°\text{C} \leq T \leq 100\,°\text{C} \\ 0.552 - 0.105\left(\frac{T}{1000}\right) + 2.72\left(\frac{T}{1000}\right)^2 - 3.28\left(\frac{T}{1000}\right)^3 & 100\,°\text{C} \leq T \leq 800\,°\text{C} \end{cases} \tag{11.62}$$

As shown in Figure 11.10, the uniaxial tensile strength of concrete (including UHPC) at high temperature decreases with increasing temperature. The uniaxial tensile strength and stiffness of concrete decrease continuously, and the volcanic ash reaction does not significantly improve the tensile properties of concrete. It is worth noting that at a given temperature, the reduction in tensile strength is greater than that in compressive strength. At 250 °C, the tensile strength obtained by Felicetti and Gambarova [43] was 45% lower than the value at ambient temperature, while the compressive strength at the same temperature was only reduced by about 10%. The test results of the uniaxial tensile test also show that, as temperature increases, the peak strain increase of the tensile strength of concrete is small and can be considered to be basically unchanged. However, plasticity and ductility after peak stress increase significantly with rising temperature.

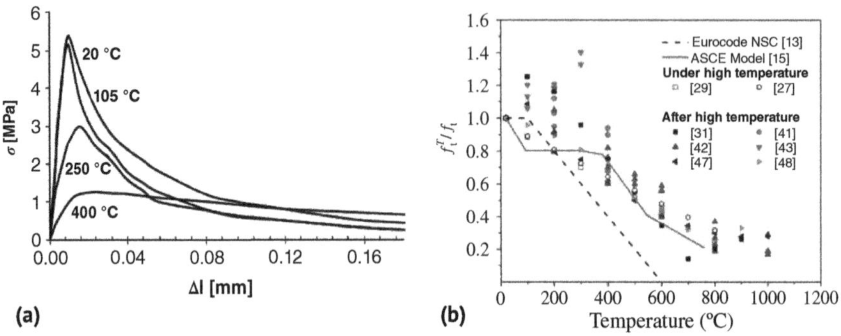

Figure 11.10 Uniaxial tensile mechanical behavior of concrete at high temperature. *Source*: [42]/with permission of Elsevier.

Aslani et al. [41] gave the tensile strength formula of ordinary concrete at high temperature:

$$f_t(T) = f_t \begin{cases} 1.02 - 0.00098\,T & 20\,°\mathrm{C} \le T \le 300\,°\mathrm{C} \\ 0.965 - 0.0001\,T - 9 \times 10^{-7}T^2 & \\ -3 \times 10^{-9}T^3 + 3.2 \times 10^{12}T^4 & 300\,°\mathrm{C} < T \le 900\,°\mathrm{C} \\ 0 & T > 900\,°\mathrm{C} \end{cases} \tag{11.63}$$

When the toughening fiber is a steel fiber, the high-temperature compressive strength formula of UHPC is as follows:

$$f_t(T) = f_t\left[0.98 - 0.925\left(\frac{T}{1000}\right)\right] \tag{11.64}$$

When the toughening fiber is polypropylene fiber, the high-temperature compressive strength formula of UHPC is as follows:

$$f_t(T) = f_t\left[0.96 - 0.958\left(\frac{T}{1000}\right)\right] \tag{11.65}$$

11.3.3 Determination of parameters

In addition to the evolution equations of concrete strength and damage variables, there are some dimensionless parameters in the high temperature plastic damage constitutive model, such as k_p^{c1}, b_c, a_t, a_c, whose values need to be determined. This section will discuss the above parameters and give corresponding value recommendations.

1. Parameter k_p^{c1}
   k_p^{c1} represents the cumulative plastic strain corresponding to the peak compressive strength of concrete $f_c(T)$. In a uniaxial compression test, the cumulative plastic strain k_p^{c1} is proportional to the plastic strain in the direction of applied stress. Gernay et al. [2] gave the relationship between k_p^{c1} and the shear expansion coefficient α_g of concrete, the strain ε_{c1} corresponding to the peak stress in uniaxial compression, and the damage $\tilde{d}_c$:

$$k_p^{c1} = \frac{\varepsilon_{c1}\left(1 - 2\tilde{d}_c\right)}{(2 - 2\tilde{d}_c)(1 - \alpha_g)} \tag{11.66}$$

where $\tilde{d}_c$ is usually taken as 0.25.

Since the accumulated plastic strain is nonnegative, the shear dilatancy coefficient α_g is less than 1. The multiaxial compression test of Kupfer et al. [44] shows that this coefficient is between 0.2 and 0.3. Similarly, the damage $\tilde{d}_c$ corresponding to the peak stress is less than 0.5. The cyclic compression test of Karsan and Jirsa [45] shows that this value is about 0.25.

2. Parameter $a_c$and a_t

a_c is a dimensionless parameter in the concrete compression damage model. To determine this parameter, Eq. (11.54) is first rearranged to isolate $a_c$on the right-hand side, yielding the following expression:

$$a_c = -ln\,(1 - d_c)\,/k_p^c \tag{11.67}$$

Then, substituting k_p^{c1} and $\tilde{d}_c$ corresponding to the peak stress in Eq. (11.67), the calculation formula of a_c can be obtained as follows:

$$a_c = -ln\left(1 - \tilde{d}_c\right)\frac{(\alpha_g - 1)(2 - 2\tilde{d}_c)}{\varepsilon_{c1}\left(1 - 2\tilde{d}_c\right)} \tag{11.68}$$

a_t is a dimensionless parameter in the concrete tensile damage model. Gernay et al. [7] introduced the concepts of equivalent grid length and plastic work, and expressed the sensitivity of the structure to the grid through the fracture energy $\bar{G}_t$. Assuming that the fracture energy is uniformly diffused, the relationship between the fracture energy per unit volume of concrete under tension and the equivalent grid length is as follows:

$$\int_0^{\infty} \tau_t\left(k_p^t\right) dk_p^t = \frac{\bar{G}_t}{l_c} \tag{11.69}$$

where $\bar{G}_t$ is the fracture energy of concrete in tension; l_c represents the characteristic length of the analysis model mesh, which depends on the type, size, shape, and integration type of the selected element.

Substituting Eqs. (11.49) and (11.54) into Eq. (11.69) and rearranging them:

$$\int_0^{\infty} \tau_t\left(k_p^t\right) dk_p^t = \frac{f_t}{2}\left(\frac{1}{a_t} + \frac{1}{6a_t}\right) = \frac{7f_t}{12a_t} \tag{11.70}$$

Combined with Eq. (11.69), a_t can be calculated by the following formula:

$$a_t = \frac{7f_t}{12\,(\bar{G}_t/l_c)} \tag{11.71}$$

3. Parameter b_c

b_c is the compressive hardening parameter of concrete. Similar to a_t, from the perspective of energy method, the fracture energy $\bar{G}_c$ of concrete under compression is sensitive to the grid, and it is assumed that the fracture energy is uniformly distributed during compression. The fracture energy per unit volume can be calculated by the following formula:

$$\int_0^{k_p^{c1}} \tau_c\left(k_p^c\right) dk_p^c + \int_0^{k_p^{c1}} \tau_c\left(k_p^c\right) dk_p^c = \frac{\bar{G}_c}{l_c} \tag{11.72}$$

where the first term represents the fracture energy dissipated before the peak f_c, and the second term represents the fracture energy dissipated after the peak. Substituting into Eq. (11.50) and integrating (11.72), the formula for b_c can be obtained as follows:

$$b_c = \frac{2f_c}{(\bar{G}_c/l_c) - f_c k_p^{c1} - (f_c - f_{c0})\, k_p^{c1} ln2} \tag{11.73}$$

where $\bar{G}_c$ is the fracture energy of concrete under compression; f_{c0} is the elastic limit of concrete under compression, which is usually taken as 0.3 f_c.

In order to avoid the negative dissipation after the peak in Eq. (11.72) and ensure that the total fracture energy is greater than the pre-peak fracture energy, the inequality $\bar{G}_c/l_c > \int_0^{k_p^{c1}} \tau_c\left(k_p^c\right) dk_p^c$ must always hold. After the formula is transformed, the inequality can be expressed as:

$$\bar{G}_c/l_c > f_c k_p^{c1} + (f_c - f_{c0})\, k_p^{c1} ln2 \tag{11.74}$$

However, in actual tests and engineering, the compressive fracture dissipation energy of concrete $\bar{G}_c$ is not easy to measure. Therefore, the parameter x_c is introduced to represent the ratio of the pre-peak fracture dissipation energy to the total fracture dissipation energy at failure under uniaxial compression:

$$x_c = \frac{f_c k_p^{c1} + (f_c - f_{c0})\, k_p^{c1} ln2}{\bar{G}_c/l_c} \tag{11.75}$$

Therefore, the value range of x_c is 0 to 1. When the material is extremely ductile, the total fracture dissipation energy is extremely large when it extends infinitely after the peak value, and the value of x_c approaches 0 infinitely. When the material is brittle and fractures at the peak stress, the fracture dissipation energy accumulated before the peak equals the total fracture dissipation energy, and x_c is equal to 1. After introducing x_c into Eq. (11.73), the result is obtained as:

$$b_c = \frac{2f_c x_c}{(1 - x_c)\left[f_c k_p^{c1} + (f_c - f_{c0})\, k_p^{c1} ln2\right]} \tag{11.76}$$

In this way, the parameter b_c can be obtained by adjusting the value of x_c according to the uniaxial compression stress–strain curve of concrete. Gernay et al. recommend x_c to be 0.19 to 0.3. CEB-FIB believes that the tensile fracture energy of concrete is related to the concrete strength and the maximum size of the aggregate. $\bar{G}_t$ is calculated using the following empirical formula:

$$\bar{G}_t = a_d\,(f_c + 8)^{0.7} \tag{11.77}$$

where a_d is the aggregate correction factor, which is related to the maximum size of aggregate. CEB-FIB gives corresponding value recommendations as shown in Table 11.1. If the value exceeds the table, a linear difference is taken.

Hilsdorf and Brameshuber [46] showed that Eq. (11.77) is in good agreement with experimental results, and that the tensile fracture energy $\bar{G}_t$ of concrete is between 50 and 150 N m^{-1}. The characteristic length l_c depends on the selected element type, size, shape, and integration method [47]. It can be approximated by the following formula:

$$l_c = \alpha_l \sqrt{A_e} \tag{11.78}$$

where, in the formula, α_l is the correction coefficient, which is 1 for linear units and 2 for quadratic units; A_e represents the unit area.

Under high temperature, the biaxial compressive strength f_b and Poisson's ratio ν of concrete will change with temperature. Ehm and Schneider [48] gave an empirical calculation formula for the biaxial compressive strength $f_b\,(T)$ at different temperatures through experiments. By introducing the parameter $\beta\,(T) = f_b\,(T)\,/f_c\,(T)$, the formula for $\beta\,(T)$ is as follows:

$$\beta\,(T) = \begin{cases} 1.16 & T \le T_{\beta 1} \\ 1.16\left(1 + 0.6 \times \left(\dfrac{T - T_{\beta 1}}{T_{\beta 2} - T_{\beta 1}}\right)\right) & T_{\beta 1} < T \le T_{\beta 2} \\ 1.86 & T_{\beta 2} < T \end{cases} \tag{11.79}$$

where $T\beta_1$ is the first transition temperature, $T\beta_1 = 350\,°\mathrm{C}$; $T\beta_2$ is the second transition temperature, $T\beta_2 = 750\,°\mathrm{C}$.

Poisson's ratio is correlated with temperature. Luccioni et al. [49] conducted relevant research on the Poisson's ratio of concrete under high temperature and found that the Poisson's ratio and temperature have a bilinear approximate relationship. The specific formula is as follows:

$$\begin{cases} \upsilon\,(T) = \upsilon_{20}\left(0.2 + 0.8 \times \dfrac{T_\upsilon - T}{T_\upsilon - 20}\right) & T \le T_\upsilon \\ \upsilon\,(T) = 0.2 \times \upsilon_{20} & T > T_\upsilon \end{cases} \tag{11.80}$$

Table 11.1 Values of coefficient a_d considering maximum aggregate size.

d[mm]	a_d
8	4
16	6
32	10

11.4 Concrete High Temperature Transient Creep Model

11.4.1 High Temperature Transient Creep Strain Model

The temperature-stress path under high temperature has an important influence on the deformation of concrete. As shown in Figure 11.11, the temperature-free expansion strain of the free specimen (external load is zero) is ε_{th} (segment *oa* in the figure). For the specimen of the constant load-heating path, stress is applied in advance, and an immediate strain $\varepsilon\sigma$ is generated (segment *oo'* in the figure). The total strain after heating is ε_T (segment *o'b'* in the figure). The curve *o'b'* is translated so that it starts from the origin o, and the curve ob is obtained. Under the action of stress σ, the thermal strain of concrete in the two different heating processes is quite different. This difference is called transient creep strain, also called transient thermal strain. Transient creep strain is the main part of the high-temperature deformation of concrete. It has a great influence on the deformation and stress redistribution of the structure and must be considered in the high-temperature analysis of the structure.

Different scholars have proposed corresponding calculation methods for transient creep strain. Anderberg and Thelandsson [14] proposed a high-temperature deformation model for concrete based on a large number of experiments, which is decomposed into four parts: thermal strain, stress–strain, transient strain, and creep strain. This model has been widely recognized and applied. Gernay [3] proposed an explicit transient creep (ETC) model, which believes that transient creep is nonlinearly proportional to applied stress and temperature, and gives the recommended value of the proportional function φ (T) in a table. Khoury [51] and Terro [18] proposed the concept of "load induced thermal strain" (LITS) of concrete at high temperature, and believed that the total strain of concrete at high temperature is equal to the free thermal strain (FTS) minus the load induced thermal strain LITS, including transitional thermal creep, drying creep, basic creep and elastic strain. They also pointed out that transient creep comprises transitional thermal creep and drying creep, and drying creep, and is the main component of LITS. Through experiments, Terro established an empirical formula for LITS, considering LITS as a function of temperature T, load level σ, and aggregate volume ratio v_a, and calculating LITS using the test data regression curve

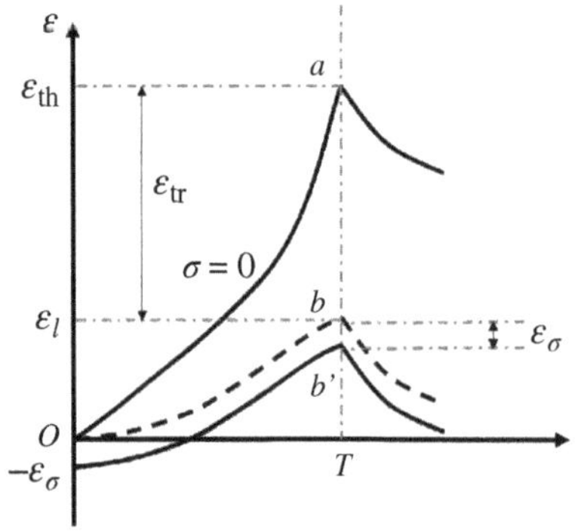

Figure 11.11 Temperature deformation and transient creep strain of concrete. *Source*: Adapted from [50].

with a load ratio of 0.3 as the "benchmark." For convenience, a dimensionless transient strain calculation parameter $\beta_{td}(T)$ can be introduced and set the transient creep strain ε_{td} equal to:

$$\varepsilon_{td} = \frac{\sigma}{f_c}\beta_{td}(T) \tag{11.81}$$

$$\beta_{td}(T) = \left[72 \times \left(\frac{T}{1000}\right)^2 - \left(\frac{T}{1000}\right)\right] \times 10^{-3} \tag{11.82}$$

However, it can be seen from the above models that the load ratio σ/f_c is the most important factor affecting transient creep, and all models take it into consideration. At the same time, most transient creep calculation formulas are proposed based on data regression methods. This empirical formula does not fully explain the generation of transient creep, and the physical meaning is not clear enough. Therefore, this chapter proposes a new calculation model based on the processes that affect the generation of transient creep, such as free water evaporation, bound water dehydration, and chemical decomposition.

The generation of transient creep is related to the internal chemical changes in the cement matrix and the escape of free water from the internal pores. As shown in Figure 11.12, when the temperature is heated from room temperature to 105 °C, transient creep is mainly caused by the evaporation and escape of free water in the pores inside the cement matrix. When the temperature exceeds 105 °C, a portion of the bound water in the hydrated calcium silicate hydrate (C–S–H) within the cement matrix becomes reactive and undergoes dehydration. At temperatures above 400 °C, calcium hydroxide begins to decompose, forming calcium oxide. The decomposition of calcium carbonate begins at 500 °C and decomposes into calcium oxide and carbon dioxide. According to the model proposed by Sabeur et al. [50, 53], the transient creep strain in this chapter takes the following form:

$$\dot{\varepsilon}_{td} = \dot{\phi}\left(\dot{h}_r, m_{dehy}, T\right) \mathbb{Q} : \frac{\sigma}{f_c} \tag{11.83}$$

where $\dot{\phi}$ represents the function describing the transient creep rate of concrete, which is related to temperature, relative humidity, and dehydration; h_r is the relative humidity; m_{dehy} is the dehydration mass; $\mathbb{Q}$ is a fourth-order tensor $\mathbb{Q} = (1+\upsilon)\,\delta\bar{\otimes}\delta - \upsilon\delta\otimes\delta$.

According to Eq. (11.83), the transient creep strain of concrete at high temperatures can be obtained by determining the function $\dot{\phi}\left(\dot{h}_r, m_{dehy}, T\right)$. Sabeur et al. [53] proposed a calculation model for transient creep only below 400 °C, and its formula is as follows:

$$\dot{\phi}\left(\dot{h}_r, m_{dehy}, T\right) = \alpha \mathrm{dc}\left|\dot{h}_r\right| + \alpha_{hc}\left(m_{dehy}\right) H(T-105)\,\dot{m}_{dehy} \tag{11.84}$$

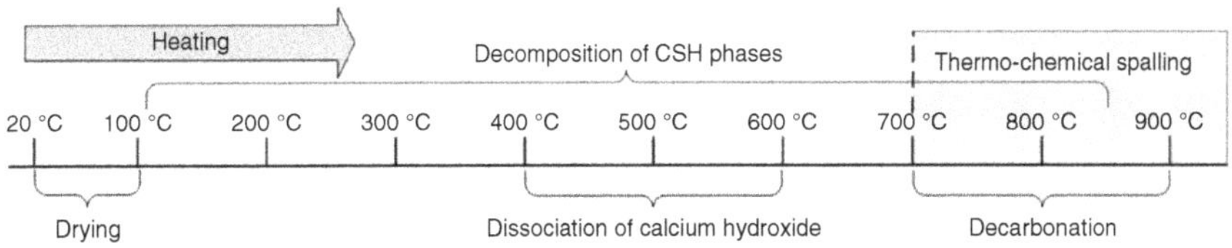

Figure 11.12 Physical and chemical reaction process of concrete at high temperature. *Source*: [52] / MDPI / CC BY 4.0.

where $\alpha_{dc}\left|\dot{h}_r\right|$ is used to describe the deformation caused by evaporation of free water below 105 °C, and $\alpha_{hc}\left(m_{dehy}\right)H\left(T-105\right)\dot{m}_{dehy}$ is used to describe the deformation caused by dehydration of bound water when the temperature is higher than 105 °C. However, when the temperature is higher than 400 °C, the accuracy of the model decreases due to the lack of corresponding terms to describe the decomposition of hydrates such as calcium hydroxide. Therefore, Eq. (11.84) is modified, and a term is introduced to describe the decomposition of hydrates:

$$\dot{\phi}\left(\dot{h}_r, m_{dehy}, T\right)=\alpha_{dc}\left|\dot{h}_r\right|+\alpha_{hc}\left(m_{dehy}\right)H\left(T-105\right)\dot{m}_{dehy} + \alpha_{chem}\left(m_{eq}\right)H\left(T-400\right) \tag{11.85}$$

where αdc is the dry creep coefficient; αhc is the dehydration creep coefficient; α_{chem} is the hydrate decomposition creep coefficient.

11.4.2 Drying Deformation

Dry creep is induced by the evaporation of free water in the microscopic pores of the cement matrix, and mainly occurs below 105 °C. As shown in Figure 11.13, Hassen et al. [53] gave the experimental data of dry creep and temperature. As temperature increases, the growth rate of dry creep gradually decreases and eventually stabilizes. Since dry creep strain accounts for a small proportion of transient creep, the mathematical model of relative humidity involves the mass conservation, energy conservation, permeation and diffusion of the three phases of liquid, dry air and water vapor, and the calculation process is very complicated. To simplify the calculation, this chapter proposes an empirical calculation formula for the average relative humidity h_r of concrete specimens and temperature:

$$h_r=\begin{cases}2.333-0.48\times ln\left(T\right) & 20\,°C\le T<105\,°C\\ 0.06 & T>105\,°C\end{cases} \tag{11.86}$$

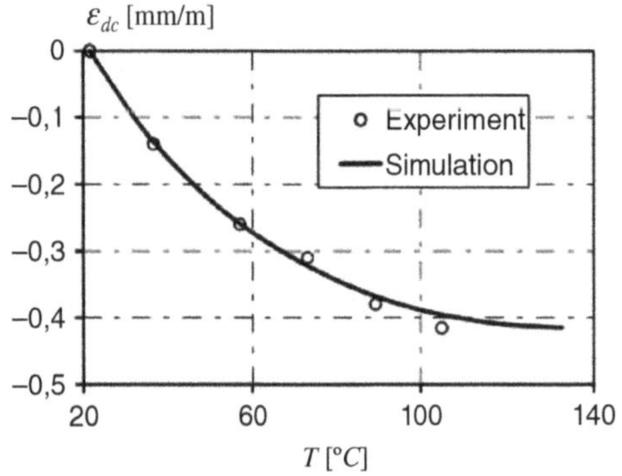

Figure 11.13 Concrete drying creep strain.

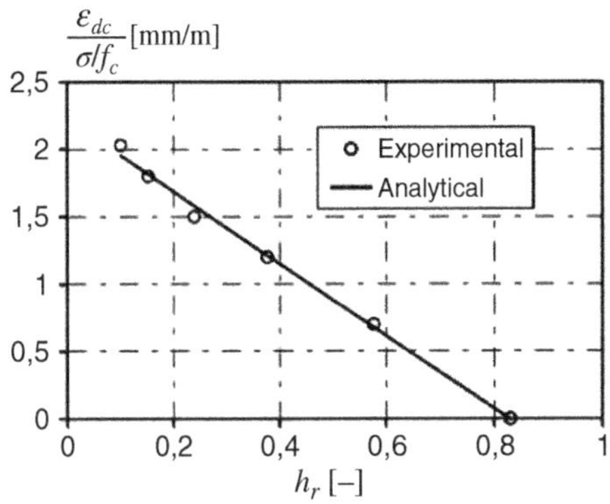

Figure 11.14 Normalized drying creep versus relative humidity.

As shown in Figure 11.14, through experimental research, it is found that after normalizing the creep deformation caused by drying, $\varepsilon_{dc}/\sigma f_c$ and relative humidity h_r approximately present a linear relationship, and the recommended value of the coefficient α_{dc} is 2.7×10^{-3}.

11.4.3 Dehydration Deformation

The creep deformation caused by the dehydration reaction is mainly described by the term $\alpha_{hc}\left(m_{dehy}\right) H\left(T-105\right) \dot{m}_{dehy}$. In order to describe the dehydration reaction of the cement matrix during the heating process, the formula for defining the dehydration rate $\dot{m}_{dehy}$ is:

$$\dot{m}_{dehy} = -\frac{m_{dehy}(t) - m_{eq}(T)}{\tau_{dehy}} \tag{11.87}$$

where $m_{dehy}(t)$ represents the dehydrated mass of concrete per unit volume at time t; $m_{eq}(T)$ represents the dehydrated mass produced when the concrete per unit volume reaches dehydration equilibrium at temperature T; τ_{dehy} is the dehydration relaxation time, indicating that the dehydration process of concrete requires a certain amount of time. Sabeur et al. [53] gave recommended values for ordinary and high-strength concrete, taking an average value of 120 minutes below 500 °C and 60 minutes above 500 °C.

$m_{eq}(T)$can be measured experimentally. By performing TGA analysis on concrete specimens at a rate of 0.2 °C min^{-1}, the residual mass of concrete specimens at high temperature can be obtained. Through data fitting, this chapter gives the empirical formula of $m_{eq}(T)$:

$$\begin{aligned} m_{eq}(T) = \Big[& \exp\left(-6.9678 \times (T-105)^{-0.2319}\right) H(T-105) \\ & + \exp\left(-9.4576 \times (T-232)^{-0.2232}\right) H(T-232) \\ & + \exp\left(-6.479 \times (T-392)^{-0.2484}\right) H(T-392) \\ & + \exp\left(-8.7276 \times (T-412)^{-0.2371}\right) H(T-412)\Big] \\ & \times 10^{3} \end{aligned} \tag{11.88}$$

where $H(x)$ represents the Heaviside step function:

$$H(x) = 0.5\,[1 + sgn(x)] \tag{11.89}$$

Dal Pont and Ehrlacher [54] found through experiments that the difference between the dehydrated mass of concrete per unit volume $m_{\mathrm{dehy}}(t)$ and the dehydrated mass at dehydration equilibrium $m_{\mathrm{eq}}(T)$ decreases with time, and the curve shows a negative logarithmic relationship, as shown in Figure 11.15.

When the temperature is determined, $m_{\mathrm{eq}}(T)$ can be regarded as a constant, and the relationship between $m_{\mathrm{eq}}(T) - m_{\mathrm{dehy}}(t)$ and time is as follows:

$$ln\left(m_{\mathrm{eq}}(T) - m_{\mathrm{dehy}}(t)\right) = (1.2 \times 10^{-5}T - 0.0095)\,t - 3.345 \tag{11.90}$$

At the same time, Dal Pont and Ehrlacher [54] believed that the dehydration mass of unit volume of concrete $m_{\mathrm{dehy}}(t)$ and the dehydration mass generated when dehydration is in equilibrium $m_{\mathrm{eq}}(T)$ can be described by the following relationship:

$$ln\left(m_{\mathrm{eq}}(T) - m_{\mathrm{dehy}}(t)\right) = ln\left(m_{\mathrm{eq}}(T) - m_{\mathrm{dehy}}(0)\right) - \frac{t}{\tau_{\mathrm{dehy}}} \tag{11.91}$$

where $m_{\mathrm{dehy}}(0)$ is the dehydrated mass of concrete per unit volume at the beginning after reaching the target temperature. It can be determined by Eq. (11.87)–(11.91).

Sabeur et al. [53] proposed a dehydration creep parameter $\alpha_{\mathrm{hc}}\left(m_{\mathrm{dehy}}\right)$ of $7 \times 10^{-7}\, m_{\mathrm{dehy}}$, but this value did not take into account the effect of heating rate on dehydration creep strain. Therefore, α_{hc} was corrected as follows

$$\alpha_{\mathrm{hc}} = 7V_{\mathrm{heating}}^{-\frac{1}{2}} \times 10^{-7} m_{\mathrm{dehy}}(t) \tag{11.92}$$

where $V_{heating}$ represents the heating rate.

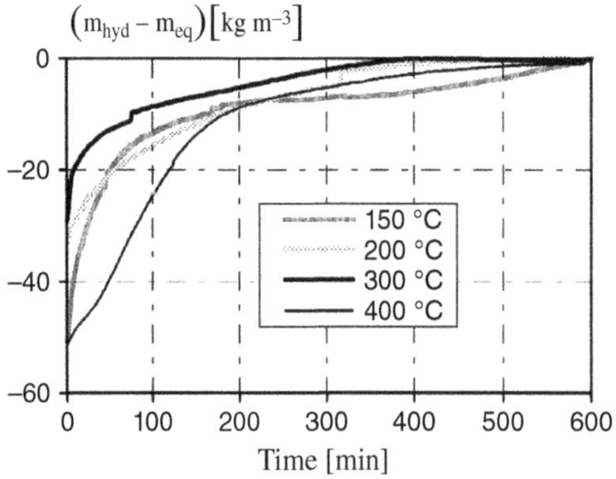

Figure 11.15 The relationship between the difference between $m_{\mathrm{dehy}}(t)$ and $m_{\mathrm{eq}}(T)$ with time.

11.4.4 Chemical Decomposition

The decomposition of CSH occurs in the temperature range above 105 °C, and the decomposition of calcium hydroxide CH occurs in the temperature range of 400~600 °C. When the temperature exceeds 400 °C, the chemical reaction rate accelerates, leading to increased transient creep strain rates due to the extensive decomposition of calcium hydroxide. Beyond 600 °C, numerous cracks develop within the microstructure of UHPC, chemical decomposition intensifies, and a substantial number of coarse pores are generated. Most of the calcium hydroxide hydrate (CH) has been decomposed, the decomposition rate of hydrates such as CSH has been greatly increased, and the strain rate has increased significantly. Based on this, a factor α_{chem} that describes the relationship between the dehydration quality and the hydrate decomposition strain is introduced into the transient creep model of this chapter, and the hydrate decomposition creep coefficient related to the dehydration quality is defined as:

$$\alpha_{\text{chem}} = \beta_1 T + \beta_2 \, \text{EXP}\,(b_c\, T) \tag{11.93}$$

where β_1 is the coefficient of decomposition creep of calcium hydroxide, which is $-0.002V_{\text{heating}}$; β_2 is the coefficient of decomposition creep of CSH; b_c is the index of decomposition creep of CSH, which is 0.0196. Through data fitting, the coefficient β_2 of decomposition creep of CSH is recommended to be calculated using the following formula:

$$\beta_2 = -0.2\dot{T} \times 10^{-9}\, \text{EXP}\,(9.19 \times \sigma / f_c) \tag{11.94}$$

11.5 Model Validation

11.5.1 Numerical Implementation of the Constitutive Model

The numerical implementation of the high-temperature plastic damage constitutive model of concrete mainly considers the implementation of the elastoplastic theory and the update and iteration of the damage variable, and on this basis, the influence of temperature is also added. The commercial finite element software ABAQUS offers a user-defined material subroutine UMAT, which can flexibly write user developed material constitutive models to supplement the deficiencies of the built-in material library. Based on the writing rules of UMAT and the above-mentioned high-temperature plastic damage constitutive model of concrete, this chapter develops a corresponding numerical model and verifies the accuracy of the constitutive model. The specific iterative process is shown in Figure 11.16.

11.5.2 Numerical Model Establishment and Verification

The verification of the high-temperature plastic damage constitutive model of concrete proposed in this chapter is divided into two parts: verification of the transient creep strain and verification of the stress–strain curve. As shown in Figure 11.17, this chapter established a 10 mm × 10 mm × 10 mm three-dimensional eight-node regular octahedral element (C3D8), applied a displacement load directly above the model, and output the transient creep strain as the state variable STATEV. The numerical calculation results were compared with the experimental data of Sabeur et al. [50, 53], Anderberg et al. [14], and Abid [52].

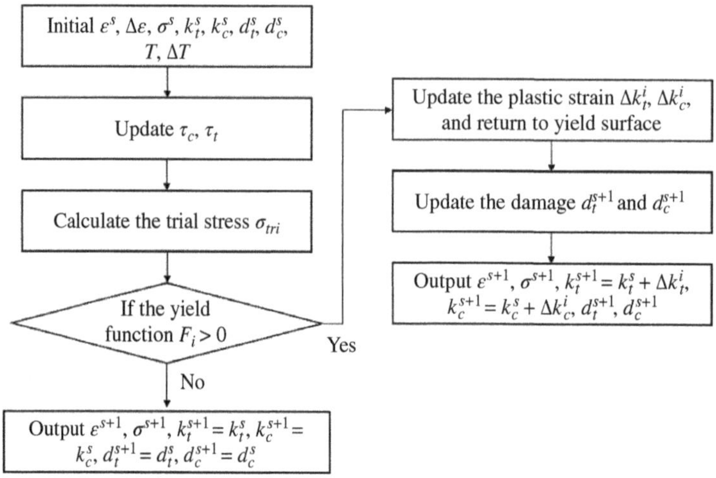

Figure 11.16 Numerical implementation process of the plastic damage constitutive model.

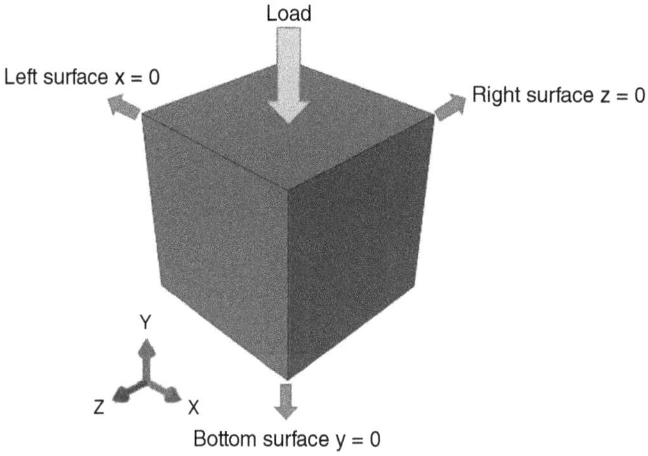

Figure 11.17 Boundary conditions of the finite element model.

Sabeur et al. [50, 53] divided concrete into four groups based on the transient creep test recommendations given by RILEM, with strengths ranging from ordinary concrete to ultra-high-strength concrete. The specific material parameters are shown in Table 11.2. The test heating rate was 1°C min^{-1}, and the load rates were 20% and 40%, respectively. The test required first applying the load to the test block, and then heating it at a uniform rate. The strain measured after the start of heating is the transient creep strain.

The fitting results are shown in Figure 11.18a–d. When the temperature is less than 400 °C, the total transient creep strain of ordinary concrete and ultra-high performance concrete remains small (less than 0.005). When the temperature is less than 105 °C, the drying strain is the main component of transient creep, and this part of the strain is usually less than 0.001. As the temperature increases, a small part of the calcium silicate hydrate (CSH) and bound water begin to dehydrate, the drying strain increases more slowly, gradually tends to be stable, and the overall development of transient creep accelerates.

Table 11.2 Test material parameters given by Sabeur et al.

Test piece number	M30C	M75C	M75SC	M100C
f_c [MPa]	39.3	99.8	89.4	120.7
E [GPa]	36.1	47.5	48.8	50.8

Source: Adapted from [50, 53].

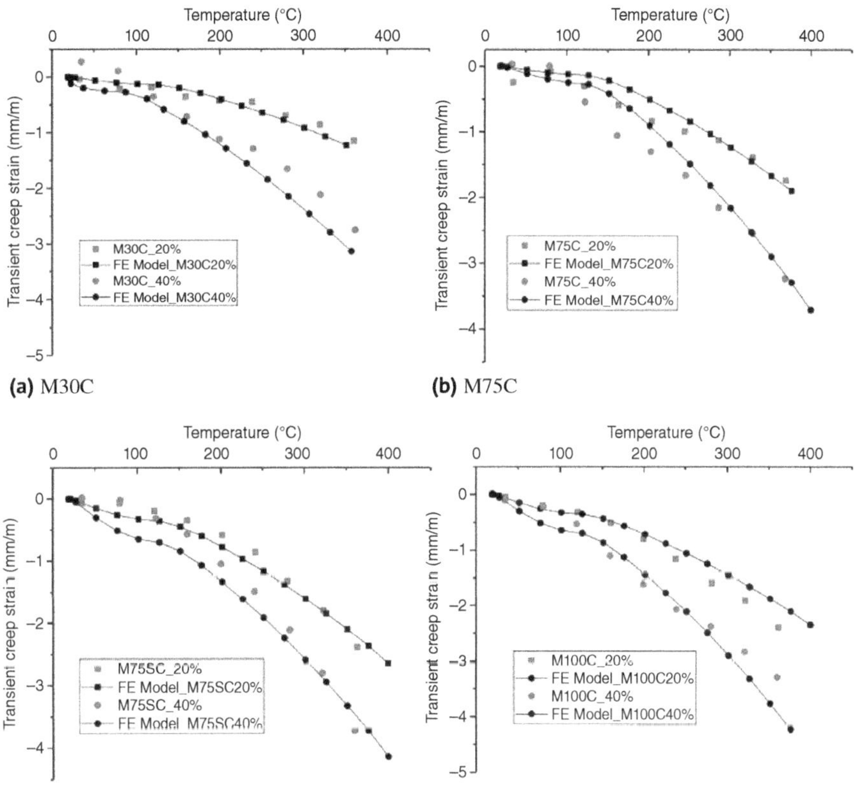

(a) M30C **(b)** M75C **(c)** M75SC **(d)** M100C

Figure 11.18 Transient creep strain test and numerical fitting results of M30C, M75C, M75SC, and M100C.

Compared with the test results, it shows that the transient creep strain calculation model considering the dehydration effect has a high calculation accuracy and a good prediction effect.

Anderberg and Thelandersson [14] also conducted an experimental study on the transient creep strain of concrete at high temperatures, using an adapted aggregate and a water–cement ratio (w/c) of 0.6. The specimens were divided into two groups, with load rates of 45% and 67.5% applied, respectively. The heating rates of each group of specimens were also divided into two types: 1 and 5 °C min^{-1}. The material parameters are shown in Table 11.3.

Table 11.3 Test material parameters given by Anderberg and Thelandersson.

Test piece number	A3	A4	A8	A9
f_c[MPa]	64	58.4	64	58.4
Curing conditions	Standard curing	Standard curing	Standard curing	Standard curing
Heating rate	1 °C min^{-1}	1 °C min^{-1}	5 °C min^{-1}	5 °C min^{-1}
Load rate	45%	67.5%	45%	67.5%

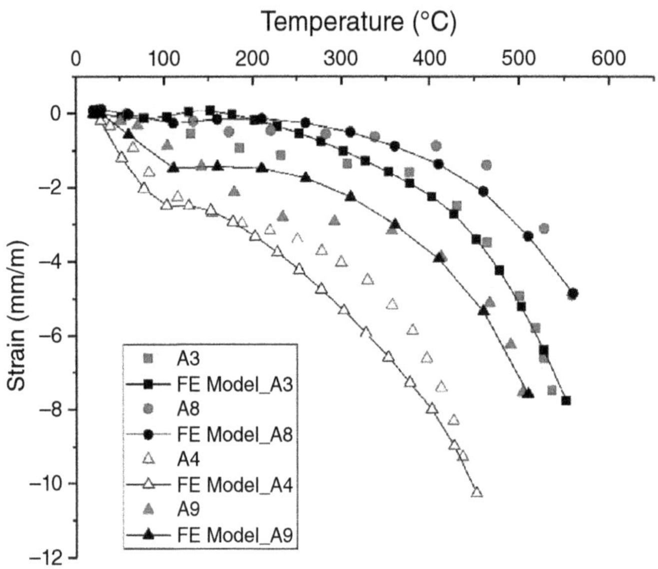

Figure 11.19 Anderberg and Thelandersson's (*Source*: Adapted from [14]) transient creep strain test and numerical fitting results.

The prediction results are shown in Figure 11.19. When the test temperature exceeds 400 °C, the hydroxide begins to decompose and the creep strain rate of change with time further increases. The calculation results of the transient creep prediction model considering the decomposition of hydrates are consistent with the test. Abid et al. [52] conducted a transient creep strain measurement test on reactive powder concrete (RPC) at high temperature. The test used two heating rates: 3 and 5 °C min^{-1}, with a material strength of 150.8 MPa, a maximum temperature of nearly 900 °C, and a load ratio of 20–60%.

As shown in Figure 11.20, the transient creep strain generated after 400 °C is much larger than the strain generated before 400 °C, and the transient creep strain generated before 400 °C can be almost ignored. After 600 °C, with the intensification of hydrate decomposition, the transient creep further increases. The transient creep model considering hydrate decomposition has good accuracy in describing the deformation after 600 °C.

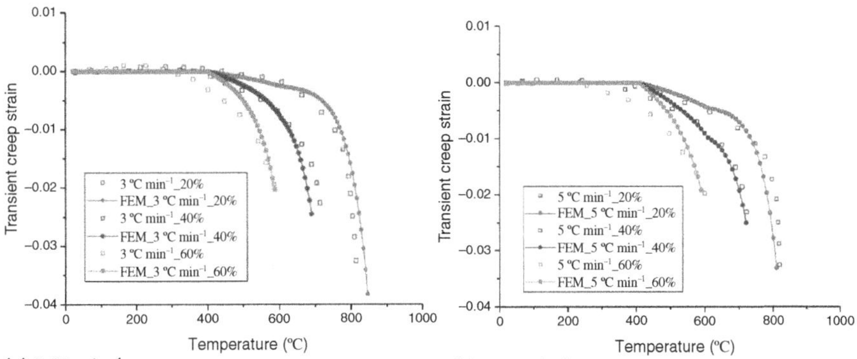

(a) 3 °C min^{-1} **(b)** 5 °C min^{-1}

Figure 11.20 Transient creep strain test and numerical fitting results by Abid et al. *Source*: [52] / MDPI / CC BY 4.0.

Table 11.4 Main parameter values.

Symbol	Parameter	Unit	Value
α_g	Shear expansion coefficient	/	0.25
υ	Poisson's ratio	/	0.2
$\tilde{d}_c$	Damage corresponding to peak stress	/	0.25
$\overline{G}_t$	Tensile fracture energy	N m^{-1}	50
l_c	Characteristic length of the grid	/	$100\sqrt{2}$

On this basis, the tensile and compressive constitutive models of concrete at high temperature are verified. The verification of the compressive constitutive model uses the measurement results of Chang et al., and the specimen is heated to the test temperature at a heating rate of 20 °C min ^{1} and kept constant for 1.5–2.5 hours. The average compressive strength at room temperature is 40 MPa, the elastic modulus is 30GPa, the temperature range is 20–503 °C, and the other constitutive parameters are listed in Table 11.4.

As shown in Figure 11.21, the compressive strength of concrete at high temperature hardly decreases below 200 °C, and begins to decrease above 300 °C. After 400 °C, the calcium hydroxide hydrate in the concrete decomposes, and its microscopic volume expansion produces a loose porous structure, which leads to a significant decrease in the strength of the concrete. As the temperature increases, the decomposition of concrete hydrates further intensifies, the strength continues to decrease, the peak strain continues to increase, and the plastic section continues to extend. The developed constitutive model can accurately describe the hardening and softening behavior of concrete at high temperatures.

The constitutive model proposed in this chapter considers the influence of steel fiber toughening on the stress–strain relationship of UHPC under high-temperature compression, and modifies the damage model. Zheng et al. gave the results of high-temperature compression tests, with a steel fiber content of 2%. As shown in Figure 11.22, this chapter

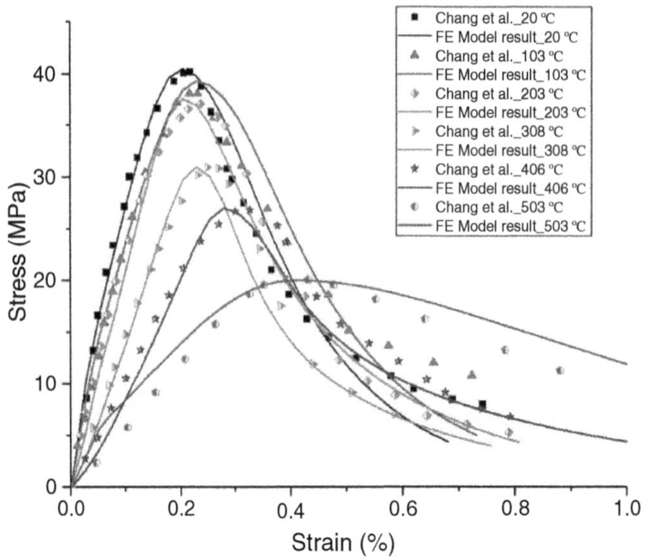

Figure 11.21 Chang et al.'s high temperature compression test and numerical results of concrete.

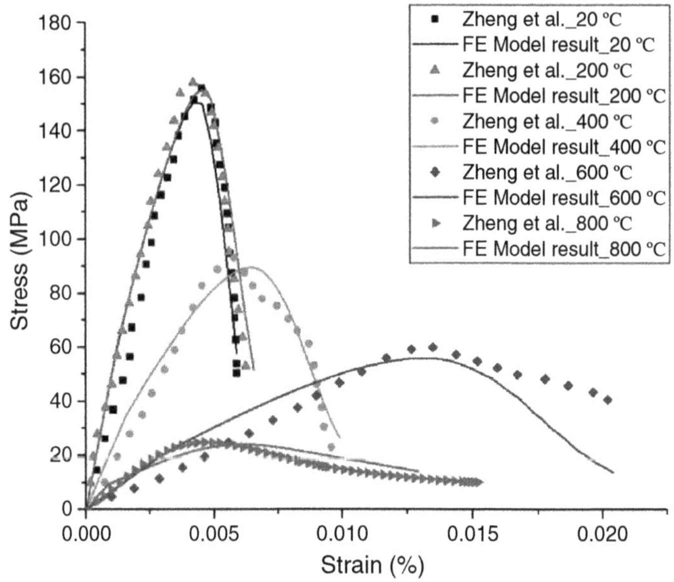

Figure 11.22 Zheng et al.'s UHPC high temperature pressure test and numerical results.

verifies the test results given by Zheng et al., and the prediction effect is good. The compressive strength of UHPC at 200 °C exhibits a slight increase, and its stress–strain behavior closely resembles that of UHPC at room temperature.

There is limited experimental data on tensile stress–strain of concrete under high temperature. This chapter uses the experimental results of Aslani et al. to verify the constitutive

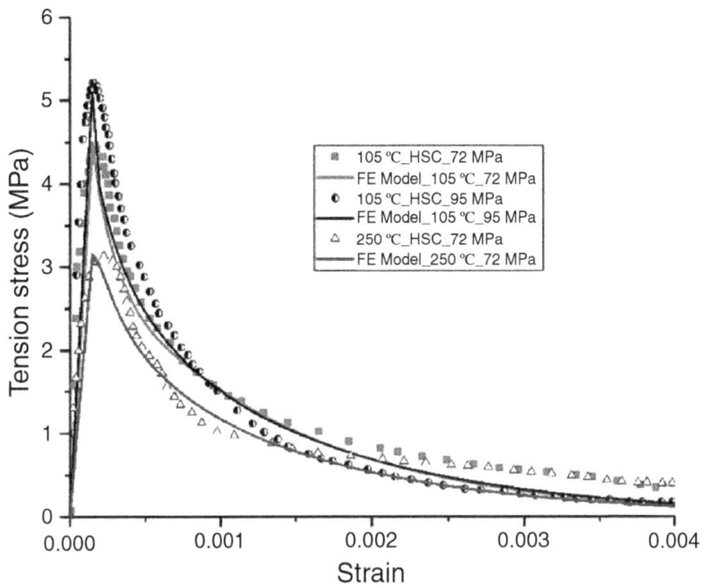

Figure 11.23 Aslani et al.'s concrete high-temperature tensile test and numerical results.

model. The test was divided into two groups, and the compressive strength of the specimens was 72 and 95 MPa, respectively. The 72 MPa specimen was heated at 105 °C and 250 °C, and the 95 MPa specimen was heated at 105 °C. The numerical prediction results are shown in Figure 11.23. The tensile stress–strain curve of concrete is very different from the compression curve. Given that concrete's tensile strength is substantially lower than its compressive strength, the tensile curve rapidly reaches its peak stress, with the corresponding peak strain being very small. Compared with the experimental data, the prediction of the established constitutive model shows reasonable accuracy.

11.6 Conclusion

This chapter mainly studies the plastic damage constitutive model of concrete under high temperature. Based on the thermodynamic theory, the plastic flow and damage evolution law of concrete are derived. On this basis, the concrete constitutive model is modified again. Considering the influence of dehydration reaction and compound decomposition, a high-temperature transient creep stress calculation model is established. The main work is summarized as follows:

1. Concrete exhibits pronounced tension-compression anisotropy, which remains significant even at elevated temperatures. Under compression, concrete demonstrates more evident plasticity and hardening, while under tension, after reaching the peak stress, the stress of concrete drops sharply, and the material softens rapidly until it is destroyed. At the same time, the strain corresponding to the peak stress of concrete

under compression is greater than that under tension, and the peak strain under compression rises with temperature. In contrast, the peak tensile strain shows little variation with temperature. Therefore, different yield criteria are employed to describe concrete under tension and compression.

2. UHPC generally incorporates steel fibers as toughening materials, which produce a hoop effect on the matrix to improve the compressive strength, thus affecting the evolution of damage variables. This chapter proposes a temperature-dependent correction formula based on the ratio of the elastic modulus of fiber-reinforced UHPC to the elastic modulus of plain UHPC.
3. Different combinations of high temperature and load of concrete will lead to different mechanical behaviors. The difference between the free thermal expansion strain and the strain caused by "first applying load and then heating" is termed high-temperature transient creep. The development of high-temperature transient creep is closely related to the load ratio and the physical and chemical changes of the matrix. Free water evaporation, bound water dehydration, and hydrate decomposition can all cause transient creep. As temperature increases, creep induced by hydrate decomposition becomes the dominant component of total transient creep. Beyond 600 °C, the transient creep caused by chemical decomposition is much greater than the free water evaporation and dehydration. Comparison with experimental results demonstrates that the model achieves good predictive accuracy.
4. The constitutive model proposed in this chapter was implemented and validated using the user subroutine platform of the commercial finite element software ABAQUS. Transient creep is considered only under compressive loading conditions. The modified high-temperature constitutive model of concrete is compared with the compressive and tensile stress–strain data under high temperature, proving that the constitutive model proposed in this chapter can better predict the mechanical behavior of high-temperature concrete.

References

1 Nechnech, W., Meftah, F., and Reynouard, J.M. (2002). An elasto-plastic damage model for plain concrete subjected to high temperatures. *Engineering Structures* 24: 597–611.

2 Gernay, T. and Franssen, J.M. (2012). A formulation of the Eurocode 2 concrete model at elevated temperature that includes an explicit term for transient creep. *Fire Safety Journal* 51: 1–9.

3 Gernay, T., Millard, A., and Franssen, J-M. (2013). A multiaxial constitutive model for concrete in the fire situation: Theoretical formulation. *International Journal of Solids and Structures* 50: 3659–3673.

4 Gernay, T. and Franssen, J-M. (2015). A plastic-damage model for concrete in fire: applications in structural fire engineering. *Fire Safety Journal* 71: 268–278.

5 Hughes, B.P. and Chapman, G.P. (1966). The deformation of concrete and micro-concrete in compression and tension with particular reference to aggregate size. *Magazine of Concrete Research* 18: 19–24.

6 Gopalaratnam, V. (1985). Softening response of plain concrete in direct tension. *Journal of the American Concrete Institute* 82: 310–323.

7 Gernay, T. and Franssen, J-M. (2011). A Comparison Between Explicit and Implicit Modelling of Transient Creep Strain in Concrete Uniaxial Constitutive Relationships. Interscience Communications Ltd, London, United Kingdom, 2011.

8 Yao, Y. and Wang, K. (2017). Elastic-plastic damage model to predict pore-pressure effect on concrete behavior at elevated temperatures. *Journal of Engineering Mechanics* 143: 04017122.

9 Purkiss, J.A. and Li, L. (2013). *Fire Safety Engineering Design of Structures[M].* CRC press.

10 Youssef, M.A. and Moftah, M. (2007). General stress–strain relationship for concrete at elevated temperatures. *Engineering Structures* 29: 2618–2634.

11 Schneider, U. (1988). Concrete at high temperatures — A general review. *Fire Safety Journal* 13: 55–68.

12 Li, L-y. and Purkiss, J. (2005). Stress–strain constitutive equations of concrete material at elevated temperatures. *Fire Safety Journal* 40: 669–686.

13 Khoury, G.A., Dias, W.P., and Sullivan, P.J.E. (1986). Deformation of concrete and cement paste loaded at constant temperatures from 140 to 724 °C. *Materials and Structures* 19: 97–104.

14 Anderberg, Y. and Thelandersson, S. (1976). *Stress and Deformation Characteristics of Concrete at High Temperatures: Experimental Investigation and Material behaviour model[M].* Sweden: Lund institute of technology Lund.

15 de Borst, R. and Peeters, P.P.J.M. (1989). Analysis of concrete structures under thermal loading. *Computer Methods in Applied Mechanics and Engineering* 77: 293–310.

16 Torelli, G., Gillie, M., Mandal, P., and Tran, V. (2017). A multiaxial load-induced thermal strain constitutive model for concrete [J]. *International Journal of Solids and Structures* 108: 115–125.

17 Nielsen, C.V., Pearce, C.J., and Bićanić, N. (2002). Theoretical model of high temperature effects on uniaxial concrete member under elastic restraint. *Magazine of Concrete Research* 54: 239–249.

18 Terro, M. (1998). Numerical modeling of the behavior of concrete structures in fire. *ACI Structural Journal* 95: 183–193.

19 Grassl, P. and Jirásek, M. (2006). Damage-plastic model for concrete failure. *International Journal of Solids and Structures* 43: 7166–7196.

20 Richard, B., Ragueneau, F., Cremona, C., and Adelaide, L. (2010). Isotropic continuum damage mechanics for concrete under cyclic loading: Stiffness recovery, inelastic strains and frictional sliding. *Engineering Fracture Mechanics* 77: 1203–1223.

21 Wu, J.Y., Li, J., and Faria, R. (2006). An energy release rate-based plastic-damage model for concrete. *International Journal of Solids and Structures* 43: 583–612.

22 Abu Al-Rub, R.K. and Voyiadjis, G.Z. (2009). Gradient-enhanced Coupled Plasticity-anisotropic Damage Model for Concrete Fracture: Computational Aspects and Applications. *International Journal of Damage Mechanics* 18: 115–154.

23 Desmorat, R., Gatuingt, F., and Ragueneau, F. (2007). Nonlocal anisotropic damage model and related computational aspects for quasi-brittle materials. *Engineering Fracture Mechanics* 74: 1539–1560.

24 Ortiz, M. (1985). A constitutive theory for the inelastic behavior of concrete. *Mechanics of Materials* 4: 67–93.

25 Yao, Y., Fang, H., and Guo, H. (2022). Unified damage constitutive model for fiber-reinforced concrete at high temperature. *Journal of Engineering Mechanics* 148.

26 Yao, Y., Guo, H., and Tan, K. (2020). An elastoplastic damage constitutive model of concrete considering the effects of dehydration and pore pressure at high temperatures. *Materials and Structures* 53.

27 Yao, Y., Wang, K., and Hu, X. (2017). Thermodynamic-based elastoplasticity multiaxial constitutive model for concrete at elevated temperatures. *Journal of Engineering Mechanics* 143.

28 Gawin, D., Baggio, P., and Schrefler, B.A. (1995). Coupled heat, water and gas flow in deformable porous media. *International Journal for Numerical Methods in Fluids* 20: 969–987.

29 Gawin, D., Majorana, C., Pesavento, F., and Schrefler, B. (1998). A fully coupled multiphase FE model of hygro-thermo-mechanical behaviour of concrete at high temperature. Computational Mechanics., Onate, E. & Idelsohn, SR (eds.), New Trends and Applications Proc. of the 4th World Congress on Computational Mechanics, Buenos Aires. 1998: 1–19.

30 Gawin, D., Majorana, C.E., and Schrefler, B.A. (1999). Numerical analysis of hygro-thermal behaviour and damage of concrete at high temperature. *Mechanics of Cohesive-frictional Materials* 4: 37–74.

31 Gawin, D., Pesavento, F., and Schrefler, B.A. (2002). Modelling of hygro-thermal behaviour and damage of concrete at temperature above the critical point of water. *International Journal for Numerical and Analytical Methods in Geomechanics* 26: 537–562.

32 Gawin, D., Schrefler, A.B., and Galindo, M. (1996). Thermo-hydro-mechanical analysis of partially saturated porous materials. *Engineering Computations: International Journal for Computer-Aided Engineering and Software* 13: 113–143.

33 Ulm, F-J., Coussy, O., Bažant, Z.P. (1999). The "Chunnel" Fire. I: chemoplastic softening in rapidly heated concrete. *Journal of Engineering Mechanics* 125: 272–282.

34 Ulm, F-J., Acker, P., and Lévy, M. (1999). The "Chunnel" Fire. II: analysis of concrete damage. *Journal of Engineering Mechanics* 125: 283–289.

35 Faria, R., Oliver, J., and Cervera, M. (1998). A strain-based plastic viscous-damage model for massive concrete structures. *International Journal of Solids and Structures* 35: 1533–1558.

36 Lubliner, J., Oliver, J., Oller, S., and Oñate, E. (1989). A plastic-damage model for concrete. *International Journal of Solids and Structures* 25: 299–326.

37 Oliver, J., Cervera, M., Oller, S., and Lubliner, J. (1990). *Isotropic Damage Models and Smeared Crack Analysis of Concrete*. Swansea: Pineridge Press.

38 Oller, S., Orate, E., Oliver, J., and Lubliners. J. (1990). Finite element nonlinear analysis of concrete structures using a "plastic-damage model." *Engineering Fracture Mechanics* 35: 219–231.

39 Aslani, F. and Bastami, M. (2011). Constitutive relationships for normal- and high-strength concrete at elevated temperatures. *Aci Materials Journal* 108: 355–364.

40 Aslani, F. and Natoori, M. (2013). Stress-strain relationships for steel fiber reinforced self-compacting concrete. *Structural Engineering & Mechanics* 46: 295–322.

41 Zheng, W., Li, H., and Wang, Y. (2012). Compressive stress–strain relationship of steel fiber-reinforced reactive powder concrete after exposure to elevated temperatures. *Construction and Building Materials* 35: 931–940.

42 Abid, M., Hou, X., Zheng, W., and Hussain, R.R. (2017). High temperature and residual properties of reactive powder concrete – A review. *Construction and Building Materials* 147: 339–351.

43 Felicetti, R. and Gambarova, P. (1998). Effects of high temperature on the residual compressive strength of high-strength siliceous concretes. *ACI Materials Journal* 95: 395–405.

44 Kupfer, H.B., Gerstle, K.H. (1973). Behavior of Concrete under biaxial stresses. *Journal of the Engineering Mechanics Division* 99: 853–866.

45 Karsan, I.D. Jirsa, J.O. (1969). Behavior of concrete under compressive loadings. *Journal of the Structural Division* 95: 2543–2564.

46 Hilsdorf, H.K. and Brameshuber, W. (1991). Code-type formulation of fracture mechanics concepts for concrete. *International Journal of Fracture* 51: 61–72.

47 Feenstra, P.H. and De Borst, R. (1996). A composite plasticity model for concrete. *International Journal of Solids and Structures* 33: 707–730.

48 Ehm, C. and Schneider, U. (1985). The high temperature behaviour of concrete under biaxial conditions. *Cement and Concrete Research* 15: 27–34.

49 Luccioni, B.M., Figueroa, M.I., and Danesi, R.F. (2003). Thermo-mechanic model for concrete exposed to elevated temperatures. *Engineering Structures* 25: 729–742.

50 Sabeur, H. (2011). On the modeling of the dehydration induced transient creep during a heating–cooling cycle of concrete. *Materials and Structures*. 44: 1609–1627.

51 Khoury, G.A. (1992). Compressive strength of concrete at high temperatures: a reassessment. *Magazine of Concrete Research* 44: 291–309.

52 Abid, M., Hou, X., Zheng, W., and Hussain, R.R. (2019). Effect of fibers on high-temperature mechanical behavior and microstructure of reactive powder concrete. *Materials (Basel)* 12.

53 Sabeur, H. and Meftah, F. (2008). Dehydration creep of concrete at high temperatures. *Materials and Structures* 41: 17–30.

54 Pont, S.D. and Ehrlacher, A. (2004). Numerical and experimental analysis of chemical dehydration, heat and mass transfers in a concrete hollow cylinder submitted to high temperatures. *International Journal of Heat and Mass Transfer* 47: 135–147.

Appendix

A.1 Machine Learning Case Studies (for Chapter 7)

```
import keras.optimizers
#load data and data bases
import pandas as pd
import numpy as np
from matplotlib import pyplot as plt
from keras.models import Sequential
from keras.layers import Dense, SimpleRNN, Activation

#Loading Data
dataset = pd.read_csv('CompressiveStressAndUHPC.csv')
pd.set_option('display.max_columns', None)   # show the
whole columns
pd.set_option('display.max_rows', None)  # show the whole rows
pd.set_option('display.expand_frame_repr', False)  # do not
fold the data
pd.set_option('display.max_colwidth', 100)
print(dataset.head())
print(dataset.shape)

Output = dataset.iloc[:, [8,9,10]]
Output = np.array(Output)
print(Output[0])
print(type(Output))
print(Output.shape)

#train-test split(for the whole dataset)
from sklearn.model_selection import train_test_split
Input_training, Input_testing, Output_training, Output_
testing = train_test_split(Input, Output, test_size=0.15,
random_state=11)
print(Input_training.shape, Input_testing.shape,
Output_training.shape)

#split training Input set
for i in range (0, 3, 1):
    locals()['Input_SubSet'+str(i)] = []
```

```
    locals()['Input_SubSet'+str(i)] = np.array(locals()
['Input_SubSet'+str(i)])
    for j in range(1, 16, 1):
        locals()['Input_SubSet'+str(i)] = np.append(locals()
['Input_SubSet'+str(i)],Input_training[(15*i+j-1)], axis=-1)
    locals()['Input_SubSet'+str(i)] = locals()['Input_SubSet'
+str(i)].reshape(15,7)
print(Input_SubSet2.shape)
print(type(Input_SubSet2))

#split testing set
for i in range (0, 3, 1):
    locals()['Output_SubSet'+str(i)] = []
    locals()['Output_SubSet'+str(i)] = np.array(locals()
['Output_SubSet'+str(i)])
    for j in range(1, 16, 1):
        locals()['Output_SubSet'+str(i)] = np.append(locals()
['Output_SubSet'+str(i)],Output_training[(15*i+j-1)], axis=-1)
    locals()['Output_SubSet'+str(i)] = locals()['Output_SubSet'+str(i)].
reshape(15,3)
print(Output_SubSet2.shape)

'''
i stands for the number of folds.
(j-1) stands for number of folds minus 1.
The array of the sequence is from 0 to (i-1).
The size of the array is ((i-1)*(number of sample),
(item within one fold)).
Input Dimension is 6.
Input Dimension is 5.
'''

#forming new training and testing dataset (from 0 to 9)
Seq_subset = np.array([0,1,2])
for i in range (0, 3, 1):
    temp_Seq = Seq_subset
    temp_Seq = np.delete(temp_Seq, i)
    locals()['training_input'+str(i)] = []
    locals()['training_input'+str(i)] = np.array(locals()
['training_input'+str(i)])
    for j in range (0, 2, 1):
        locals()['training_input'+str(i)] = np.append(locals()
['training_input'+str(i)], locals()['Input_SubSet'+str(temp_Seq[j])])
    locals()['training_input'+str(i)] = locals()['training_input'
+str(i)].reshape(30,7)
for i in range (0, 3, 1):
    temp_Seq = Seq_subset
    temp_Seq = np.delete(temp_Seq, i)
    locals()['training_output'+str(i)] = []
    locals()['training_output'+str(i)] = np.array(locals()
['trainin_output'+str(i)])
    for j in range (0, 2, 1):
        locals()['training_output'+str(i)] = np.append(locals()
['training_output'+str(i)], locals()['Output_SubSet'+str
(temp_Seq[j])]
    locals()['training_output'+str(i)] = locals()
```

```
['training output'+str(i)]reshape(30,3)

'''
i stands for the number of folds.
'''

#testing dataset
for i in range (0, 3, 1):
    locals()['testing_input'+str(i)] = locals()['Input_SubSet'
+str(i)]
for i in range (0, 3, 1):
    locals()['testing_output'+str(i)] = locals()['Output_SubSet'
+str(i)]
print(training_input2.shape)
print(type(training_input0))
print(training_output2.shape)
print(type(training_output0))
print(testing_input0.shape)
print(type(testing_input0))
print(testing_output0.shape)
print(type(testing_output0))

'''
i stands for the n-th of trained out.
'''

#creat model for each parameter
FWT_Predict_model = Sequential()
FWT_Predict_model.add(Dense(units=15, activation='relu',
input_dim=7))
FWT_Predict_model.add(Dense(units=15, activation='relu'))
FWT_Predict_model.add(Dense(units=15, activation='relu'))
FWT_Predict_model.add(Dense(units=15, activation='relu'))
FWT_Predict_model.add(Dense(units=15, activation='relu'))
FWT_Predict_model.add(Dense(units=15, activation='relu'))
FWT_Predict_model.add(Dense(units=15, activation='relu'))
FWT_Predict_model.add(Dense(units=15, activation='relu'))
FWT_Predict_model.add(Dense(units=15, activation='relu'))
FWT_Predict_model.add(Dense(units=15, activation='relu'))
FWT_Predict_model.add(Dense(units=1, activation='linear'))
FWT_Predict_model.summary()
FST_Predict_model = Sequential()
FST_Predict_model.add(Dense(units=15, activation='relu',
input_dim=7))
FST_Predict_model.add(Dense(units=15, activation='relu'))
FST_Predict_model.add(Dense(units=15, activation='relu'))
FST_Predict_model.add(Dense(units=15, activation='relu'))
FST_Predict_model.add(Dense(units=15, activation='relu'))
FST_Predict_model.add(Dense(units=15, activation='relu'))
FST_Predict_model.add(Dense(units=15, activation='relu'))
FST_Predict_model.add(Dense(units=15, activation='relu'))
FST_Predict_model.add(Dense(units=15, activation='relu'))
FST_Predict_model.add(Dense(units=15, activation='relu'))
FST_Predict_model.add(Dense(units=1, activation='linear'))
FST_Predict_model.summary()
SL_Predict_model = Sequential()
```

```
SL_Predict_model.add(Dense(units=15, activation='relu',
input_dim=7))
SL_Predict_model.add(Dense(units=15, activation='relu'))
SL_Predict_model.add(Dense(units=15, activation='relu'))
SL_Predict_model.add(Dense(units=15, activation='relu'))
SL_Predict_model.add(Dense(units=15, activation='relu'))
SL_Predict_model.add(Dense(units=15, activation='relu'))
SL_Predict_model.add(Dense(units=15, activation='relu'))
SL_Predict_model.add(Dense(units=15, activation='relu'))
SL_Predict_model.add(Dense(units=15, activation='relu'))
SL_Predict_model.add(Dense(units=15, activation='relu'))
SL_Predict_model.add(Dense(units=1, activation='linear'))
SL_Predict_model.summary()

'''
Optimizer setting:
Learning Rate = 0.01
Number of Iterations = 300
Loss Function = MSE
Metrix
'''

#optimizer setting
from keras import optimizers
adam = optimizers.adam(learning_rate=0.01, beta_1=0.9,
beta_2=0.999,amsgrad=False)

#model compile
FWT_Predict_model.compile(loss='mse', optimizer='adam',
metrics=['mse'])
FST_Predict_model.compile(loss='mse', optimizer='adam',
metrics=['mse'])
SL_Predict_model.compile(loss='mse', optimizer='adam',
metrics=['mse'])

#Trainging for FWT model
for i in range (0, 3, 1):
        FWT_Predict_model.fit(locals()['training_input'+str
(i)], locals()['training_output'+str(i)][:, 0], epochs=4000)
print('finish prediction')

#Trainging for FST model
for i in range (0, 3, 1):
        FST_Predict_model.fit(locals()['training_input'+str(i)],
locals()['training_output'+str(i)][:, 1], epochs=4000)
print('finish prediction')

#Trainging for SL model
for i in range (0, 3, 1):
        SL_Predict_model.fit(locals()['training_input'+str
(i)], locals()['training_output'+str(i)][:, 2], epochs=4000)
print('finish prediction')

#training data accuracy
from sklearn.metrics import mean_squared_error, r2_score
Output_training_predicted_FWT = FWT_Predict_model.predict
```

```
(Input_training)
accuracy_train_FWT = mean_squared_error(Output_training[:, 0],
Output_training_predicted_FWT)
r2_score_train_FWT = r2_score(Output_training[:, 0], Output_
training_predicted_FWT)
print(type(Output_training[:, 0]), type(Output_training_
predicted_FWT))
print(Output_training[:, 0].shape, Output_training_predicted_
FWT.shape)
print('FWT accuracy')
print('mse='+str(accuracy_train_FWT))
print('r2='+str(r2_score_train_FWT))
#training data accuracy
from sklearn.metrics import mean_squared_error, r2_score
Output_training_predicted_FST = FST_Predict_model.predict
(Input_training)
accuracy_train_FST = mean_squared_error(Output_training[:, 1],
Output_training_predicted_FST)
r2_score_train_FST = r2_score(Output_training[:, 1],
Output_training_predicted_FST)
print(type(Output_training[:, 1]), type(Output_training_
predicted_FST))
print(Output_training[:, 1].shape, Output_training_predicted_
FST.shape)
print('FST accuracy')
print('mse='+str(accuracy_train_FST))
print('r2='+str(r2_score_train_FST))
#training data accuracy
from sklearn.metrics import mean_squared_error, r2_score
Output_training_predicted_SL = SL_Predict_model.predict
(Input_training)
accuracy_train_SL = mean_squared_error(Output_training[:, 2],
Output_training_predicted_SL)
r2_score_train_SL = r2_score(Output_training[:, 2],
Output_training_predicted_SL)
print(type(Output_training[:, 2]), type(Output_training_
predicted_SL))
print(Output_training[:, 2].shape, Output_training_predicted_
SL.shape)
print('SL accuracy')
print('mse='+str(accuracy_train_SL))
print('r2='+str(r2_score_train_SL))

#training data accuracy
from sklearn.metrics import mean_squared_error, r2_score
Output_testing_predicted_FWT = FWT_Predict_model.predict
(Input_testing)
accuracy_test_FWT = mean_squared_error(Output_testing[:, 0],
Output_testing_predicted_FWT)
r2_score_test_FWT = r2_score(Output_testing[:, 0], Output_
testing_predicted_FWT)
print(type(Output_testing[:, 0]), type(Output_testing_
predicted_FWT))
print(Output_testing[:, 0].shape, Output_testing_predicted_
FWT.shape)
print('Y1 accuracy')
```

```
print('mse='+str(accuracy_test_FWT))
print('r2='+str(r2_score_test_FWT))
#training data accuracy
from sklearn.metrics import mean_squared_error, r2_score
Output_testing_predicted_FST = FST_Predict_model.predict
(Input_testing)
accuracy_test_FST = mean_squared_error(Output_testing[:, 1],
Output_testing_predicted_FST)
r2_score_test_FST = r2_score(Output_testing[:, 1], Output_
testing_predicted_FST)
print(type(Output_testing[:, 1]), type(Output_
testing_predicted_FST))
print(Output_testing[:, 1].shape, Output_testing_
predicted_FST.shape)
print('Y2 accuracy')
print('mse='+str(accuracy_test_FST))
print('r2='+str(r2_score_test_FST))
#training data accuracy
from sklearn.metrics import mean_squared_error, r2_score
Output_testing_predicted_SL = SL_Predict_model.predict
(Input_testing)
accuracy_test_SL = mean_squared_error(Output_testing[:, 2],
Outputtesting_predicted_SL)
r2_score_test_SL = r2_score(Output_testing[:, 2], Output_
testing_predicted_SL)
print(type(Output_testing[:, 2]), type(Output_testing_
predicted_SL))
print(Output_testing[:, 2].shape, Output_testing_predicted_
SL.shape)
print('Y3 accuracy')
print('mse='+str(accuracy_test_SL))
print('r2='+str(r2_score_test_SL))

from matplotlib import rcParams
rcParams['font.family'] = 'serif'
rcParams['font.serif'] = ['Times New Roman']

#Y1 Prediction
fig1 = plt.figure(figsize=(7, 7))
x = np.linspace(0,250,250)
plt.plot(x, x, '--k')
plt.plot(x, 0.9*x, '-.c', label = 'Lower Bound')
plt.plot(x, 1.1*x, '-.r', label = 'Upper Bound')
plt.scatter(Output_training_predicted_FWT, Output_training
[:, 0],marker='x', label = 'Training dataset')
plt.scatter(Output_testing_predicted_FWT, Output_testing[:, 0],
label = 'Testing dataset')
plt.legend(fontsize=16)
plt.title('Predicted Compressive Strength V.S. Predicted
CompressiveStrength', fontsize=16)
plt.ylabel('True Compressive Strength, MPa', fontsize=16)
plt.xlabel('Predicted Compressive Strength, MPa', fontsize=16)

#Y1 Prediction
fig2 = plt.figure(figsize=(7, 7))
x = np.linspace(0,250,250)
```

```
plt.plot(x, x, '--k')
plt.plot(x, 0.9*x, '-.c', label = 'Lower Bound')
plt.plot(x, 1.1*x, '-.r', label = 'Upper Bound')
plt.scatter(Output_training_predicted_FST, Output_training
[:, 1], marker='x')
plt.scatter(Output_testing_predicted_FST, Output_testing[:, 1])
plt.title('Predicted Tensile Strength V.S. Predicted
Tensile Strength')
plt.ylabel('Ture Compressive Strength, MPa')
plt.xlabel('Predicted Compressive Strength, MPa')

#Y1 Prediction
fig3 = plt.figure(figsize=(7, 7))
x = np.linspace(0, 250, 250)
plt.plot(x, x, '--k')
plt.plot(x, 0.9 * x, '-.c', label='Lower Bound')
plt.plot(x, 1.1 * x, '-.r', label='Upper Bound')
plt.scatter(Output_training_predicted_SL, Output_training
[:, 2], marker='x')
plt.scatter(Output_testing_predicted_SL, Output_testing[:, 2])
plt.title('Predicted Slump V.S. Predicted Slump')
plt.ylabel('Ture Compressive Strength, MPa')
plt.xlabel('Predicted Compressive Strength, MPa')
```

A.2 Tensorial Notations and Operations (for Chapter 10)

The standard tensorial notations and operations are adopted throughout. Boldface letters denote tensors, the order of which depends on the context. The Einstein summation convention is used for Latin indices unless otherwise indicated.

- Dots and colons

$$\boldsymbol{u}\cdot\boldsymbol{v} = u_i v_i \quad (\boldsymbol{a} : \boldsymbol{a})_{ij} = a_{ik} b_{kj} \qquad (\boldsymbol{A} : \boldsymbol{a})_{ij} = A_{ijkl} a_{kl}$$
$$(\boldsymbol{A} : \boldsymbol{B})_{ijkl} = A_{ijmn} B_{mnkl}$$

- Double colons

$$\boldsymbol{A} :: \boldsymbol{B} = A_{ijmn} B_{mnij}$$

- The double contraction between any two fourth-order tensors is scale-invariant.
- Dyadic product

$$(\boldsymbol{u} \otimes \boldsymbol{v})_{ij} = u_i v_j \quad (\boldsymbol{a} \otimes \boldsymbol{a})_{ijkl} = a_{ij} b_{kl}$$

- Three commonly used fourth-order projection tensors: $\boldsymbol{I}, \boldsymbol{J}, \boldsymbol{K}$

$$I_{ijkl} = \frac{1}{2}\left(\delta_{ik}\delta_{jl} + \delta_{il}\delta_{jk}\right) \quad J_{ijkl} = \frac{1}{3}\delta_{ij}\delta_{kl} \quad K_{ijkl} = I_{ijkl} - J_{ijkl}$$

where δ_{ij} is the Kronecker delta tensor. For any symmetric second-order tensor $\boldsymbol{a}$, its hydrostatic and deviatoric parts could be obtained via:

$$\frac{1}{3}\mathrm{tr}\left(\boldsymbol{a}\right)\delta = \boldsymbol{J} : \boldsymbol{a} \quad \mathrm{dev}\left(\boldsymbol{a}\right) = \boldsymbol{K} : \boldsymbol{a}$$

Moreover, the symmetric second-order tensor $\boldsymbol{a}$ can be projected to himself by:

$$\boldsymbol{a} = \frac{1}{2}\left(\boldsymbol{a} + \boldsymbol{a}^{T}\right) = \boldsymbol{I} : \boldsymbol{a}$$

where the superscript T denotes the transpose operation.

A.3 Basic Framework for Calculation of Elastic-plastic Damage Constitutive Model of Concrete (for Chapter 11)

```
      SUBROUTINE UMAT (STRESS,STATEV,DDSDDE,SSE,SPD,SCD,
      1 RPL,DDSDDT,DRPLDE,DRPLDT,
      2 STRAN,DSTRAN,TIME,DTIME,TEMP,DTEMP,PREDEF,DPRED,CMNAME,
      3 NDI,NSHR,NTENS,NSTATV,PROPS,NPROPS,COORDS,DROT,PNEWDT,
      4 CELENT,DFGRD0,DFGRD1,NOEL,NPT,LAYER,KSPT,KSTEP,KINC)
C
       INCLUDE 'ABA_PARAM.INC'
C
       CHARACTER*80 CMNAME
       DIMENSION STRESS(NTENS),STATEV(NSTATV),
      1 DDSDDE(NTENS,NTENS),DDSDDT(NTENS),DRPLDE(NTENS),
      2 STRAN(NTENS),DSTRAN(NTENS),TIME(2),PREDEF(1),DPRED(1),
      3 PROPS(NPROPS),COORDS(3),DROT(3,3),DFGRD0(3,3),DFGRD1(3,3)
C     LOCAL ARRAYS
C ----------------------------------------------------------------
C     EELAS - ELASTIC STRAINS
C     EPLAS - PLASTIC STRAINS
C     FLOW - PLASTIC FLOW DIRECTIONS
C
       DIMENSION EELAS(6), EPLAS(6), FLOWC(6),
      1 DPSTRC(6), DULTIC(6), DEVI(6), ESTRESS(6), OLDS(6),
      2 ESTRAN(6), DPSTRD(6), DSTRTR(6), STRETR(6)
C
       PARAMETER(ZERO=0.0D0,ONE=1.0D0,TWO=2.0D0,THREE=3.
0D0,SIX=6.0D0,
      1 HALF=0.5D0,ENUMAX=0.4999D0,TOLER=1.0D-4,FOUR=4.0D0,
LPH=0.121D0,
      2 FT=3.5D0,ALPHG=0.25D0,CCONE=0.000933D0,FCZERO=9.9D0,
FC=33.0D0,
      3 BC=520.4D0,ALPHT=1537D0,ALPHC=308.34D0,EMOD_F=20000,
EMOD_0=20000)
C     4 RR_0=2.0D0,RR_L=2.0D0,BR=50D0,AP=1.0D0,MIU=1.0D0,OMIG=0.009D0,
C     5 ALPH=0.3235D0,D_CR=0.35D0,TRIN_CR=1.29D0,TRIN_TH=0.001D0,
C     6 VISCO=0.055D0,KK=0.00D0,KKK=1.0D0,SIG=31.89D0)
C     7 DC=0.0D0,NN=1.2D0,
C
C ----------------------------------------------------------------
C     UMAT FOR ISOTROPIC PLASTICITY-DAMAGE CONCRETE
C     WITH HARDENING AND SOFTENING - CANNOT BE USED FOR PLANE STRESS
C ----------------------------------------------------------------
C    PROPS(1) - E
C    PROPS(2) - NU
C    ELASTIC PROPERTIES
C
C       CT=STATEV(15)
       CC=STATEV(16)
C       DT=STATEV(17)          ÷
```

```
      DC=STATEV(58)
      EMOD=PROPS(1)
      ENU=MIN(PROPS(2), ENUMAX)
      EBULK3=EMOD/(ONE-TWO*ENU)
      EG2=EMOD/(ONE+ENU)
      EG=EG2/TWO
      EG3=THREE*EG
      ELAM=(EBULK3-EG2)/THREE
C
C   SET INITIAL STIFFNESS
C
      DO K1=1,NTENS
         DO K2=1,NTENS
           DDSDDE(K2,K1)=ZERO
         END DO
      END DO

C  ELASTIC STIFFNESS
      DO K1=1, NDI
        DO K2=1, NDI
          DDSDDE(K2,K1)=ELAM
        END DO
        DDSDDE(K1,K1)=(EG2+ELAM)
      END DO
      DO K1=NDI+1, NTENS
        DDSDDE(K1,K1)=EG
      END DO
C
C      RECOVER ELASTIC STRAIN, PLASTIC STRAIN AND SHIFT TENSOR
 AND ROTATE no use
C      NOTE: USE CODE 1 FOR (TENSOR) STRESS, CODE 2 FOR
 (ENGINEERING) STRAIN no use
C
C   SAVE STRESS AND PLASTIC STRAINS AND
C   CALCULATE PREDICTOR STRESS AND ELASTIC STRAIN
C
      DO K1=1, NTENS
         OLDS(K1)=STRESS(K1)
         EELAS(K1)=STATEV(K1)
        EPLAS(K1)=STATEV(K1+NTENS)
        EELAS(K1)=EELAS(K1)+DSTRAN(K1)
      END DO
C
      DO K1=1,NTENS
       DSTRTR(K1)=DSTRAN(K1)
       DSTRAN(K1)=(ONE-DC)*DSTRTR(K1)
      END DO
C     NOMINAL STRESS
      DO K1=1, NTENS
          DO K2=1, NTENS
          STRESS(K2)=STRESS(K2)+DDSDDE(K2, K1)*DSTRAN(K1)
          END DO
       END DO
C
C     COMPRESSIVE YEILD PARAMETER
      TONE=(STRESS(1)+STRESS(2)+STRESS(3))
```

```
       SJTWO=(ONE/SIX*((STRESS(1)-STRESS(2))**2
      1 +(STRESS(1)-STRESS(3))**2+(STRESS(2)-STRESS(3))**2))
      2 +STRESS(4)**2+STRESS(5)**2+STRESS(6)**2
       DEVI(1)=ONE/THREE*(TWO*STRESS(1)-STRESS(2)-STRESS(3))
       DEVI(2)=ONE/THREE*(TWO*STRESS(2)-STRESS(1)-STRESS(3))
       DEVI(3)=ONE/THREE*(TWO*STRESS(3)-STRESS(1)-STRESS(2))
       DEVI(4)=STRESS(4)
       DEVI(5)=STRESS(5)
       DEVI(6)=STRESS(6)
       THEGML=STRESS(3)
       THEMAX=STRESS(1)
       STATEV(64)=THEGML
       STATEV(65)=THEMAX
C
C      IF YEILD
       IF(KINC.EQ.ONE) THEN
        RC=FCZERO
       ELSE
        RC=STATEV(14)
       END IF
C
C      COMPRESSIVE YEILD   THEGML
       SC=ABS(THEGML)-RC
       STATEV(63)=SC
       IF(SC.GT.ZERO) THEN
         DO K1=1,NTENS
         DULTIC(K1)=ALPHG+DEVI(K1)/TWO/SQRT(SJTWO)
         END DO
         DO K1=NDI+1,NTENS
         DULTIC(K1)=DEVI(K1)/TWO/SQRT(SJTWO)
         END DO
         DO K1=1,NTENS
         FLOWC(K1)=STATEV(K1+5*NTENS)
         END DO
        RCC=(ABS(STRESS(1)*FLOWC(1)+STRESS(2)*FLOWC(2)+STRESS(3)*FLOWC(3)
      1  +TWO*(STRESS(4)*FLOWC(4)+STRESS(5)*FLOWC(5)+STRESS(6)*FLOWC(6))
      2  +TOLER/SIX)/(ALPHG*TONE+SQRT(SJTWO))+0.00001)
         DO K1=1,NTENS
         FLOWC(K1)=RCC*DULTIC(K1)
         END DO
         DO K1=1,NTENS
         DPSTRC(K1)=FLOWC(K1)*DTIME
         END DO
         CC=CC+RCC*DTIME
C
        IF(CC.GE.CCONE) THEN
C        RC=FC-6512*(CC-CCONE)
        RC=FC*(ONE+BC*(CC-CCONE))*EXP(-BC*(CC-CCONE))
        DC=ONE-EXP(-ALPHC*CC)
C        DO K1=1,NTENS
C          STRETR(K1)=STRESS(K1)
C          STRESS(K1)=(ONE-DC)*STRETR(K1)
C        END DO
C
        ELSE
        RC=FCZERO+(TWO*(FC-FCZERO)*CC)/(CCONE*(ONE+(CC/CCONE)**2))
```

```
      DC=ONE-EXP(-ALPHC*CC)
      DO K1=1,NTENS
       STRETR(K1)=STRESS(K1)
       STRESS(K1)=(ONE-DC)*STRETR(K1)
      END DO
      END IF
C
      DO K1=1,NTENS
       DPSTRD(K1)=(ONE-DC)*DPSTRC(K1)
      END DO
      DO K1=1,NTENS
       DO K2=1,NTENS
         STRESS(K2)=STRESS(K2)-DDSDDE(K2, K1)*DPSTRD(K1)
       END DO
      END DO
     ELSE
       DO K1=1,NTENS
       DPSTRC(K1)=ZERO
       END DO
     END IF
C
     DO K1=1,NDI
      EPLAS(K1)=EPLAS(K1)+DPSTRC(K1)
      EELAS(K1)=EELAS(K1)-DPSTRC(K1)
     END DO
     DO K1=NDI+1,NTENS
      EELAS(K1)=EELAS(K1)-TWO*DPSTRC(K1)
      EPLAS(K1)=EPLAS(K1)+TWO*DPSTRC(K1)
     END DO
C     IF(CC.GE.CCONE) THEN
C     DO K1=1,NTENS
C       DO K2=1,NTENS
C         STRESS(K2)=-RC-DDSDDE(K2, K1)*DPSTRD(K1)
C       END DO
C     END DO
C     ELSE
C     END IF
C
     DO K1=1,NTENS
       STATEV(K1)=EELAS(K1)
       STATEV(K1+NTENS)=EPLAS(K1)
       STATEV(13)=RCC
       STATEV(15)=ALPHG*TONE+SQRT(SJTWO)
       STATEV(17)=SCTRY
       STATEV(14)=RC
       STATEV(16)=CC
       STATEV(58)=DC
       STATEV(K1+3*NTENS)=DULTIC(K1)
       STATEV(K1+4*NTENS)=DEVI(K1)
       STATEV(K1+5*NTENS)=FLOWC(K1)
       STATEV(K1+6*NTENS)=DPSTRC(K1)
     END DO
C
     RETURN
     END
```

Index

Page numbers in *italics* and **bold** refers to figures and tables respectively.